Lecture Notes in Computer Scien

Edited by G. Goos, J. Hartmanis, and J. van Leeuwen

T0254050

Springer

Berlin
Heidelberg
New York
Barcelona
Hong Kong
London
Milan
Paris
Tokyo

Andreas Brandstädt   Van Bang Le (Eds.)

Graph-Theoretic Concepts in Computer Science

27th International Workshop, WG 2001
Boltenhagen, Germany, June 14-16, 2001
Proceedings

Springer

Series Editors

Gerhard Goos, Karlsruhe University, Germany
Juris Hartmanis, Cornell University, NY, USA
Jan van Leeuwen, Utrecht University, The Netherlands

Volume Editors

Andreas Brandstädt
Van Bang Le
University of Rostock, Department of Computer Science
18051 Rostock, Germany
E-mail: {ab/le}@informatik.uni-rostock.de

Cataloging-in-Publication Data applied for

Cataloging-in-Publication Data applied for

Die Deutsche Bibliothek - CIP-Einheitsaufnahme

Graph theoretic concepts in computer science : 27th international workshop ;
proceedings / WG 2001, Boltenhagen, Germany, June 14 - 16, 2001. Andreas
Brandstädt ; Van Bang Le (ed.). - Berlin ; Heidelberg ; New York ; Barcelona ;
Hong Kong ; London ; Milan ; Paris ; Tokyo : Springer, 2001
   (Lecture notes in computer science ; Vol. 2204)
   ISBN 3-540-42707-4

CR Subject Classification (1998): F.2, G.1.2, G.1.6, G.2, G.3, E.1, I.3.5

ISSN 0302-9743
ISBN 3-540-42707-4 Springer-Verlag Berlin Heidelberg New York

This work is subject to copyright. All rights are reserved, whether the whole or part of the material is
concerned, specifically the rights of translation, reprinting, re-use of illustrations, recitation, broadcasting,
reproduction on microfilms or in any other way, and storage in data banks. Duplication of this publication
or parts thereof is permitted only under the provisions of the German Copyright Law of September 9, 1965,
in its current version, and permission for use must always be obtained from Springer-Verlag. Violations are
liable for prosecution under the German Copyright Law.

Springer-Verlag Berlin Heidelberg New York
a member of BertelsmannSpringer Science+Business Media GmbH

http://www.springer.de

© Springer-Verlag Berlin Heidelberg 2001
Printed in Germany

Typesetting: Camera-ready by author, data conversion by Steingräber Satztechnik GmbH, Heidelberg
Printed on acid-free paper   SPIN: 10840737   06/3142   5 4 3 2 1 0

Preface

The 27th International Workshop on Graph-Theoretic Concepts in Computer Science (WG 2001) was held in Boltenhagen (Mecklenburg-Vorpommern) June 14–16, 2001. It was organized by the Theoretical Computer Science group of the Department of Computer Science at the University of Rostock. The organizers gratefully acknowledge the support by the German Research Community (Deutsche Forschungsgemeinschaft - DFG) and the state Mecklenburg-Vorpommern.

As was the case for previous WG workshops, this workshop was devoted to the theoretical and practical aspects of graph concepts in computer science, and its contributed talks showed how recent research results from algorithmic graph theory can be used in computer science and which graph-theoretic questions arise from new developments in computer science. Moreover, the workshop gave an impression of the impact of computer science on the efficiency of graph algorithms and on structural aspects of graphs and graph decomposition.

The workshop looks back to a remarkable tradition of 27 years; all previous WG conferences took place in the middle of Europe (Germany, The Netherlands, Italy, Switzerland, Slowakia). The participants of WG 2001 came from various countries such as Canada, China, Czech Republic, France, Great Britain, Greece, Italy, Norway, Russia, Sweden, Syria, Taiwan, The Netherlands, U.S.A., and, of course, Germany.

The program committee represented the wide scientific spectrum and the aims of the conference. In a careful reviewing process with at least four reports per submission, the program committee selected 27 papers from the submissions. All accepted papers were presented at the conference, and the referees' comments as well as the numerous fruitful discussions during the workshop have been taken into account by the authors of these conference proceedings. Moreover, there were two fascinating invited talks by

> H.-J. Bandelt (University of Hamburg) on some connections between graph concepts such as median hulls and Steiner trees on one hand and molecular biology on the other hand,

and by

> F. Meyer auf der Heide (University of Paderborn) on data management in networks, giving a survey on strategies for distributing and accessing shared objects in large parallel and distributed systems.

It is our pleasure to thank all those who contributed to the scientific success of WG 2001:
- all authors of submitted and of presented papers, and in particular the speakers
- the referees and subreferees
- the invited speakers

- the DFG
- the state of Mecklenburg-Vorpommern, and last but not least
- the organizers Katrin Erdmann, Thomas Szymczak, Ernst de Ridder, supported by Roswitha Fengler before and Tilo Klembt, Suhail Mahfud, and Hans-Jörg Schulz during the conference.

August 2001   Andreas Brandstädt and Van Bang Le

The Tradition of WG

Hosts – Location

1975   U. Pape – Berlin
1976   H. Noltemeier – Göttingen
1977   J. Mühlbacher – Linz
1978   M. Nagl, H.J. Schneider – Schloß Feuerstein, near Erlangen
1979   U. Pape – Berlin
1980   H. Noltemeier – Bad Honnef
1981   J. Mühlbacher – Linz
1982   H.J. Schneider, H. Göttler – Neunkirchen, near Erlangen
1983   M. Nagl, J. Perl – Haus Ohrbeck, near Osnabrück
1984   U. Pape – Berlin
1985   H. Noltemeier – Schloß Schwanenberg, near Würzburg
1986   G. Tinhofer, G. Schmidt – Stift Bernried, near München
1987   H. Göttler, H.J. Schneider – Schloß Banz, near Bamberg
1988   J. van Leeuwen – Amsterdam
1989   M. Nagl – Schloß Rolduc, near Aachen
1990   R.H. Möhring – Johannesstift Berlin
1991   G. Schmidt, R. Berghammer – Richterheim Fischbachau, München
1992   E.W. Mayr – Wilhelm-Kempf-Haus, Wiesbaden-Naurod
1993   J. van Leeuwen – Sports Center Papendal, near Utrecht
1994   G. Tinhofer, E.W. Mayr, G. Schmidt – Herrsching, near München
1995   M. Nagl – Haus Eich, Aachen
1996   G. Ausiello, A. Marchetti-Spaccamela – Cadenabbia
1997   R.H. Möhring – Bildungszentrum am Müggelsee, Berlin
1998   J. Hromkovič, O. Sýkora – Smolenice-Castle, near Bratislava
1999   P. Widmayer – Centro Stefano Franscini, Monte Verità, Ascona
2000   D. Wagner – Waldhaus Jakob, Konstanz
2001   A. Brandstädt – Boltenhagen, near Rostock

Program Committee

Hans Bodlaender	Utrecht University, The Netherlands
Andreas Brandstädt	University of Rostock, Germany (chair)
Victor Chepoi	Université de la Mediterranée, Marseille, France
Michel Habib	LIRMM Montpellier, France
Klaus Jansen	University of Kiel, Germany
Dieter Kratsch	Université de Metz, France
Luděk Kučera	Charles University, Prague, Czech Republic
Alberto Marchetti-Spaccamela	Università di Roma "La Sapienzia", Italy
Ernst W. Mayr	TU München, Germany
Rolf H. Möhring	TU Berlin, Germany
Manfred Nagl	RWTH Aachen, Germany
Hartmut Noltemeier	University of Würzburg, Germany
Jeremy P. Spinrad	Vanderbilt University, Nashville, USA
Ondrej Sýkora	Loughborough University, United Kingdom
Gottfried Tinhofer	TU München, Germany
Peter Widmayer	ETH Zürich, Switzerland

Additional Reviewers

Gerardo Bandera
Thomas Bayer
Hans-J. Boeckenhauer
Peter Damaschke
Feodor F. Dragan
Stefan Dobrev
Thomas Erlebach
Jens Ernst
Jiří Fiala
Stefanie Gerke
Yubao Guo
Alexander Hall
Volker Heun
Klaus Holzapfel
Juraj Hromkovič

Sandi Klavžar
Ekkehard Köhler
Christian Laforest
Jean-Marc Lanlignel
Van Bang Le
Zsuzsanna Lipták
Sonia P. Mansilla
Ross McConnell
Michal Mnuk
Haiko Müller
Takao Nishizeki
Conrad Pomm
Pascal Prea
Martin Raab
Ingo Rohloff

Udi Rotics
Mark Scharbrodt
Konrad Schlude
Heiko Schröder
Martin Skutella
Ladislav Stacho
Robert Szelepcsenyi
Thomas Szymczak
Ioan Todinca
Yann Vaxes
Ulrich Voll
Imrich Vrt'o
Karsten Weihe

Sponsoring Institutions

Deutsche Forschungsgemeinschaft (DFG)
Land Mecklenburg-Vorpommern

Table of Contents

Median Hulls as Steiner Hulls
in Rectilinear and Molecular Sequence Spaces
(Invited Presentation)

Hans-J. Bandelt

Departement of Mathematics, University of Hamburg, Germany
bandelt@math.uni-hamburg.de

1   Introduction

The *Steiner problem* is to find a shortest connection of a finite subset X in a given metric space (M, d). If the space meets some local compactness and connectivity conditions, then a solution, a *Steiner minimal tree* for X, exists [7]. More generally, considering any graph-theoretic tree T with all nodes of degree < 3 labeled by elements of X (that is, an X-*labeled tree*), we may ask for a minimal length realization of T in (M, d), that is, for an embedding of the node set of T in M which extends the identity map of X and yields the smallest possible length of the image Steiner tree (relative to the labeled tree T). The embedded unlabeled nodes of T are called *Steiner points*. A Steiner minimal tree is then that realization which attains minimum length among the (finite) collection of X-labeled trees. Given the subset X, one is interested in restricting the search for Steiner points in the metric space (M, d). Any proper subset of M that is guaranteed to harbor at least one Steiner minimal tree for X is called a *Steiner hull* for X [7,12]. Particular interest attaches to Steiner hulls that are finitely generated in that they are determined by a finite subset S of M containing X and all Steiner points of some Steiner minimal tree for X; this finite set S can trivially be turned into a Steiner hull by attaching one geodesic from M for each pair of its points. For the Steiner problem in rectilinear space, that is, a d-dimensional real space equipped with the metric associated with the 1-norm, finitely generated Steiner hulls have been described [7,12].

One of the motivations to study the Steiner problem in rectilinear space comes from biology, where one is interested in reconstructing the phylogeny or genealogy of species or populations based on molecular or morphological data. The Steiner problem is there referred to as the parsimony problem and Steiner minimal trees are then called most parsimonious reconstructions of most parsimonious trees. Actually, Farris [8] was the first to describe an algorithm for finding a minimal length realization of any X-labeled tree within a rectilinear space (an "ordered character" in the biological context refers to one coordinate of the space). This was accordingly modified for molecular sequence spaces by Fitch [9]; cf. [11]. Such a space (equipped with the Hamming distance) is, for some $d > 0$, the d-th Cartesian power of an equidistant alphabet (typically, the nucleotide alphabet). In the binary case, that is, say, for the alphabet $\{0, 1\}$, it

A. Brandstädt and V.B. Le (Eds.): WG 2001, LNCS 2204, pp. 1–7, 2001.
© Springer-Verlag Berlin Heidelberg 2001

is knowm that median hulls generate Steiner hulls [2,14]. In fact, an algebraic analysis of the algorithm of Farris and Fitch (the FFH algorithm, see below) reveals that any X-labeled tree T has a minimal length realization as a Steiner tree relative to T using only Steiner points from the median hull of X. In practical applications where this hull would still be too large, one could further reduce this set in a heuristic fashion, of course at the expense that the Steiner hull property may no longer provably be maintained [1,4].

2   Reduction to Hypercubes

Consider a tree T interconnecting a family of sequences from a finite subset X of the d-dimensional rectilinear space. Then one can realize T as a Steiner tree in the rectilinear space attaining minimal length relative to T [7]. Adjoin to X all Steiner points of this Steiner tree and extend the resulting set to a finite grid Y (of dimension at most d). This grid can be recoded by 0-1 sequences and thus be embedded into a finite-dimensional hypercube H in which every coordinate receives a certain weight. We can think of X being embedded in H such that a Steiner tree realization of T in H has the same minimal length as in the original rectilinear space. We will see below that the adjunction of a set of Steiner points was an unnecessary proviso since the median hull of X (necessarily contained in Y) turns out to harbor an ample collection of potential Steiner points to guarantee a minimal length realization relative to any prescribed tree T.

3   Farris-Fitch-Hartigan (FFH) Labeling

First, in order to describe and analyze the FFH algorithm, it is convenient to encode any given tree T in terms of its leaves $t_0, t_1, \ldots t_{n-1}$ $(n > 2)$. One leaf, say t_0, is selected as a root. In dealing with the thus rooted tree T we employ the metaphor of a maternal genealogy and think of T as being visualized in the plane such that t_0 is at the top and the other leaves are at the bottom. The rooted subtree $T' := T - t_0$ obtained by deleting t_0 (and the edge incident with it) has the (unique) daughter node u_0 of t_0 as its root. The so-called New Hampshire encoding [10] of T' is a string comprising brackets and the leaf names $t_1, \ldots, t_{n-1}$. The code for T' is obtained bottom-up: if in the first step, say, $t_1, \ldots, t_i$ $(i > 1)$ comprise all the daughter nodes of their common mother node u, then we record this by $u = (t_1, \ldots, t_i)$ where the order within the bracket is arbitrary. Then delete $t_1, \ldots, t_i$ and turn u into a leaf, which represents the subtree with original leaves $t_1, \ldots, t_i$. Continue until the bracketing stops with a single node, the root u_0 of T'. To arrive at the final code, successively replace every interior node by the associated bracket.

We now turn the New Hampshire encodings of the subtrees into operations on the hypercube. First we deal with the case of a single coordinate which may take values 0, 1. In order to handle ties , we introduce a third letter N, which stands for "not determined". For each $k > 1$ we define the k-ary majority operation on $\{0, 1, N\}$ as follows. The operation returns 0 if 0 occurs more often than 1 among the k entries, it reurns 1 if 1 is more frequent than 0, and returns N otherwise, that is , in the case of a tie between 0 and 1. The outcome is then simply written as a bracket with the entries listed as a string. By definition, this operation is commutative. For $k = 2$ we have $(00) = (0N) = 0$, $(11) = (1N) = 1$, and $(01) = (NN) = N$, and for $k = 3$ we have $(000) = (001) = (00N) = (0NN) = 0$, $(111) = (110) = (11N) = (1NN) = 1$, and $(01N) = (NNN) = N$. Restricted to $\{0, 1\}$ the latter entails the median operation on $\{0, 1\}$ [3]. The k-ary bracket operation on the set $\{0, 1, N\}^d$ of sequences of length d is then defined coordinate-wise.

The FFH algorithm makes two passes through the tree T rooted at t_0, a bottom-up pass and then a top-down pass [10,13]. Starting from the labeled leaves $t_1, \ldots, t_{n-1}$, any mother node with k children labeled by $x_1, \ldots, x_k$ receives the label $(x_1 \ldots x_k)$. Proceeding bottom-up this pass stops with the labeling of the root u_0 of T'. In the top-down pass, beginning with u_0, the un-determined ("N") coordinates are successively specified by 0 or 1: a node of T with current label x such that the mother node is already labeled by a 0-1 sequence y receives the 0-1 sequence (xxy) as its new label. This, in brief, is how the algorithm works. The (well-known) proof that this labeling yields a minimal length realization relative to T is straightforward by induction: interpret each label of the bottom-up pass as a set of 0-1 sequences obtained by all combinations when inserting 0 and 1 for N; this set describes exactly the 0-1 sequences at the interior node under consideration each leading to a minimal length realization of the subtree rooted at this node; the final top-down pass then executes the minimization link-wise by using the specified 0-1 sequence at the mother node in question.

4   Examples

First consider the set $X = \{v, w, x, y\}$ of length 7 sequences $v = 1000111$, $w = 0100100$, $x = 0010010$, $y = 0001001$. Let T be the tree with four leaves t_0, t_1, t_2, t_3 and interior edge separating t_0, t_1 from t_2, t_3. Then $T' := T - t_0 = (t_1, (t_2, t_3))$. When we label t_0, t_1, t_2, t_3 by v, w, x, y, respectively, then the root of T' receives the label $(vw(wxy)) = 0000100$ under FFH and the second interior node of T is then labeled by $(wxy) = 0000000$. When we assign to the leaves t_0, t_1, t_2, t_3 either w, v, x, y or x, y, v, w, or y, x, v, w instead, we obtain six further sequences, so that eventually all 0-1 sequences of length 7 with prefix 0000 are generated. The resulting set of 12 sequences (including X) is then closed under the median operation and constitutes the graph $\lambda(4)$ described in [5].

Next we consider the set $X = \{v, w, x, y, z\}$ of length 15 sequences where the coordinates correspond to the partitions of X into two parts such that 0 is always assigned to the majority part (for convenience). Assume that T is a tree with five leaves t_0, t_1, t_2, t_3, t_4 which are labeled by v, w, x, y, z, respectively, unless stated otherwise. Let t_0 serve as the root of T, and put $T' := T - t_0$. Since there are two binary rooted trees with four leaves, we distinguish two cases.

Case 1: $T' = ((t_1, t_2), (t_3, t_4))$. Then the root of T' receives the label

$$(vwxyz) = (v((vwx)yz)((vyz)wx))$$

by the FFH algorithm. Its daughter node that is the mother node of t_3 and t_4 then gets the label

$$(vwxyyzz) = ((vwx)yz).$$

When we permute the labels, the root label always remains the same, but the mother node of t_3 and t_4 can receive nine further possible labels.

Case 2: $T' = (((t_1, t_2), t_3), t_4)$. Then via FFH the root of T' carries the label

$$(vvwxyyzzz) = (((wxy)yz)vz),$$

its interior daughter node

$$(wxyyz) = ((wxy)yz),$$

and the interior granddaughter node

$$(wxy).$$

Reshuffling the labels will produce altogether 30 label candidates at the root, 20 at its interior daughter node, and 10 at its interior granddaughter node.

In either case and under each permutation of labels, we obtain a Steiner minimal tree (of lenghth 23), where each time the FFH algorithm selects three Steiner points from a pool of 71 sequences.

The FFH algorithm can in fact contruct further sequences from X when one prescribes a tree with more than five leaves (and thus multiple occurrences of labels), which would no longer lead to a Steiner minimal tree for X. Specifically, pick a tree with six leaves t_i $(i = 0, \ldots, 5)$ such that $T - t_0 = ((((t_1, t_2), t_3), t_4), t_5)$ and $t_0, t_1, t_2, t_3, t_4, t_5$ are labeled by z, w, x, v, y, v, respectively. Then the root of $T - t_0$ receives

$$(vvvwxyz) = (((vwx)vy)vz).$$

Permutation of the labels gives four additional sequences. The original five sequences of X together with the former 71 Steiner points and the latter five sequences form a set closed under the median operation and constitute the nodes of the graph $\lambda(5)$, see [5]. We will see below that this set exhausts all possible interior labels obtainable via FFH from any tree with leaves labeled by sequences of X.

5   Median Hull Includes FFG Hull

The median hull of a finite subset X of a rectilinear space or a hypercube is the smallest (necessarily finite) set containing X and closed under the median operation which assigns to any triplet u, v, w its median (uvw). This hull can be constructed by successively adding medians until no further new points (sequences) arise.

Now, regarding X as a subset of the k-dimensional hypercube H. One can easily check whether a particular sequence $a_1 \ldots a_k$ from H belongs to the median hull of X: a necessary and sufficient condition is that each truncation (alias projection) $a_i a_j$ of the sequence to two coordinates belongs to the corresponding truncation of X, that is, agrees at coordinates i and j with some sequence of X [6]. To see this, apply induction on the sequence length. Assume that $a_1 a_2 \ldots a_{k-1} a_k$ meets this criterion on coordinate pairs. Then by the induction hypothesis the two truncated sequences $a_1 a_2 \ldots a_{k-1}$ and $a_2 \ldots a_{k-1} a_k$ belong to the median hull of the corresponding truncation of X, and the length 2 sequence $a_1 a_k$ belongs to the truncation of X to coordinates $1, k$. Hence there exist $b_1, \ldots, b_k$ from $\{0, 1\}$ such that $a_1 a_2 \ldots a_{k-1} b_k$, $b_1 a_2 \ldots a_{k-1} a_k$, $a_1 b_2 \ldots b_{k-1} a_k$ belong to the median hull of X. Since $(a_i a_i b_i) = a_i$ for all i, we obtain $a_1 \ldots a_k$ as the median of these three sequences, as required.

In order to show that the FFH algorithm produces only labels from the median hull of X, we may therefore assume that the leaves of the leaf-rooted tree T under consideration are labeled by 0-1 sequences of length 2. If all four sequences occur at the leaves, then there is nothing left to show. So, assume that no leaf receives label 11, say. We claim that after the bottom-up phase 1N, N1 and 11 never label any interior node. Argue by induction from the bottom-upwards. Assume that the interior node under consideration has 1 as the first label coordinate. This means that more daughter nodes have 1 than 0 at the first coordinate. Then by the induction hypothesis the former daughter nodes must have 0 at the second coordinate, whereas only the latter daughter nodes may have 1 as the second coordinate. Hence the mother node necessarily receives label 10, which is fine. Finally, in the top-down phase the feasible labels 0N and N0 are specified as either 00 or 01 and 10, respectively, while NN is replaced by the label at the mother node, which is different from 11 (by employing a top-down induction argument).

We conclude that all labels erected by the FFH algorithm lie in the median hull of X.

6   FFH Hull Includes Median Hull

Now that we know that the FFH labeling stays within the median hull of the given set X of binary sequences we may ask whether it actually exhausts the latter: can every sequence in the median hull of X be obtained as a FFH label of some interior node of a suitable tree T rooted at some leaf t_0? Yes, of course. To see this, use induction on the number of median operations employed in generating a particular sequence v of the median hull of X. Assume that $v = (u_1 u_2 u_3)$ where each u_i is the FFH label of the daughter node of a leaf-root t_i of some tree T_i, independent of the label from X attached to t_i. In other words, the FFH algorithm already fixes the labels (as a 0-1 sequence) at the root of the tree $T_i' := T_i - t_i$ in the bottom-up phase. Stipulating that all rooted trees are identified with their New Hampshire codes, the rooted tree $T' := (T'_1, T'_2, T'_3)$ then does the job, no matter how we label a new leaf t attached to the root of T' in order to arrive at the required tree T. Note that this recursive construction yields a ternary tree T'. If instead a binary tree T' is desired, one simply has to modify the above definition of T' by setting $T' := (((T'_1, T'_2), (T'_1, T'_3)), (T'_2, T'_3))$. In fact, the root of T' then receives the label $(((u_1u_2)(u_1u_3))(u_2u_3)) = (u_1u_2u_3)$ as before. To verify the latter inequality, it suffices to assume u_1, u_2, u_3 are 0 or 1 and to consider the two cases where either $u_2 = u_3$ holds or $u_1 = u_2$ is different from u_3.

7   Conclusion

The rectilinear Steiner problem reduces to the Steiner problem in weighted hypercubes. The median hull, generating a universal Steiner hull, then assists in focussing the search for Steiner points. This hull is universal in the sense that, for whatever tree T is considered that interconnects a family of given sequences, one can always achieve a minimal length realization of T within this hull. In particular, the median hull also generates a Steiner hull for the restricted Steiner problem where the number of Steiner points is bounded from above by a fixed number k, thus solving a problem of [7, p. 213]. In this case one actually needs only a minor part of the median hull of the set X of sequences to be optimally interconnected. For instance, when $k = 1$, the Steiner point is always found among the $(2i + 1)$-ary medians $(x_1 \ldots x_{2i+1})$ of different sequences from X. But even if one searches for a Steiner minimal tree in the unrestricted setting, a Steiner hull may be generated from a proper subset of the median hull (when $\#X > 4$) because different nodes then necessarily carry different labels.

For non-binary sequences the FFH algorithm in its original formulation does not determine the labels uniquely, unless one specifies a tie-breaking rule. This can in fact be achieved with a priority ordering of the tree nodes ("pFFH algorithm"). With arbitrary molecular sequences the pertinent hull is the quasi-median hull [1]; one can then show that the quasi-median hull and the pFFH hull of a finite set X in a molecular sequence space coincide (Bandelt and Röhl, in preparation).

References

1. H.-J. BANDELT, P. FORSTER and A. RÖHL, Median-joining networks for inferring intraspecific phylogenies, *Mol. Biol. Evol.* 16 (1999), 37-48.

2. H.-J. BANDELT, P. FORSTER, B.C. SYKES and M.B. RICHARDS, Mitochondrial portraits of human populations using median networks, *Genetics* 141 (1995), 743-753.

3. H.-J. BANDELT and J. HEDLÍKOVÁ, Median algebras, *Discrete Math.* 45 (1983), 1-30.

4. H.-J. BANDELT, V. MACAULAY and M.B. RICHARDS, Median networks: speedy construction and greedy reduction, one simulation, and two case studies from human mtDNA, *Molec. Phylogen. Evol.* 16 (2000), 8-28.

5. H.-J. BANDELT and M. VAN DE VEL, Superextensions and the depth of median graphs, *J. Combin. Theory Ser.* A 57 (1991), 187-202.
6. G. BERGMAN, On the existence of subalgebras of direct products with prescribed *d*-fold projections, *Algebra Universalis* 7 (1977), 341-356.
7. D. CIESLIK, Steiner minimal trees, *Kluwer Acad. Publ.*, Dordrecht, 1998.
8. J. FARRIS, Methods for computing Wagner trees, *Syst. Zool.* 19 (1970), 83-92.
9. W.M. FITCH, Toward defining the course of evolution: minimal change for a specific tree topology, *Syst. Zool.* 20 (1971), 406-416.
10. D. GUSFIELD, Algorithms on strings, trees, and sequences: computer science and computational biology, *Cambridge University Press*, Cambridge, 1997.
11. J.A. HARTIGAN, Minimum mutation fits to a given tree, *Biometrics* 29 (1973), 53-65.
12. F.K. HWANG, D.S. RICHARDS and P. WINTER, The Steiner tree problem, *North-Holland*, Amsterdam, 1992.
13. D.L. SWOFFORD, G.J. OLSEN, P.J. WADDELL and D.M. HILLIS, Phylogenetic inference, In: *Molecular systematics*, Sinauer, Sunderland (D.M. Hillis, C. Moritz, and B.K. Marble, eds.), 1996.
14. M.L.J. VAN DE VEL, Theory of convex structures, *North-Holland*, Amsterdam, 1993

Data Management in Networks
(Invited Presentation Abstract)

Friedhelm Meyer auf der Heide

Heinz Nixdorf Institute and Department of Mathematics and Computer Science
University of Paderborn, Germany
`fmadh@upb.de`

We survey strategies for distributing and accessing shared objects in large parallel and distributed systems. Examples for such objects are, e.g., global variables in a parallel program, pages or cach lines in a virtual shared memory system, or shared files in a distributed system, for example in a distributed data server. We focus on strategies for distributing, accessing, and (consistently) updating such objects, which are provably efficient with respect to various cost measures. Our focus is on presenting strategies that are tailored to situations where the bandwidth of the network is the bottleneck, so we aim to organise the shared objects in such a way that congestion is minimized.

First we present schemes that are efficient w.r.t. information about read- and write-frequencies. The main part of the talk will deal with online, dynamic, redundant data mamnagement strategies that have good competitive ratio, i.e., that are efficient compared to an optimal dynamic offline strategy that is constructed using full knowledge of the dynamic access pattern. Especially the case of memory restrictions in the processors will be discussed.

A. Brandstädt and V.B. Le (Eds.): WG 2001, LNCS 2204, p. 8, 2001.
© Springer-Verlag Berlin Heidelberg 2001

Edge-Isoperimetric Problems
for Cartesian Powers of Regular Graphs [*]

Sergei L. Bezrukov[1] and Robert Elsässer[2]

[1] Departament of Math. and Comp. Sci.
University of Wisconsin-Superior, USA
[2] Departament of Math. and Comp. Sci.
University of Paderborn, Germany

Abstract. We consider an edge-isoperimetric problem (EIP) on the cartesian powers of graphs. One of our objectives is to extend the list of graphs for whose cartesian powers the lexicographic order provides nested solutions for the EIP. We present several new classes of such graphs that include as special cases all presently known graphs with this property. Our new results are applied to derive best possible edge-isoperimetric inequalities for the cartesian powers of arbitrary regular, resp. regular bipartite, graphs with a high density.

1 Introduction

Let $G = (V_G, E_G)$ be a graph and $A, B \subseteq V_G$. Denote

$$I_G(A, B) = \{(u, v) \in E_G \mid u \in A, \ v \in B\},$$
$$I_G(A) = I_G(A, A),$$
$$I_G(m) = \max_{A \subseteq V_G, \ |A| = m} |I_G(A)|.$$

We will often omit the index G. Our subject is the following version of the *edge-isoperimetric problem* (EIP): for a fixed m, $1 \leq m \leq |V_G|$, find a set $A \subseteq V_G$ such that $|A| = m$ and $|I(A)| = I(m)$. We call such a set A *optimal*. This problem is known to be NP-complete in general and has many applications in various fields of knowledge, for instance see a survey [4].

We emphasize on graphs representable as cartesian products of other graphs. Given two graphs $G = (V_G, E_G)$ and $H = (V_H, E_H)$, their *cartesian product* is defined as a graph $G \times H$ with the vertex-set $V_G \times V_H$ whose two vertices (x, y) and (u, v) are adjacent iff either $x = u$ and $(y, v) \in E_H$, or $(x, u) \in E_G$ and $y = v$. The graph $G^n = G \times G \times \cdots \times G$ (n times) is called the nth *cartesian power* of G.

The EIP for the cartesian powers of a graph G have been well studied for different graph classes. To present these results we need the definition of the

---

[*] This work was partially supported by the German Research Association (DFG) within the SFB 376 "Massive Parallelität: Algorithmen, Entwurfsmethoden, Anwendungen"

lexicographic order on a set of n-tuples with integral entries. For that we say that $(x_1, \ldots, x_n)$ is greater than $(y_1, \ldots, y_n)$ iff there exists an index i, $1 \leq i \leq n$, such that $x_j = y_j$ for $1 \leq j < i$ and $x_i > y_i$. Assuming that the vertices of a graph G are totally ordered, denote by $\mathcal{F}^n(m)$ the collection of the first m vertices of G^n in the lexicographic order. We call such a set an *initial segment*. Let G be the cartesian product of n complete graphs. Then the following Theorem holds.

Theorem 1. (Harper [7] for $p_1 = \ldots = p_n = 2$, Lindsey [8] in general) *Let $n \geq 2$ and $p_1 \geq p_2 \geq \cdots \geq p_n$. Then $|I(\mathcal{F}^n(m))| \geq |I(A)|$ for any $A \subseteq K_{p_1} \times \cdots \times K_{p_n}$, $|A| = m$, and any $m \geq 1$.*

In other words, what is more commonly said, for the products of cliques there exists a *nested structure of solutions*, i.e., a family of optimal subsets $A_1, A_2, \ldots, A_m, \ldots$ such that $|A_m| = m$ for any m and $A_1 \subseteq A_2 \subseteq \cdots \subseteq A_m \subseteq \cdots$. We call an order providing the nested structure of solutions an *optimal order* (if such exists). Existence of a nested structure of solutions provides as an immediate consequence solutions to such important problems as cutwidth and wirelength, and often allows to construct good k-partitioning of graphs and their embedding to some other graphs [4].

The classical results presented in Theorem 1 can be extended in various directions. For instance, taking an (infinite) path instead of a clique leads to a grid. In this case the EIP also has nested structure of solutions [1,6], although the optimal order $\mathcal{G}$ is not the lexicographic one. It is further shown in [4] that the order $\mathcal{G}$ is also optimal for the products of arbitrary trees. For the definition of this order and further details, readers are referred to [1,4,6]. It is also known that no optimal order exists for the powers of cycles of length 5 and larger.

The order $\mathcal{G}$ and the lexicographic order were for a long time the only examples of optimal orders with respect to the EIP. However, as shown in [4], the order $\mathcal{G}$ is optimal for products of trees only. Therefore, two natural questions arise: *(i) for what other graphs is the lexicographic order on the power optimal with respect to the edge isoperimetric problems; and (ii) what other optimal orders can one expect?* In [5], we studied the cartesian powers of the Petersen graph and presented a new optimal order for them. However, it looks like this order is optimal for a rather narrow class of graphs. This, in the light of the results presented above, motivates to study further the lexicographic order.

Following this direction, Ahlswede and Cai proved in [2] that the lexicographic order is optimal for the products of complete bipartite graphs $K_{p,p}$. They also showed in [2] that this order yields the so-called *local-global principle*. That is, if the lexicographic order is optimal for G^2 then it is optimal for G^n for any $n \geq 3$ (see [4] for the local-global principles for some other orders). The main difficulty in applying this powerful result is to establish the optimality of the lexicographic order for the second cartesian power of a graph. Unfortunately, no efficient methods have been developed and the known approaches are very specific. In our paper we present new ideas in this direction and apply them for three new graph families. These new graphs include, in particular, all presently known graphs such that the lexicographic order is optimal on their cartesian powers.

The paper is organized as follows. In Section 2 we introduce new graph classes $H_{p,i}$ and $B_{p,i}$ and present some basic definitions and lemmas which are used in the sequel. A motivation for studying these graph classes is provided by Lemma 1, which establishes best-possible isoperimetric inequalities for powers of arbitrary regular, resp. regular bipartite, graphs. In Section 3 we prove the optimality of the lexicographic order for the cartesian powers of $H_{p,i}$. The approach used in this proof also works well for the powers of complete bipartite graphs and complete t-partite graphs and generalize the result of Ahlswede and Cai [2] concerning the complete bipartite graphs. Concluding remarks and open questions in Section 5 complete the paper.

2   Definitions and Auxiliary Lemmas

Here we introduce three new families of graphs, for whose cartesian powers the lexicographic order is optimal. The graph $H_{p,i}$, can be constructed from K_{2p} by partitioning its vertex set into two parts V' and V'' with $|V'| = |V''| = p$ and removing i disjoint perfect matchings between these sets. More exactly,

$$V_{H_{p,i}} = V' \cup V'', \quad V' = \{0, 1, \ldots, p-1\}, \quad V'' = \{p, p+1, \ldots, 2p-1\},$$

$$E_{H_{p,i}} = \{(u,v) \mid u, v \in V', \text{ or } u, v \in V''\} \cup$$

$$\{(u,v) \mid u \in V', \ v \in V'', \ u+v \neq (2p-j) \bmod 2p, \ j = 1, \ldots, i\}.$$

An example of $H_{3,1}$ is shown in Fig. 1(a). The vertices belonging to V' and V'' are put into ovals. Note that various ways of deleting i perfect matchings from K_{2p} may lead to non-isomorphic graphs. However, we do not worry much about that, because the function $I(m)$ for all these graphs is the same, as it follows from the proof of Lemma 3. This function is mostly important for our approach rather than the structural properties of the underlying graphs. Another graph we study in our paper, $B_{p,i}$, is the one obtained from $K_{2p,2p}$ for an even p by removing i disjoint perfect matchings (cf. Fig. 1(b) and the conclusion for a precise definition). The lexicographic order is also optimal for $B_{p,i}^n$ for any $n \geq 2$ and $i \leq \frac{p}{4}$. This can be proved by following the lines of the proof of Theorem 2 in the next section. The interest to these graphs is provided by the following lemma, whose proof is presented at the end of this section after establishing some auxiliary results.

Lemma 1.

a. Let G be a regular graph with $|V_G| = V_{H_{p,i}}$ and $\deg(G) = \deg(H_{p,i})$, that admits an optimal order in the EIP. Then $I_{H_{p,i}^n}(m) \geq I_{G^n}(m)$ for any $n \geq 1$ and $m = 1, \ldots, (2p)^n$.

b. Let G be a regular bipartite graph with both independent sets of size p (p even) and $\deg(G) = \deg(B_{p,i})$, that admits an optimal order in the EIP. Then $I_{B_{p,i}^n}(m) \geq I_{G^n}(m)$ for any $n \geq 1$ and $m = 1, \ldots, (2p)^n$.

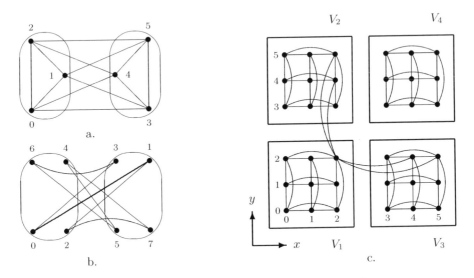

Fig. 1. Representations of $H_{3,1}$ (a), $B_{4,2}$ (b), and $H_{3,1}^2$ (c)

This provides best-possible isoperimetric inequalities for powers of regular, resp. regular bipartite, graphs. The class of regular graphs is important, because in the light of (1) the isoperimetric problem for regular graphs is equivalent to the problem of minimization of $|\theta(A)|$, where $\theta(A) = I(A, \overline{A})$ and $\overline{A} = V_G \setminus A$. The function $\theta(A)$ appears in most applications of the edge-isoperimetric inequalities [4]. For a non-regular graph G a solution to the EIP on G provides a lower bound for this function following from the identity $I(A) \cup I(\overline{A}) \cup \theta(A) = E_G$. Moreover, we conjecture that the regularity of G is important for the lexicographic order to be optimal for G^n (cf. the conclusion).

The third graph class we consider here is given by the complete t-partite graphs $K_{p,p,...,p}$. It can be also shown that the lexicographic order is optimal for its cartesian powers too. This extends a result of [2] for complete bipartite graphs. We also develop a new method for solving the EIP for the product of two graphs. This method works well for all graphs mentioned above.

We will need the following two auxiliary propositions for the sequel.

Lemma 2. Let $G = (V_G, E_G)$ be a regular graph and $A \subseteq V_G$. Then A is an optimal set iff $\overline{A} = V_G \setminus A$ is optimal.

Proof. Let $\deg(v) = k$ for any $v \in V_G$. Then

$$2|I(A)| + |I(A, \overline{A})| = k|A| \tag{1}$$
$$2|I(\overline{A})| + |I(A, \overline{A})| = k|\overline{A}| = k(|V_G| - |A|)$$

The above equations imply $2|I(\overline{A})| = k(|V_G| - 2|A|) + 2|I(A)|$. Thus, $|I(A)|$ and $|I(\overline{A})|$ differ on an additive constant for any $A \subseteq V_G$. Hence, if one of these sets is optimal, then the other one is optimal too.   □

Denote $V = V_{H_{p,i}}$ for brevity and let $\mathcal{F}(m) = \{0, 1, \ldots, m-1\}$ for $m = 0, 1, \ldots, 2p$.

Lemma 3. $|I(\mathcal{F}(m))| \geq |I(A)|$ *for any* $A \subseteq V$ *with* $|A| = m$.

Proof. The assertion is obvious if $m \leq p$, because $H_{p,i}$ contains K_p, whose vertices are numbered first. Now if $m > p$ then the assertion follows from Lemma 2. $\qquad\square$

Let $n \geq 2$ and $A \subseteq V^n$. For $1 \leq i \leq n$ and $\mathbf{x} = (x_1, \ldots, x_n) \in V^n$ denote

$$A_i(\mathbf{x}) = \{y \mid (x_1, \ldots, x_{i-1}, y, x_{i+1}, \ldots, x_n) \in A\}.$$

The set $A \subseteq V^n$ is called *compressed* if $A_i(\mathbf{x}) = \{0, 1, \ldots, |A_i(\mathbf{x})|\}$ for any $i = 1, \ldots, n$ and any $\mathbf{x} \in A$. Standard arguments (see [2,4]) provide that if a graph G admits an optimal order, then for any $m \geq 1$ there exist compressed optimal sets in V_{G^n}. This observation is a key point for the proof of Lemma 1.

Proof of Lemma 1.
For a graph G from the statement and $j = 1, \ldots, |V_G|$, let us denote

$$\Delta_G(j) = I_G(j) - I_G(j-1) \quad \text{with} \quad \Delta_G(0) = 0.$$

Taking into account that G admits an optimal order, denote the vertices of G by $0, 1, \ldots, 2p-1$ according to their positions in this order. This provides $V_G = V$.

We prove the first proposition of the lemma by induction on n. For $n = 1$ we can use the same arguments as in the proof of Lemma 3. Note that $I_G(m) = \sum_{j=0}^{m-1} \Delta_G(j)$ for any graph G that admits an optimal order.

Now for $n \geq 2$ let $A \subseteq V_{G^n}$ with $|A| = m$ be an optimal compressed set. Then A can be viewed as a subset of V^n. Note (see e.g. [2] for (2)) that for any compressed set $A \subseteq V_{G^n} = V^n$ one has

$$|I_{G^n}(A)| = \sum_{(x_1, \ldots, x_n) \in A} \sum_{i=1}^{n} \Delta_G(x_i) \tag{2}$$

$$= \sum_{j=0}^{p-1} \left[\sum_{(j, x_2, \ldots, x_n) \in A} \sum_{i=2}^{n} \Delta_G(x_i) \right] + \sum_{(x_1, \ldots, x_n) \in A} \Delta_G(x_1). \tag{3}$$

Similar formulas are also valid for compressed subsets $B \subseteq V_{H_{p,i}^n} = V^n$. Note that A and B are the subsets of the same vertex set. Set $B = A$. Now for any fixed j in (3) the induction hypothesis implies

$$\sum_{(j, x_2, \ldots, x_n) \in A} \sum_{i=2}^{n} \Delta_G(x_i) \leq \sum_{(j, x_2, \ldots, x_n) \in B} \sum_{i=2}^{n} \Delta_{H_{p,i}}(x_i). \tag{4}$$

Furthermore, for a fixed $\mathbf{x} \in V$ and a compressed set $A = B$ (with $A_1(\mathbf{x}) = B_1(\mathbf{x})$) Lemma 3 implies

$$\sum_{x_1 \in A_1(\mathbf{x})} \Delta_G(x_1) = I_G(|A_1(\mathbf{x})|) \leq I_{H_{p,i}}(|B_1(\mathbf{x})|) = \sum_{x_1 \in B_1(\mathbf{x})} \Delta_{H_{p,i}}(x_1) \tag{5}$$

Summing up (4) for $j = 0, \ldots, p - 1$ and (5) for $x_2, \ldots, x_n$ such that $\mathbf{x} = (x_1, x_2, \ldots, x_n) \in A$ and using (3) one gets

$$I_{G^n}(m) = |I_{G^n}(A)| \leq |I_{H_{p,i}^n}(B)| \leq \max_{B \subseteq V^n, \, |B| = m} |I_{H_{p,i}^n}(B)| = I_{H_{p,i}^n}(m),$$

and the induction goes through.

The second proposition can be proved similarly by taking into account that an analog of Lemma 3 is valid for the graph $B_{p,i}$.   □

By using a similar argument the following result can be proved, that can be considered as a best-possible edge-isoperimetric inequality for the cartesian powers of general graphs.

Corollary 1. *Let G be a graph with $|V_G| = p$ that admits nested solutions in the EIP. Then $I_{K_p^n} \geq I_{G^n}(m)$ for any $n \geq 1$ and $m = 1, \ldots, p^n$.*

3   EIP for the Cartesian Powers of $H_{p,i}$

One of the main results of the paper is formulated in the following theorem.

Theorem 2. *If $p \geq 3$ and $1 \leq i \leq p - i$ then $|I(\mathcal{F}^2(|A|))| \geq |I(A)|$ for any $A \subseteq V \times V$.*

Proof. Let $A \subseteq V \times V$ be an optimal compressed set non-isomorphic to $\mathcal{F}^2(|A|)$ (if such exists). Denote

$$V_1 = \{(x, y) \mid 0 \leq x < p, \ 0 \leq y < p\},$$
$$V_2 = \{(x, y) \mid 0 \leq x < p, \ p \leq y < 2p\},$$
$$V_3 = \{(x, y) \mid p \leq x < 2p, \ 0 \leq y < p\},$$
$$V_4 = \{(x, y) \mid p \leq x < 2p, \ p \leq y < 2p\},$$

and let $A_i = A \cap V_i$, $i = 1, 2, 3, 4$. Obviously, for any i the subgraph induced by the vertex set V_i is isomorphic to $K_p \times K_p$. These sets are shown in Fig. 1(c) for the graph $H_{3,1}^2$. The graphs induced by vertices located in one row (or in one column) in this figure are isomorphic to $H_{p,i}$. We will often use the fact that any vertex of V_2 (and V_3) is connected with $p - i$ vertices of V_1 and versa. In Fig. 1(c) such connections are shown for the vertex $(2, 2) \in V_1$.

Case 1. Assume $A \subseteq V_1 \cup V_2$. Since A is compressed, then $|A_1| \geq |A_2|$. Thus, due to Theorem 1 we can assume that A_1 and A_2 are initial segments of V_1 and V_2 respectively (see Fig. 2). Let $|A_1| = k_1 p + \delta_1$ and $|A_2| = k_2 p + \delta_2$ for some integer $k_1, k_2, \delta_1, \delta_2$ with $0 \leq k_1, k_2 \leq p - 1$ and $0 \leq \delta_1, \delta_2 < p$. If $|A_1| = p^2$, we assume $k_1 = p - 1$ and $\delta_1 = p$. Now $|A_1| \geq |A_2|$ implies $k_1 \geq k_2$. Moreover, if $k_1 = k_2$ then $\delta_1 = p$ or $\delta_2 = 0$ due to the compression, hence $A = \mathcal{F}^2(|A|)$. So we can assume $k_1 > k_2$. *Case 1a.* Assume $k_1 - k_2$ is even (and positive). First consider the case $\delta_1 + \delta_2 \leq p$. We transform A into $B = \mathcal{F}^2(|A|)$ in two steps. First, we construct a set C by moving $(k_2, y) \in A_2$ for $p \leq y < p + \delta_2$ to $(k_1, y + \delta_1 - p) \in V_1 \setminus A_1$. It is easily shown that

$$|I(C)| - |I(A)| = \delta_2(\delta_1 + k_1 - k_2) - \delta_2(p - i). \tag{6}$$

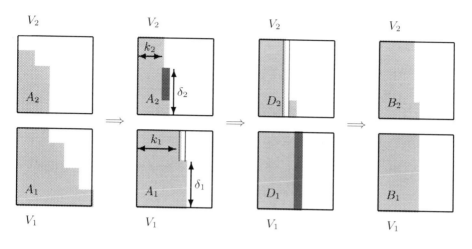

Fig. 2. Compression of A in case 1 and transformation for $\delta_1 + \delta_2 > p$ in case 1a.

Denote $C_i = C \cap V_i$, $D_i = D \cap V_i$ and $B_i = B \cap V_i$, for $i = 1, 2, 3, 4$. Now we move $(x, y) \in C_1$ with $\frac{k_1 + k_2}{2} \le x < k_1$, $0 \le y < p$ to $(x - (k_1 - k_2)/2, y + p) \in V_2 \setminus C_2$ and $(x, y) \in C_1$ with $x = k_1$, $0 \le y < \delta_1 + \delta_2$ to $((k_1 + k_1)/2, y) \in V_1 \setminus C_1$. For the resulting set D one has

$$|I(D)| - |I(C)| = \frac{p(k_1 - k_2)}{2}\left(p - i - \frac{k_1 - k_2}{2}\right) - (\delta_1 + \delta_2)\frac{k_1 - k_2}{2}. \qquad (7)$$

Note that $B_2 = D_2$ and $|B_1| = |D_1|$. Since the subgraphs induced by V_1 and V_2 are isomorphic to $K_2 \times K_2$, then $|I(D)| = |I(B)|$. Therefore, from (6) and (7) for $\Delta = |I(B)| - |I(A)|$ one gets

$$\Delta = \frac{p(k_1 - k_2)}{2}\left(p - i - \frac{k_1 - k_2}{2}\right) + \delta_2\left(\delta_1 + \frac{k_1 - k_2}{2}\right) - \delta_1\frac{k_1 - k_2}{2} - \delta_2(p - i). \qquad (8)$$

Since $i \le p - i$, then $p/2 \ge i$. Furthermore, $k_2 \ge 0$ and $k_1 \le p - 1$ imply $k_1 - k_2 \le p - 1$. These inequalities imply $p - i \ge p/2 > (p - 1)/2 \ge (k_1 - k_2)/2$. Therefore, $p - i - (k_1 - k_2)/2 \ge 1$, and from (8) it follows

$$\Delta \ge \frac{(p - \delta_1)(k_1 - k_2)}{2} - \delta_2 + \delta_1\delta_2. \qquad (9)$$

Since $p \ge \delta_1$ and $\delta_2\delta_1 \ge \delta_2$ for $\delta_1 \ne 0$, then $\Delta \ge 0$. If $\delta_1 = 0$, then taking into account $k_1 - k_2 \ge 2$ and (9), we get

$$\Delta \ge \frac{p(k_1 - k_2)}{2} - \delta_2 \ge p - \delta_2 > 0.$$

Now consider the case $\delta_1 + \delta_2 > p$. Then $\delta_2 > 0$. We transform A into $B = \mathcal{F}^2(|A|)$ in three steps. First construct the set C by moving $p - \delta_1$ vertices of the form $(k_2, y) \in A_2$ for $\delta_1 + \delta_2 \le y < \delta_2 + p$ to $(k_1, y - \delta_2) \in V_1 \setminus A_1$. One has

$$|I(C)| - |I(A)| = (p - \delta_1)(\delta_1 + k_1 - k_2 - (\delta_1 + \delta_2 - p) - (p - i)). \qquad (10)$$

Next, we move the vertices $(k_2, y) \in C_2$ for $p \leq y < \delta_1 + \delta_2$ to $((k_1 + k_2)/2, y)$. For the resulting set D one has $|I(D)| = |I(C)|$. Finally, we move $(x, y) \in D_1$ with $\frac{k_1 + k_2}{2} < x \leq k_1$ and $0 \leq y < p$ to $(x + 1 - (k_1 - k_2)/2, y + p) \in V_2 \setminus D_2$. The resulting set is B. One has

$$|I(B)| - |I(D)| = \frac{p(k_1 - k_2)}{2}\left(p - i - \frac{k_1 - k_2}{2} - 1\right) + (\delta_1 + \delta_2 - p)\frac{k_1 - k_2}{2}. \quad (11)$$

It follows from (10) and (11) that for $\Delta = |I(B)| - |I(A)|$ one has

$$\Delta = \frac{p(k_1 - k_2)}{2}\left(p - i - \frac{k_1 - k_2}{2}\right) - \delta_1 \frac{k_1 - k_2}{2}$$
$$+ (p - \delta_1)(\delta_1 - \delta_2 + (p - \delta_1) - (p - i)) + \delta_2 \frac{k_1 - k_2}{2}$$
$$= \frac{p(k_1 - k_2)}{2}\left(p - i - \frac{k_1 - k_2}{2}\right) - \delta_1 \frac{k_1 - k_2}{2}$$
$$+ (p - \delta_1)(i - \delta_2) + \delta_2 \frac{k_1 - k_2}{2}. \quad (12)$$

Similarly as above, $p - 1 - (k_1 - k_2)/2 \geq 1$. Now if $i \geq \delta_2$, then

$$\frac{p(k_1 - k_2)}{2}\left(p - i - \frac{k_1 - k_2}{2}\right) - \delta_1 \frac{k_1 - k_2}{2} \geq \frac{(p - \delta_1)(k_1 - k_2)}{2} \geq 0, \quad (13)$$

which implies $\Delta > 0$.

Note that $p - \delta_1 < \delta_2$. Thus, if $i < \delta_2$ and $(k_1 - k_2)/2 > \delta_2 - i$, then (12) and (13) imply

$$\Delta > \frac{p(k_1 - k_2)}{2}\left(p - i - \frac{k_1 - k_2}{2}\right) - \delta_1 \frac{k_1 - k_2}{2} - \delta_2(\delta_2 - i) + \delta_2 \frac{k_1 - k_2}{2}$$
$$\geq \frac{p(k_1 - k_2)}{2}\left(p - i - \frac{k_1 - k_2}{2}\right) - \delta_1 \frac{k_1 - k_2}{2} \geq 0.$$

Now if $i < \delta_2$ and $\delta_2 - i \geq (k_1 - k_2)/2$, then (12) implies

$$\Delta = \frac{p(k_1 - k_2)}{2}\left(p - i - \frac{k_1 - k_2}{2}\right) + \delta_1 \left(\delta_2 - i - \frac{k_1 - k_2}{2}\right) - p(\delta_2 - i) + \delta_2 \frac{k_1 - k_2}{2}$$
$$= \left[\frac{p(k_1 - k_2)}{2}\left(p - i - \frac{k_1 - k_2}{2}\right) - \delta_2 \left(p - i - \frac{k_1 - k_2}{2}\right)\right]$$
$$+ \delta_1 \left(\delta_2 - i - \frac{k_1 - k_2}{2}\right) + (p - \delta_2)i. \quad (14)$$

Since $p > \delta_2$ and $k_1 - k_2 \geq 2$, then the last term of (14) and the term in brackets are positive. Therefore, $\Delta > 0$.

Case 1b. In this case we assume that $k_1 - k_2 - 1$ is even. The proof of this case is similar to case 1a. and it is omitted because of space limitations.

Case 2. Assume $A \subseteq V_1 \cup V_2 \cup V_3$ and $A_2 \neq \emptyset$ and $A_3 \neq \emptyset$. Without loss of generality we can assume that A_2 is an initial segment (in "columns" of V_2) and so is A_3 (in "rows" of V_3) (cf. Fig. 3). Indeed, after such a transformation the number of inner edges in A_2 and A_3 cannot decrease by Theorem 1 and

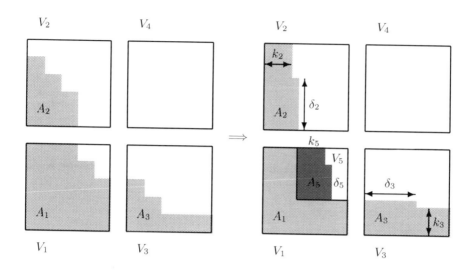

Fig. 3. Compression of A in case 2

the number of edges between A_1 and A_2 (also between A_1 and A_3) remains the same. Let $|A_2| = k_2 p + \delta_2$ and $|A_3| = k_3 p + \delta_3$ with $0 \le \delta_2, \delta_3 < p$. We can assume $k_2 \ge k_3$. Now we consider an $(p - k_2) \times (p - k_3)$ rectangle $V_5 \subseteq V_1$ with $V_5 = \{(x, y) \in V_1 \mid k_2 \le x < p, \ k_3 \le y < p\}$ and replace $A_5 = A \cap V_5$ with the initial segment in V_5 of the same size. This initial segment is shown by the dark gray area in Fig. 3. By Theorem 1 (applied for V_5), the resulting set will be optimal too. This is true because the number of inner edges within A_5 will not decrease, and the number of edges between A_5 and $V_1 \setminus V_5$ remains the same. Let $|A_5| = (p - k_3) k_5 + \delta_5$ with $0 \le \delta_5 < p - k_3$.

Without loss of generality we can assume $\delta_2 = 0$. Indeed, otherwise if $\delta_3 > p - \delta_2$ then we move $(x, k_3) \in A_3$ for $\delta_2 + \delta_3 \le x < p + \delta_3$ to $(k_2, x - \delta_2 + p) \in V_2 \setminus A_2$. For the resulting set C one has

$$|I(C)| - |I(A)| = (p - \delta_2)(\delta_2 - ((\delta_2 + \delta_3 - p) + k_2 - k_3)) > 0.$$

If $0 < \delta_3 \le p - \delta_2$ then we move $(x, k_3) \in A_3$ for $p \le x < p + \delta_3$ to $(k_2, x + \delta_2) \in V_2 \setminus A_2$. For the resulting set C one has

$$|I(C)| - |I(A)| = \delta_3(\delta_2 + k_2 - k_3) > 0.$$

If $\delta_3 = 0$ then $A_3 \ne \emptyset$ implies $k_3 > 0$. We move $(x, k_3 - 1) \in A_3$ for $p + \delta_2 \le x < 2p$ to $(k_2, x) \in V_2 \setminus A_2$. For the resulting set C one has

$$|I(C)| - |I(A)| = (p - \delta_2)(k_2 - k_3 + 1) > 0.$$

In all cases we get a contradiction with the optimality of A.

Case 2a. Assume $k_5 > 0$. If $k_5 > k_3$ or $k_5 = k_3$ and $\delta_3 = 0$, then $|A| \le 2p^2$. We move $(x, y) \in A_3$ to $(y + k_2, x) \in V_2 \setminus A_2$. This results in a set $C \subseteq V_1 \cup V_2$

with $|I(C)| \geq |I(A)|$, and we apply the arguments of case 1 to further transform C to $\mathcal{F}^2(|A|)$.

If $1 \leq k_5 < k_3$ or $1 \leq k_5 = k_3$ and $\delta_3 > 0$, then we move $(x,y) \in A_3$ for $p \leq x < 2p$ and $0 \leq y < k_5$ to $(y + k_2, x) \in V_2 \setminus A_2$. For the resulting set C one has
$$|I(C)| - |I(A)| = k_5 p(k_2 - (k_3 - k_5)) - \delta_3 k_5 \geq k_5(p - \delta_3) \geq 0.$$
Moreover, $|C_3| < |A_3|$, and C satisfies the assumptions of case 2. Applying similar arguments to C we obtain a set D either with $D_3 = \emptyset$ or with $k_5 = 0$. In the first case we apply to D the arguments of case 1 and transform it to $\mathcal{F}^2(|A|)$.

Case 2b. Assume $k_5 = 0$ and $k_3 > 0$. We show that in this case A is not optimal. We start with interchanging $(x,0) \in A_3$ for $p + \delta_3 \leq x < 2p$ and $(x,k_3) \in V_3 \setminus A_3$. This transformation does not affect the number of inner edges in a set. First assume $k_2 < p-1$. If $\delta_5 > 0$ we move $(k_2,y) \in A_1$ for $k_3 < y \leq k_3 + \delta_5$

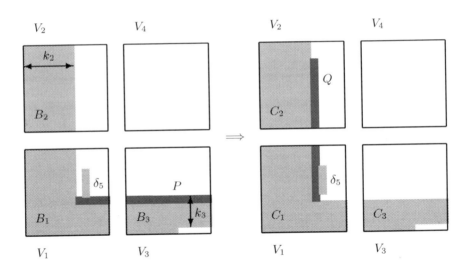

Fig. 4. Transformation of A into B and C in case 2b

to $(k_2 + 1, y)$. The resulting set B has the same number of inner edges. If $\delta_5 = 0$ we set $B = A$. The set B is shown in the left part of Fig. 4. Denote
$$P = \{(x, k_3 - \epsilon) \in A \mid k_2 \leq x < 2p\},$$
$$Q = \{(k_2, y) \notin A \mid k_3 - \epsilon \leq y < 2p + k_3 - k_2 - \epsilon\}$$
where $\epsilon = 0$ if $\delta_3 > 0$ and $\epsilon = 1$ if $\delta_3 = 0$. Now $|P| = |Q| = 2p - k_2$ and we replace P with Q (cf. Fig 4). We show that for the obtained set C one has $|I(C)| > |I(B)|$. Indeed, if $\delta_3 > 0$ then for $\Delta = |I(C)| - |I(B)|$ one has
$$\Delta = [|I(Q)| + |I(Q,A)|] - [|I(P)| + |I(P, A \setminus P)|]$$
$$= \left[\binom{p}{2} - \binom{k_3}{2} + (2p - k_2)k_2 + \binom{p + k_3 - k_2}{2} + (p + k_3 - k_2)(p - i) \right]$$

$$-\left[(p-k_2)(k_3+k_2) + \binom{p-k_2}{2} + p(p-i) + \binom{p}{2} + k_3\delta_3 + (p-\delta_3)(k_3-1)\right]$$
$$= pk_2 + (k_3-k_2)(p-i) - pk_3 - p + \delta_3 = i(k_2-k_3) + (p-\delta_3) > 0.$$

Similarly, if $\delta_3 = 0$ then $\Delta = i(k_2 - k_3 + 1) > 0$.

Now assume $k_2 = p - 1$. In this case we construct the set C by replacing P with the set $R = \{(p-1, y) \notin A \mid k_3 + \delta_5 - \epsilon < y < p + k_3 + \delta_5 - \epsilon\}$ with ϵ defined above. For $\Delta = |I(C)| - |I(B)|$ one has

$$\Delta = [|I(R)| + |I(R, A)|] - [|I(P)| + |I(P, A \setminus P)|]$$
$$= \binom{p-\delta_5-k_3-1}{2} + \binom{k_3+1+\delta_5}{2}$$
$$+ p(p-1) + (p-\delta_5-k_3-1)(k_3+1+\delta_5) + (k_3+1+\delta_5)(p-i)$$
$$- \left[\binom{p}{2} + k_3\delta_3 + (p-\delta_3)(k_3-1) + p(p-i)\right]$$
$$= 2(p-\delta_5-k_3-1)(k_3+1+\delta_5+i/2) + (p-\delta_3) + p\delta_5 \geq p - \delta_3 > 0.$$

In all cases we get a contradiction with the optimality of A.

Case 2c. We assume that $k_5 = 0$, $k_3 = 0$ and $\delta_3 > 0$. Similar arguments to the case 2b. can be used to show contradiction to the optimality in this case. We omit the proof because of space limitations.

Case 3. Assume $V_1 \subset A$. Then $\overline{A} = V \setminus A \subseteq V_2 \cup V_3 \cup V_4$, and we can apply the arguments of case 1 and case 2 to $\overline{A}$ and transform it to $\mathcal{F}^2(|\overline{A}|)$. Since, by Lemma 3, both A and $\overline{A}$ are optimal sets then the proof is completed.   □

Due to the local-global principle for the lexicographic order [2], we have the following result:

Corollary 2. *If $1 \leq i \leq p - i$, then $|I(\mathcal{F}^n(|A|))| \geq |I(A)|$ for any $A \subset V^n$ and $n \geq 1$.*

The next theorem shows that the assumption concerning i is important for the existence of an optimal order.

Theorem 3. *There exist no optimal order for $H_{p,i} \times H_{p,i}$ if $p \geq 3$ and $p - i < i < p - 1$.*

We omit the proof of this Theorem due to space limitations. Note, that Theorem 3 is not true for $i = p - 1$. In this case $H_{p,i}$ (resp. $H^2_{p,i}$) is isomorphic to $K_p \times K_2$ (resp. $K_p \times K_p \times K_2 \times K_2$) and the existence of an optimal order follows from Theorem 1. Furthermore, Theorem 3 is also not true for $i = p$. In this case $H_{p,i}$ (resp. $H^2_{p,i}$) is non-connected and is isomorphic to two disjoint copies of K_p (resp. to four disjoint copies of $K_p \times K_p$). In this case the lexicographic order is also optimal as it follows from [3].

4   Concluding Remarks

Results similar to Theorems 2 and 3 can be obtained for the graph $B_{p,i} = (V_{B_{p,i}}, E_{B_{p,i}})$, where

$$V_{B_{p,i}} = V' \cup V'', \quad V' = \{0, 2, \ldots, 2p - 2\}, \ V'' = \{1, 3, \ldots, 2p - 1\},$$
$$E_{B_{p,i}} = \{(u, v) \mid u \in V', \ v \in V'', \quad u + v \neq (2p - j) \bmod 2p, \ j = 1, \ldots, i\}.$$

It is easily shown that for any m the set $\{0, 1, \ldots, m - 1\}$ is a solution to the EIP on $B_{p,i}$. With this numbering the lexicographic order on $V = V_{B_{p,i}} \times V_{B_{p,i}}$ is well-defined.

Theorem 4. *Let $i \leq p/2 - i$ and let $p \geq 4$ be even. Then $|I(\mathcal{F}^2(|A|))| \geq |I(A)|$ for any $A \subseteq V \times V$.*

The local-global principle [2] implies that a similar result holds for any cartesian power of $B_{p,i}$. We also have some preliminary results concerning the EIP for the cartesian powers of $K_{p,\ldots,p}$ (t parts) with up to two t-factors deleted. In these cases the lexicographic order is also optimal.

References

1. Ahlswede R., Bezrukov S.L., *Edge Isoperimetric Theorems for Integer Point Arrays*, Appl. Math. Lett. **8** (1995), No. 2, 75–80.
2. Ahlswede R., Cai N., *General Edge-isoperimetric Inequalities, Part II: a Local-Global Principle for Lexicographic Solution*, Europ. J. Combin. **18** (1997), 479–489.
3. Bezrukov S.L., *Encoding of analog signals for discrete binary channel*, in: Proc. Int. Conf. Algebraic and Combinatorial Coding Theory, Varna 1988, 12–16.
4. Bezrukov S.L., *Edge Isoperimetric Problems on Graphs*, in: Graph Theory and Combinatorial Biology, Bolyai Soc. Math. Stud. **7**, L. Lovasz, A. Gyarfas, G.O.H. Katona, A. Recski, L. Szekely eds., Budapest 1999, 157–197.
5. Bezrukov S.L., Das S., Elsässer R., *An Edge-Isoperimetric Problem for Powers of the Petersen Graph*, Annals of Combinatorics, **4** (2000), 153–169.
6. Bollobás B., Leader I., *Edge-isoperimetric Inequalities in the Grid*, Combinatorica **11** (1991), 299–314.
7. Harper L.H., *Optimal Assignment of Numbers to Vertices*, J. Sos. Ind. Appl. Math. **12** (1964), 131–135.
8. Lindsey II J.H., *Assignment of Numbers to Vertices*, Amer. Math. Monthly **7** (1964), 508–516.

Approximate Constrained Bipartite Edge Coloring[*]

Ioannis Caragiannis[1], Afonso Ferreira[2], Christos Kaklamanis[1],
Stéphane Pérennes[2], Pino Persiano[3], and Hervé Rivano[2]

[1] Computer Technology Institute and
Dept. of Computer Engineering and Informatics,
University of Patras, 26500 Rio, Greece
{caragian,kakl}@cti.gr
[2] MASCOTTE Project, INRIA Sophia Antipolis,
2004 route des Lucioles, B.P. 93,
06902 Sophia Antipolis Cedex, France
{Afonso.Ferreira,Stephane.Perennes,Herve.Rivano}@sophia.inria.fr
[3] Dipartimento di Informatica ed Applicazioni,
Università di Salerno, 84081 Baronissi, Italy
giuper@dia.unisa.it

Abstract. We study the following *Constrained Bipartite Edge Coloring
(CBEC)* problem: We are given a bipartite graph $G(U, V, E)$ of maximum
degree l with n vertices, in which some of the edges have been legally
colored with c colors. We wish to complete the coloring of the edges of G
minimizing the total number of colors used. The problem has been proved
to be NP–hard even for bipartite graphs of maximum degree three [5].
In previous work Caragiannis et al. [2] consider two special cases of the
problem and proved tight bounds on the optimal number of colors by
decomposing the bipartite graph into matchings which are colored into
pairs using detailed potential and averaging arguments. Their techniques
lead to 3/2–aproximation algorithms for both problems. In this paper
we present a randomized $(1.37 + o(1))$–approximation algorithm for the
general problem in the case where $\max\{l, c\} = \omega(\ln n)$. Our techniques
are motivated by recent work of Kumar [11] on the *Circular Arc Coloring*
problem and are essentially different and simpler than those presented
in [2].

1   Introduction

König's classical result from graph theory [10], states that the edges of a bipartite
graph with maximum degree l can be colored using exactly l colors so that
edges that share an endpoint are assigned different colors (see also [1]). We call
such edge colorings *legal* colorings. König's proof [10] is constructive, yielding a
polynomial–time algorithm for finding optimal bipartite edge colorings. Faster

---

[*] This work was partially funded by the European Communities under IST FET
Project ALCOM–FT and RTN Project ARACNE.

A. Brandstädt and V.B. Le (Eds.): WG 2001, LNCS 2204, pp. 21–31, 2001.
© Springer-Verlag Berlin Heidelberg 2001

algorithms have been presented in [3,6,7,15]. These algorithms usually use as a subroutine an algorithm that finds perfect matchings in bipartite graphs [8,15].

Bipartite edge coloring can be used to model scheduling problems such as timetabling. An instance of timetabling consists of a set of teachers, a set of classes, and a list of pairs (t, c) indicating that the teacher t has to teach class c during a time slot within the time span of the schedule [15]. A timetable is an assignment of the pairs to time slots in such a way that no teacher t and no class c occurs in two pairs that are assigned to the same time slot. This problem can be modelled as an edge coloring problem on a bipartite graph.

In real–life situations, the problem is made somehow harder due to additional constraints that are imposed on the solutions. This is a general feature of practical optimization problems and it is due to the fact that an optimization problem at hand is most of the time just a subproblem of a larger–scale optimazition that one seeks to obtain. In the example of scheduling classes and teachers, it is sometimes the case that some teachers have been assigned to a timeslot because of some other duties that they have to attend during other time slots; thus, assignments will have to take into account this extra restriction. So, usually, additional constraints that are put on a timetable make the problem NP–complete [4].

Caragiannis et al. in [2] study two special cases of problem CBEC that arise from algorithmic problems in optical networks (see [12,9]). Their results can be summarized as follows:

- **Problem A:** Some of the edges adjacent to a specific pair of opposite vertices of an l–regular bipartite graph are already colored with S colors that appear only on one edge (*single* colors) and D colors that appear on two edges (*double* colors). They show that the rest of the edges can be colored using at most $\max\{\min\{l + D, \frac{3l}{2}\}, l + \frac{S+D}{2}\}$ total colors. They also show that this bound is tight by constructing instances in which $\max\{\min\{l + D, \frac{3l}{2}\}, l + \frac{S+D}{2}\}$ colors are indeed necessary.

- **Problem B:** Some of the edges of an l–regular bipartite graph are already colored with S colors that appear only on one edge. They show that the rest of the edges can be colored using at most $\max\{l + S/2, S\}$ total colors. They also show that this bound is tight by constructing instances in which $\max\{l + S/2, S\}$ total colors are necessary.

Their techniques are based on the decomposition of the bipartite graph into matchings which are colored into pairs using detailed potential and averaging arguments. Their results imply 3/2–aproximation algorithms for both problems.

The original proofs in [2] consider l–regular bipartite graphs $G(U, V, E)$ with $|U| = |V| = n/2$. However, these results extend to bipartite graphs of maximum degree l with n vertices using a simple observation presented in Section 2. Note that CBEC has been proved to be NP–hard even for bipartite graphs of maximum degree three [5].

Our approach. In this paper, motivated by recent work of Kumar [11] on the circular arc coloring problem, the steps we follow to obtain a provably good approximation to problem CBEC are summarized below:

– Given a bipartite graph of maximum degree l in which some of the edges are legally colored with c colors, we reduce the problem to an integral multicommodity flow problem with constraints.
– We formulate the multicommodity flow problem as a 0–1 integer linear program.
– We relax the integrality constraint, and solve the linear programming relaxation obtaining an optimal fractional solution.
– We use randomized rounding to obtain a provably good integer solution of the integral multicommodity flow problem which corresponds to a partial edge coloring.
– We extend the edge coloring by assigning extra colors to uncolored edges.

In this way we extend the coloring of the edges of G using a total number of colors which is provably close to the optimal one. Our algorithm is randomized and works with high probability provided that the optimal number of colors is large (i.e., $\omega(\log n)$).

Roadmap. The rest of the paper is structured as follows. We present the reduction from the costrained bipartite edge coloring problem to an integral multicommodity flow problem in Section 2. In Section 3 we demostrate how to approximate the solution of the integral multicommodity flow problem and prove that this solution corresponds to an approximate edge coloring. An improvement to our approach is presented in Section 4.

2   Bipartite Edge Coloring and Multicommodity Flows

In this section we describe the reduction of an instance of problem CBEC to an instance of an integral multicommodity flow problem with constraints. We first present a reduction of the initial instance of the CBEC problem to the following one.

Let $G = (U, V, E)$ be a bipartite graph with $n = n_1 + n_2$ vertices, with $U = \{u_1, ..., u_{n_1}\}$, $V = \{v_1, ..., v_{n_2}\}$, and with maximum degree l, in which some of the edges in E are already legally colored. For any integer $k \geq 0$, we construct the bipartite graph $G_k = (A, B, E(G_k))$ where the sets of vertices A and B are defined as

$$A = \{x_i | u_i \in U\} \cup \{y'_i | v_i \in V\},$$

and

$$B = \{y_i | v_i \in V\} \cup \{x'_i | u_i \in U\}.$$

For graph G_0, the set of edges $E(G_0)$ is defined as follows. For any edge $(u_i, v_j) \in E(G)$ with $u_i \in U$ and $v_j \in V$, $E(G_0)$ contains two edges: (x_i, y_j) and (x'_i, y'_j). We call these edges *regular* edges. Also, let l be the maximum degree of G and let $d(u_i)$ (resp. $d(v_i)$) be the degree of a vertex $u_i \in U$ (resp. $v_i \in V$) in G. The edge set $E(G_0)$ also contains $l - d(u_i)$ copies of (x_i, x_i') for $i = 1, ..., n_1$, and $l - d(v_i)$ copies of (y_i, y_i') for $i = 1, ..., n_2$. These edges are called *cross* edges. Graph G_k for $k \geq 0$ is obtained from G_0, by adding k copies of the edges (x_i, x_i') for $i = 1, ..., n_1$, and k copies of the edges (y_i, y_i'), for $i = 1, ..., n_2$. An example for the construction of graph G_k from G is depicted in Figure 1.

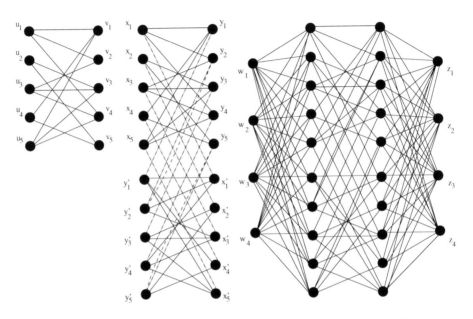

Fig. 1. The graph G, the graph G_1, and the corresponding multicommodity network H_1.

Lemma 1. *G can be edge colored with $l + k$ colors iff G_k can be edge colored with $l + k$ colors.*

Proof. Since G is a subgraph of G_k, any legal edge coloring of G_k trivially yields a legal edge coloring of G.

Assume that we have a legal edge coloring of G with $l + k$ colors. Then, the edges of G_k can be colored with $l + k$ colors as follows. For any edge $(u_i, v_j) \in E(G)$ colored with a color χ, we color the edges (x_i, y_j) and (x_i', y_j') of $E(G_k)$ with χ. This gives a partial edge coloring of G_k in which the cross edges are uncolored. Let $u_i \in U$ (resp. $v_i \in V$) and let $d(u_i)$ (resp. $d(v_i)$) be the degree of u_i (resp. v_i) in G. The cross edges between x_i and x_i' (resp. between y_i and y_i') are now constrained by $d(u_i)$ (resp. $d(v_i)$) colors. Thus, we can use the $l + k - d(u_i)$ (resp. $l + k - d(v_i)$) colors not used by edges adjacent to u_i (resp. v_i) to color the cross edges between x_i and x_i' (resp. y_i and y_i'). This completes the coloring of G_k with $l + k$ colors. □

Now, for any integer $k \geq 0$, consider the multicommodity network $H_k = (W, A, B, Z, E(H_k))$ constructed as follows. Sets of vertices A and B are the same with those of graph G_k. Also,

$$W = \{w_1, ..., w_{l+k}\}$$

and

$$Z = \{z_1, ..., z_{l+k}\}.$$

The set $E(H_k)$ is defined as

$$E(H_k) = E(G_k)$$
$$\cup \{(w_i, x_j) | 1 \leq i \leq l + k, 1 \leq j \leq n_1\}$$
$$\cup \{(w_i, y'_j) | 1 \leq i \leq l + k, 1 \leq j \leq n_2\}$$
$$\cup \{(y_j, z_i) | 1 \leq i \leq l + k, 1 \leq j \leq n_2\}$$
$$\cup \{(x'_j, z_i) | 1 \leq i \leq l + k, 1 \leq j \leq n_1\}$$

All the edges in $E(H_k)$ have unit capacity, and an edge can carry only an integral amount of flow for each commodity. The source for the i–th commodity is located at w_i, while the corresponding sink is located at z_i. An example for the construction of network H_k from graph G_k is depicted in Figure 1.

Intuitively, an integral flow of the $l+k$ commodities corresponds to a (partial) legal coloring of the edges of G_k: an edge between A and B carrying one unit of flow for commodity i in H_k corresponds to an edge colored with color i in G_k.

Since some of the the edges of the graph G_k are precolored, our multicommodity flow problem has some additional constraints. If an edge is precolored with color i in G_k, it is constrained to carry a unit amount of flow for commodity i in H_k. So, we can reduce an instance of CBEC to multicommodity flow with constraints using the following observation.

Lemma 2. G_k can be edge colored with $l + k$ colors iff there is an integral (constrained) flow of value $n(l + k)$ for commodities $1, ..., l + k$ in network H_k.

In the next section we show how to approximate the corresponding integral constrained multicommodity flow problem, and, using the reduction above, we obtain a provably good solution for the initial instance of CBEC.

3 Approximating the Multicommodity Flow Problem

In general, integral multicommodity flow (without constraints) is NP–complete [4]. However, it is straightforward to formulate the constrained multicommodity flow problem as a 0–1 integer linear program and solve its linear programming relaxation by setting aside the integrality constraint. In this way, we obtain an optimal fractional solution.

Clearly, $\max\{l, c\}$ is a lower bound on the minimum number of colors sufficient for extending the partial edge–coloring of G. We begin with network

$H_{\max\{l,c\}-l}$, solving the corresponding linear program $\text{LP}_{\max\{l,c\}-l}$. If the maximum flow is smaller than $n\max\{l,c\}$, this means that the integer linear program has no flow with value $n\max\{l,c\}$, meaning (by Lemma 2) that there exist no coloring of $G_{\max\{l,c\}-l}$ with $\max\{l,c\}$ colors. We continue with networks $H_{\max\{l,c\}-l+1}, H_{\max\{l,c\}-l+2}, ...$, until we find some L such that the solution of LP_{L-l} gives a fractional (constrained) multicommodity flow of value nL. Clearly, L is a lower bound for the minimum number of colors sufficient for coloring the edges of G_{L-l}.

Now, we will use the fractional solution of the linear program LP_{L-l} in order to obtain a solution for the corresponding integer linear program ILP_{L-l} which is provably close to the optimal one. We will use the randomized rounding technique proposed by Raghavan [14].

Let f be the flow obtained by solving LP_{L-l}. Flow f can be decomposed into L flows $f_1, f_2, ..., f_L$; one for each commodity. Each f_i can be further broken up into sets $P_{i,1}$, $P_{i,2}$, ..., of n vertex–disjoint paths from w_i to z_i (i.e., the edges between A and B in each set of vertex–disjoint paths forms a perfect matching) each carrying an amount $m_{i,j}$ of flow for commodity i, such that $\sum_{j=1}^{t_i} m_{i,j} = f_i = 1$. We call the procedure of decomposing flow *matching stripping* (since it is similar in spirit to the path stripping technique proposed in [14]).

Lemma 3. *Matching stripping can be done in polynomial time.*

Proof. Matching stripping can be performed as follows. Consider a solution to LP_k and the associated flows for commodity i in network H_k. Set $j = 1$. Let e_j be the edge carrying the smallest non–zero amount $m_{i,j}$ of flow for commodity i. Find a set $P_{i,j}$ of n vertex–disjoint paths from w_i to z_i containing edges that carry non–zero amount of flow for commodity i including e_j. Associate amount $m_{i,j}$ with $P_{i,j}$ and subtract amount $m_{i,j}$ from the flow for commodity i carried by each edge in $P_{i,j}$. Repeat this process for $j = 2, 3, ...$, until no flow remains. This will decompose the flow f_i into sets of n vertex–disjoint paths $P_{i,j}$ between w_i and z_i each carrying amount $m_{i,j}$ of flow for commodity i.

We first inductively prove that a set of n vertex–disjoint paths $P_{i,j}$ from w_i to z_i can be found at any execution of the above process. Let e_1 be the edge carrying the smallest non–zero amount $m_{i,1}$ of flow for commodity i in the beginning of the first execution. Assume that any set of vertex–disjoint paths from w_i to z_i containing edges that carry non–zero amount of flow for commodity i including e_j has size at most $n-1$. This means that there is no perfect matching containing e_1 in the subgraph of H_k containing the vertex sets A and B and the edges between them that carry non–zero amount of flow for commodity i. By Hall's Matching Theorem (see [1]), we obtain that there exists a set $S \subseteq A$ (such that e_1 is incident to one of its vertices) with neighborhood $N(S) \subseteq B$ of size $|N(S)| \leq |S| - 1$. Observe that since the solution of LP_k is optimal the edges incident to a vertex of A carry unit total amount of flow for commodity i. Thus, the edges incident to S carry a total amount $|S|$ of flow for commodity i and, since $|N(S)| \leq |S| - 1$, the capacity constraints for some of the edges incident to $N(S)$ are violated. Thus, a perfect matching $M_{i,1}$ containing edges between

A and B including e_1 exists. The set $P_{i,1}$ of vertex–disjoint paths is constructed by adding all edges between w_i and A and all edges between B and z_i to $M_{i,1}$.

Assume now that $j - 1$ sets of n vertex disjoint paths $P_{i,1}$, $P_{i,2}$, ..., $P_{i,j-1}$ between w_i and z_i have been constructed in the beginning of the j–th execution of the above process and let $m_{i,1}$, $m_{i,2}$, ..., $m_{i,j-1}$ be the associated flows for commodity i. Furthermore, assume that there still exists an edge which carries non–zero amount of flow for commodity i. Note that an amount of $\sum_{t=1}^{j-1} m_{i,t}$ of flow for commodity i has been subtracted from each edge between w_i and A, from each edge between B and z_i, from the edges between A and B incident to each vertex of A, and, similarly, from the edges between A and B incident to each vertex of B. Following the same reasoning as above, we consider the edge e_j carrying the smallest non–zero amount of flow and we obtain that there exists a perfect matching $M_{i,j}$ between A and B containing edges that carry non–zero amount of flow for commodity i including e_j (otherwise, some the edge capacity contraints in the original solution of LP_k would have been violated). Again, the set $P_{i,j}$ of vertex–disjoint paths is constructed by adding all edges between w_i and A and all edges between B and z_i to $M_{i,j}$.

We now easily prove that the number t_i of executions of the above process is polynomial. Observe that after the j–th execution, there exists at least one edge (e_j) which carry zero amount of flow, and, thus, it will not be considered in the construction of paths $P_{i,t}$ for $t > j$. Thus, the number of executions of the process is at most the number of edges between A and B, i.e., $t_i \leq n(l + k)$.

The lemma follows since maximum bipartite matching can be solved in polynomial time. □

In order to obtain an integer solution for ILP_{L-l}, for each commodity i, we will select one out of the t_i sets of vertex–disjoint paths, and use its edges to route commodity i. To select a set of vertex–disjoint paths for commodity i, we cast a t_i–faced die (one face per each of the t_i sets of vertex–disjoint paths) where $m_{i,j}$ are the probabilities associated with the faces. The selection is performed independently for each commodity. Performing this procedure for each commodity, we obtain L sets of n vertex–disjoint paths to route the L commodities.

However, these sets of n vertex–disjoint paths may not constitute a feasible integer solution to ILP_{L-l} since some edge capacities may be violated. Since in the fractional solution an edge between A and B may carry more than one commodity, it is possible that, during the rounding procedure, more than one commodities may select sets of vertex–disjoint paths that contain that edge.

Next, in each edge between A and B that was selected by more than one commodities we arbitrarily select one commodity that will use this edge. In this way, we obtain a feasible integer solution for ILP_{L-l}.

Note that the feasible solution of the integral multicommodity flow problem in H_{L-l} corresponds to a partial edge coloring of G_{L-l} with L colors. We also have to assign extra colors to the edges that do not belong to the sets of vertex–disjoint paths that were selected by the rounding procedure. Let G'_{L-l} be the (random) subgraph of G_{L-l} that contains all vertices of G_{L-l} and the edges that do not correspond to edges of H_{L-l} that were selected by the rounding procedure.

Next, in Lemma 5, we will provide an upper bound on the maximum degree of graph G'_{L-l}. Our proof is based on the following technical lemma on a well–known occupancy problem. A proof can be found in Kumar [11] (see also [13]).

Lemma 4. *Consider the process of randomly throwing m_1 balls into m_2 bins such that the expectation of the number of balls thrown into any bin is at most one. For the random variable Z denoting the number of empty bins, it holds that*

$$\Pr[Z \geq m_2 - m_1 + m_1/e + \lambda\sqrt{m_1}] \leq 2\exp(-\lambda^2/2).$$

Lemma 5. *The maximum degree of G'_{L-l} is at most $L/e + 2\sqrt{L\ln n}$, with probability at least $1 - 4/n$.*

Proof. During randomized rounding, each commodity i randomly selects a set of n vertex–disjoint paths between the source w_i and destination z_i in H_{L-l}. Thus, for each node u of G_{L-l}, one of the L edges incident to u is selected to carry unit flow for a specific commodity. Intuitively, we can think of the integral flow for each commodity as a ball and the edges incident to a vertex u as bins. The randomized rounding procedure can be modelled by the classical occupancy problem where L balls are to be randomly and independently thrown into L bins with the restriction that the expectation of the number of balls thrown into any bin is at most one (this is due to the edge capacity constraints of the flow problem). Using Lemma 4 with $m_1 = m_2 = L$ and $\lambda = 2\sqrt{\ln n}$, we obtain that the random variable denoting the number of empty bins, i.e., the number of edges incident to u which are not selected for carrying flow for any commodity, is at most $L/e + 2\sqrt{L\ln n}$ with probability at least $1 - 2/n^2$.

Thus, the probability that more than $L/e + 2\sqrt{L\ln n}$ edges incident to some of the $2n$ vertices of G_k have not been selected after the execution of the randomized rounding procedure is at most $2n \cdot 2/n^2 = 4/n$. The lemma follows. $\square$

By Lemma 5, the edges of G'_{L-l} can be colored with at most $L/e + 2\sqrt{L\ln n}$ extra colors, with high probability. Thus, in the case where L is large (i.e., $L = \omega(\ln n)$), we have proved the following theorem.

Theorem 1. *With very high probability, the algorithm uses at most $(1+1/e)L + o(L)$ total colors.*

Since L is a lower bound to the optimal number of colors sufficient for coloring the edges of the bipartite graph, we obtain that our algorithm has approximation ratio $1 + 1/e + o(1) = 1.37 + o(1)$.

4   Decreasing the Number of Colors

In this section we discuss some modifications of our algorithm which lead to a better upper bound on the total number of colors sufficient for solving instances of CBEC. Note that this improved result does not imply an approximation ratio better than the one obtained in Section 3.

We slightly modify the reduction described in Section 2. Consider again the bipartite graphs $G_k = (A, B, E(G_k))$ (for integer $k \geq 0$) defined in Section 2. For any integer $k \geq \max\{l, c\} - l$, we construct the multicommodity flow network $H'_k = (W', A, B, Z', E(H'_k))$ where now

$$W' = \{w_1, ..., w_{\max\{l,c\}}\}$$

and

$$Z' = \{z_1, ..., z_{\max\{l,c\}}\}.$$

The set $E(H'_k)$ is defined as

$$E(H'_k) = E(G_k)$$
$$\cup \{(w_i, x_j) | 1 \leq i \leq \max\{l, c\}, 1 \leq j \leq n_1\}$$
$$\cup \{(w_i, y'_j) | 1 \leq i \leq \max\{l, c\}, 1 \leq j \leq n_2\}$$
$$\cup \{(y_j, z_i) | 1 \leq i \leq \max\{l, c\}, 1 \leq j \leq n_2\}$$
$$\cup \{(x'_j, z_i) | 1 \leq i \leq \max\{l, c\}, 1 \leq j \leq n_1\}$$

Our reduction is now based on the following lemma.

Lemma 6. G_k can be edge colored with $l + k$ colors iff there is an integral (constrained) flow of value $n \max\{l, c\}$ for commodities in network H'_k.

Proof. A coloring of G_k with $l + k$ colors can be reduced to an integral (constrained) flow of value $n \max\{l, c\}$ for commodities in network H'_k by making each edge between A and B colored with some color i in G_k (for $1 \leq i \leq \max\{l, c\}$) carry a unit amount of flow for commodity i.

Given an integral (constrained) flow of value $n \max\{l, c\}$ for commodities in network H'_k, we can achieve a partial coloring of G_k with $\max\{l, c\}$ colors by using color i (for $1 \leq i \leq \max\{l, c\}$) to color an edge which carries a unit amount of flow for commodity i. We observe that the vertex–induced subgraph of G_k which contains the edges of G_k left uncolored is $(l + k - \max\{l, c\})$–regular. Thus, $l + k - \max\{l, c\}$ colors can be used to complete the coloring of the edges of G_k with $l + k$ colors in total.   □

The general structure of our approach is the same with the one described in Section 3. We begin with network $H'_{\max\{l,c\}-l}$, solving the corresponding linear program $LP_{\max\{l,c\}-l}$. If the maximum flow is smaller than $n\max\{l, c\}$, this means that the integer linear program has no flow with value $n\max\{l, c\}$, meaning (by Lemma 6) that there exist no coloring of $G_{\max\{l,c\}-l}$ with $\max\{l, c\}$ colors. We continue with networks $H'_{\max\{l,c\}-l+1}, H'_{\max\{l,c\}-l+2}, ...,$ until we find some L such that the solution of LP_{L-l} gives a fractional (constrained) multicommodity flow of value $n\max\{l, c\}$. By Lemma 6, L is a lower bound for the minimum number of colors sufficient for coloring the edges of G_{L-l}.

Then, we use the fractional solution of LP_{L-l} to obtain a feasible solution of ILP_{L-l} using randomized rounding. Again, we can prove that matching stripping can be correctly performed in polynomial time; however, some minor modifications are needed in the proof of Lemma 3.

In order to obtain an upper bound on the degree of the graph G'_{L-l} (the subgraph of G_{L-l} containing edges of G_{L-l} left uncolored after the application of the rounding procedure), we again use Lemma 4 (with $m_1 = \max\{l, c\}$ and $m_2 = L$) to show that G'_{L-l} can be edge colored with $L - \max\{l, c\} + \max\{l, c\}/e + o(\max\{l, c\})$ additional colors. In this way, when $\max\{l, c\} = \omega(\ln n)$, we obtain the following.

Theorem 2. *With very high probability, the algorithm uses at most $L + \frac{\max\{l,c\}}{e} + o(\max\{l, c\})$ total colors.*

References

1. C. Berge. Graphs and Hypergraphs. *North Holland*, 1973.
2. I. Caragiannis, C. Kaklamanis, and P. Persiano. Edge Coloring of Bipartite Graphs with Constraints. In *Proc. of the 24th International Symposium on Mathematical Foundations of Computer Science (MFCS 99)*, LNCS 1672, Springer, pp. 771–779, 1999.
3. R. Cole and J. Hopcroft. On Edge Coloring Bipartite Graphs. *SIAM Journal on Computing*, 11 (1982), pp. 540–546.
4. S. Even, A. Itai, and A. Shamir. On the Complexity of Timetable and Multicommodity Flow Problems. *SIAM Journal on Computing*, 5 (1976), pp. 691–703.
5. J. Fiala. NP–Completeness of the Edge Precoloring Extension Problem on Bipartite Graphs. Manuscript, 2000.
6. H.N. Gabow. Using Eulerian Partitions to Edge Color Bipartite Graphs. *Internat. J. Comput. Inform. Sci.*, 5 (1976), pp. 345–355.
7. H.N. Gabow and O. Kariv. Algorithms for Edge Coloring Bipartite Graphs and Multigraphs. *SIAM Journal on Computing*, 11 (1982), pp. 117–129.
8. J. Hopcroft and R. Karp. An $n^{5/2}$ Algorithm for Maximum Matchings in Bipartite Graphs. *SIAM Journal on Computing*, 2 (1973), pp. 225–231.
9. C. Kaklamanis, P. Persiano, T. Erlebach, K. Jansen. Constrained Bipartite Edge Coloring with Applications to Wavelength Routing. In *Proc. of the 24th International Colloquium on Automata, Languages, and Programming (ICALP '97)*, LNCS 1256, Springer Verlag, 1997, pp. 493–504.
10. D. König. Graphok és alkalmazásuk a determinások és a halmazok elméletére. *Mathematikai és Természettudományi Értesitö*, 34 (1916), pp. 104–119 (in Hungarian).
11. V. Kumar. Approximating Circular Arc Colouring and Bandwidth Allocation in All–Optical Ring Networks. In *Proc. of the 1st International Workshop on Approximation Algorithms for Combinatorial Optimization Problems (APPROX '98)*, LNCS, Springer, 1998. Journal version: An Approximation Algorithm for Circular Arc Coloring. Algorithmica, to appear.
12. M. Mihail, C. Kaklamanis, S. Rao. Efficient Access to Optical Bandwidth. In *Proc. of the 36th Annual Symposium on Foundations of Computer Science (FOCS '95)*, 1995, pp. 548–557.

13. R. Motwani and P. Raghavan. Randomized Algorithms. Cambridge University Press, 1995.
14. P. Raghavan. Randomized Rounding and Discrete Ham–Sandwiches: Provably Good Algorithms for Routing and Packing Problems. PhD Thesis, UC Berkeley, 1986.
15. A. Schrijver. Bipartite Edge Coloring in $O(\Delta m)$ Time. *SIAM Journal on Computing*, 28(3) (1998), pp. 841–846.

Maximum Clique Transversals

Maw-Shang Chang[1], Ton Kloks[2,*], and Chuan-Min Lee[1]

[1] Department of Computer Science and Information Engineering
Chung Cheng University
Ming-Shiun, Chiayi 621, Taiwan
mschang,cmlee@cs.ccu.edu.tw
[2] Department of Computer Science
Royal Holloway, University of London
Egham, Surrey TW20 OEX, United Kingdom
ton.kloks@cs.rhul.ac.uk

Abstract. A *maximum clique transversal set* in a graph G is a set S of vertices such that every maximum clique of G contains at least a vertex in S. Clearly, removing a maximum clique transversal set reduces the clique number of a graph. We study algorithmic aspects of the problem, given a graph, to find a maximum clique transversal set of minimum cardinality. We consider the problem for planar graphs and present fixed parameter and approximation results.

We also examine some other graph classes: subclasses of chordal graphs such as k-trees, strongly chordal graphs, etc., graphs with few P_4s, comparability graphs, and distance hereditary graphs.

1 Introduction

A *maximum clique transversal set* in a graph G is a set S of vertices such that every maximum clique of G contains at least one vertex in S. The *clique transversal set* is the same, but with "maximum clique" replaced by "maximal clique" [5,8,14]. The maximum clique transversal set problem is, given a graph G, to find a maximum clique transversal set of minimum cardinality of G.

In this paper, we study algorithmic aspects of the maximum clique transversal set problem. One of the main objectives for this research is the placement of transmitters for mobile telephones. One common method (mostly) used to place transmitter for mobile telephones is to place them at the corner points of a hexagonal grid that is placed over the area of the country.

A common solution is the placement of another transmitter close to it when a transmitter can no longer handle all communication requirements. The drawback is that this new transmitter needs another frequency to avoid interference. At some instance the number of these transmitters that are placed close to each other grows so large, that it is desirable to replace them by a more efficient (but

---

* Most of this work was done while this author visited Chung Cheng University. During the initial part of the research for this project this author was supported by an Earmarked Research Grant from the Research Grants Council of Hong Kong.

A. Brandstädt and V.B. Le (Eds.): WG 2001, LNCS 2204, pp. 32–43, 2001.
© Springer-Verlag Berlin Heidelberg 2001

also much more expensive) big transmitter tower. These towers are quite large, and contain also very costly hardware to switch between different frequencies between transmitters contained in it in some optimal way.

The problem stated as a graph algorithmic problem is to find the minimum number of vertices (transmitters) such that every maximum clique contains at least one element of this set. We emphasize on planar graphs since for this class of graphs the practical applications seem of most importance. Unfortunately, the problem remains NP-complete for planar graphs in general, but it turns out to be fixed parameter tractable. We present fixed parameter results for planar graphs in this paper and show that the optimal solution can be approximated within an arbitrary constant.

For theoretical reasons we also investigate some other common graph classes. We consider subclasses of chordal graphs such as k-trees, strongly chordal graphs etc., graphs with few P_4's, comparability graphs, and distance hereditary graphs. It seems that the problem remains NP-complete in a class of graphs in which the clique transversal problem is NP-complete, and is polynomially solvable in a class of graphs in which the clique transversal problem is polynomially solvable.

2   Preliminaries

Let G be a graph. The maximum clique number of G is denoted by $\omega(G)$ and the domination number of G is denoted by $\gamma(G)$. For a subset S of vertices we let $G - S$ be the graph induced by $V \setminus S$. For every $x \in V$, we use $S - x$ and $S + x$ to represent $S - \{x\}$ and $S \cup \{x\}$.

Definition 1. *A* maximum clique transversal set *in a graph* $G = (V, E)$ *is a subset* $S \subseteq V$ *of vertices such that every maximum clique of* G *contains at least a vertex in* S.

The maximum clique transversal problem *is the problem of finding a maximum clique transversal set in* G *of minimum cardinality. We write* $h(G)$ *for the minimum size of a maximum clique transversal set in graph* G.

Definition 2. *Let* G *be a graph and consider the union of all maximum cliques in* G. *Call the subgraph induced by this union the subgraph induced by all maximum cliques, denoted by* G^*.

We start with a very basic observation.

Lemma 1. *There is a maximium clique transversal set of cardinality* k *in graph* G *if and only if it is a clique transversal set in* G^*.

The following lemma is also quite obvious. Notice that every maximum clique transversal set in a graph G is a dominating set of G^*, since every vertex of G^* is contained in at least one maximum clique.

Lemma 2. *For any graph* G, $\gamma(G^*) \leq h(G)$.

3   Fixed Parameter Results for Planar Graphs

We here use two widely-applicable techniques, search-tree method and reduction to a problem kernel to achieve our fixed parameter results for planar graphs. Before concerning them first notice that the maximum clique transversal problem remains NP-complete when restricted to planar graphs:

Lemma 3. *The maximum cique-transversal problem is NP-complete in planar graphs.*

Proof. This follows from the NP-completeness of the vertex cover problem when restricted to triangle-free planar graphs. Notice that the independent set problem remains NP-complete when restricted to triangle-free, 3-connected, cubic planar graphs [17].   □

Recall the definition of a layer-decomposition $L_1, \ldots, L_t$ of a planar graph G, where L_1 is the outerface and t an upperbound for the outerplanarity of G.

For further definitions and notions on special graph classes not defined here, we refer to [6].

Remark 1. Notice that the number of 4-cliques in a planar graph is linear: Every 4-clique in a plane embedding has a vertex "in the middle" which is unique for this 4-clique. It can also be proved that the number of triangles in a planar graph is linear. The clique number of a planar graph is less than 5. Hence the number of all maximum cliques in a planar graph is linear.

3.1   Search-Tree Method

From [18], the clique number of a planar graph can be found in linear time and we can have the following lemma by [10,15,18].

Lemma 4. *There exists a linear time algorithm to list all maximum cliques of a planar graph.*

Following the above lemma, all maximum cliques of a planar graph can be generated in linear time. We now construct a search tree as follows. Consider all maximum cliques of G in arbitrary order $C_1, \ldots, C_\ell$. Label the root R of search tree with the empty set and then build the search tree that consists of R with children which are the vertices of an arbitrarily selected maximum clique C_i, $1 \leq i \leq \ell$. Each child v gets children that correspond with the vertices of another maximum clique C that does not contain v. In general, each node v of the search tree gets children that correspond with the vertices of a maximum clique C that does not contain any vertex in the path from node v to the root of the search tree.

Notice that for a maximum clique transversal set of size at most k, we have to consider only paths of this search tree from the root with at most k vertices.

Since each node has degree at most $\omega(G)$, the total number of these paths is bounded by ω^k where $\omega = \omega(G)$. We search the tree by depth-first traversal to determine for each path if it indeed hits every maximum clique in $O(n)$ time. Hence the total time needed for our search tree algorithm is $O(\omega^k n)$.

Theorem 1. *For each constant k there exists an algorithm which runs in time $O(\omega^k n)$ to determine if there is a maximum clique transversal set in a given planar graph G with n vertices and $\omega(G) = \omega$.*

For more weighty k the following algorithm can be beneficial. Recall the result of [1]:

Theorem 2. *Let G be a planar graph and let k be a constant. There exists an algorithm which runs in time $O(c^{\sqrt{k}} n)$ to find a dominating set in G if it exists. Here c is a moderate constant.*

Our algorithm computes an optimal maximum clique transversal set for planar graphs in linear time works as follows. Since $\gamma(G) \le h(G) \le k$ for a constant k, we know that the treewidth is at most $O(\sqrt{k})$ (see [1]). Furthermore, we can construct an approximate tree-decomposition of width at most $O(\sqrt{k})$ in time $O(\sqrt{k}n)$.

We just use the customary way for finding a solution using dynamic programming on the tree-decomposition.

Due to space limitations, we confine to mentioning the final result.

Theorem 3. *For every fixed constant k there exists a linear time algorithm to compute a maximum clique transversal set of size at most k (if it exists) in time $O(c^{\sqrt{k}} n)$.*

3.2   Reduction to a Problem Kernel

In this subsection we show that we can either reduce the graph to a subgraph with $O(k^2)$ vertices, or decide that no maximum clique transversal set can exist of order at most k. The original graph has a maximum clique transversal set of order at most k if and only if the kernel has a maximum clique transversal set of a small enough order. Combining this result with the algorithm of the previous section we obtain an algorithm that runs in time $O(\omega^k k^2 + n)$ (or, alternatively, $O(c^{\sqrt{k}} k^2 + n)$). For simplicity we assume that in the following cases the clique number of a planar graph is 4. It is easy to see how to obtain corresponding solutions for the cases with smaller clique number.

Notice that we may assume that every vertex of the graph G is contained in at least one 4-clique, since otherwise we could remove it from the graph together with all it's incident edges. If some edge is not contained in a 4-clique we can remove the edge (without the two endpoints) from the graph. These changes can be performed in linear time (using the list of 4-cliques) and do not change the problem. Henceforth we assume that every vertex and every edge of G is contained in at least one 4-clique.

Lemma 5. *If there exists a vertex of degree larger than $3(k+1)$ then it must be contained in any maximum clique transversal set of size at most k.*

Proof. Let x be a vertex of degree larger than $3(k + 1)$. Assume the degree of x is at least $3k + 4$.

Recall that each edge is contained in at least one 4-clique.
Since the graph is planar, at least $k + 1$ 4-cliques have only x in common and do not share any edge from each other. If we do not pick x as an element in the maximum clique transversal set, the size of any maximum clique transversal set is at least $k + 1$. Hence, x must be in the maximum clique transversal set of size at most k.   □

Remove all vertices of degree larger than $3(k + 1)$ in G and lower the desired maximum clique transversal set cardinality by one for every removed vertex. Henceforth we assume that the degree of all vertices is at most $3(k+1)$. Consider this kernel $G^* = (V^*, E^*)$. Let n^* be the number of vertices in this kernel.

Lemma 6. $n^* = |V^*| = O(k^2)$.

Proof. Let c be the number of 4-cliques in G^*. Since every vertex is contained in a 4-clique $c \geq \frac{n^*}{4}$. We claim there are at least $\frac{c}{12k}$ vertex disjoint 4-cliques in G^*. This is justly to prove by induction on c.

Hence, $c \leq 12k^2$ and $n^* \leq 4c = O(k^2)$.   □

Theorem 4. *There exists an algorithm that runs in time $O(4^k k^2 + n)$ to determine if a planar graph has a maximum clique transversal set of size at most k. Alternatively, there exists an algorithm that runs in time $O(c^{\sqrt{k}} k^2 + n)$ for some constant c.*

4   Approximations for Planar Graphs

In this section we show that we can approximate the size of a maximum clique transversal set for planar graph arbitrarily close.

Recall that if a graph is d-outerplanar, the treewidth is at most $3d - 1$, and for graphs with bounded treewidth the maximum clique transversal problem can easily be solved in linear time.

We divide all nodes into k classes; each class corresponds to level congruent to $i \pmod{k}$ for $i = 0, 1, \ldots, k - 1$. For each i, let G_j be a $(k + 1)$-outerplanar graph induced by the levels $jk + i$ to $(j + 1)k + i$, $j \geq 0$. For every G_j, find its optimal maximum clique transversal set by using tree-decomposition. Let H_i denote the union of these solutions for class i. Among $H_0, H_1, \ldots, H_{k-1}$, pick the minimum size of them as our approximation solution. By using the similar analysis to Baker [4], we have the approximation at most $\frac{k+1}{k}$ optimal as follows.

Consider any optimal solution H. For some s, $0 \leq s \leq k - 1$, at most $\frac{|H|}{k}$ nodes in H are in levels congruent to $s \pmod{k}$. For V_j be the set of nodes in H in levels $jk + s$ through $(j + 1)k + s$, and V_j' the optimal solution for the subgraph induced by these levels. Obviously, $|V_j'| \leq |V_j|$ for each j. Since only the nodes in levels congruent to $s \pmod{k}$ are computed twice, the sum over j of the $|V_j|$'s is at most $\frac{k+1}{k}|H|$. However, according to our algorithm, our approximation solution is not larger than the sum of the $|V'_j|$'s. Hence for fixed k, we have a linear time algorithm to find a maximum clique transversal set of size at most $\frac{k+1}{k}$ optimal. We obtain:

Theorem 5. *For every constant ϵ there exists an efficient algorithm which finds a maximum clique transversal set of size at most $h(1+\epsilon)$ where h is the optimal maximum clique transversal set size of G.*

Proof. Fix k such that $\frac{1}{k} \le \epsilon$. $h\frac{k+1}{k} = h(1 + \frac{1}{k}) \le h(1 + \epsilon)$.   □

5   Chordal Graphs of Diameter Two and k-Trees

Consider an instance of the ordinary parameterised hitting set problem. Let S be a finite set and let $\mathcal{C}$ be a collection of subsets of S. Let k be the parameter of the problem. Then the (ordinary) hitting set problem asks for a subset $S' \subseteq S$ with $|S'| \le k$ such that S' contains at least one element of each subset of $\mathcal{C}$.

We can reduce this instance to the problem of finding a maximum clique transversal set as follows. Take as a central clique the clique with vertex set S, and define for each element $c \in \mathcal{C}$ a clique $Q(c)$ of cardinality $w = |S| + 1$ with $Q(c) \cap S = c$ and with unique extra vertices added to each $Q(c)$ in order to obtain the cardinality w for each $Q(c)$. Hence, the size of the maximum clique transversal set for this graph is at most k if and only if the original parameterised hitting set problem has a yes-instance for parameter at most k.

The parameterised hitting set problem was proven to be $W[2]$-complete by Wareham, as reported (personal communication) in the book on parameterised complexity of Downey and Fellows.

Theorem 6. *The parameterised hitting set problem is $W[2]$-complete for chordal graphs.*

k-Trees.   We show that the maximum clique transversal problem for maximum cliques remains NP-complete for k-trees with k unbounded. Due to space limitations we mention the following without proof.

Lemma 7. *For k-trees the maximum clique transversal number $h(G)$ is equal to the domination number $\gamma(G)$.*

Proof. If G is a $k + 1$-clique then $h(G) = \gamma(G)$. Otherwise, consider a simplicial vertex x adjacent to a k-clique ω. Consider a dominating set D. If $x \in D$ we can replace x by another vertex from ω and obtain a dominating set of the same cardinality. By induction every maximal clique in $G - x$ contains at least one element of D. Hence D is a maximum clique transversal set.

Now consider a maximum clique transversal set $\mathcal{H}$ of minimum cardinality. If $x \in \mathcal{H}$ replace it by another vertex of ω and thus obtain a maximum clique transversal set $\mathcal{H}^*$ of the same cardinality. By induction, $\mathcal{H}^*$ has a vertex in every maximal clique of $G - x$. Hence also in G.

Theorem 7. *The hitting set problem remains NP-complete for k-trees with k unbounded.*

Proof. The domination problem is NP-complete for k-trees (arbitrary k) [11]. □

Since G^* of a strongly chordal graph G can be computed in linear time and G^* is also a strongly chordal graph, following Lemma 1, the linear time algorithm for the clique transversal problem in strongly chordal graphs can be used to find a minimum hitting set in G by finding a minimum clique transversal set in G^*.

6   Cographs and Graphs with Few P_4s

An algorithm for the clique transversal problem for cographs was already described in [14].

Cographs are properly contained in the set of distance hereditary graphs for which we describe an efficient algorithm in the next section. In this section we briefly give the formulas for cographs, and the extension to graphs with few P_4s, i.e., $(q, q - 4)$-graphs with fixed parameter q. Let G be a cograph which is the join of two cographs G_1 and G_1 (i.e., every vertex of G_1 is made adjacent to every vertex of G_2). Let $h_i = h(G_i)$ be the maximum clique transversal numbers for G_i ($i = 1, 2$). Then $h(G) = \min(h_1, h_2)$.

Let G be a cograph which is the disjoint union of two cographs G_1 and G_2 (i.e., no vertex of G_1 is adjacent to any vertex of G_2). Let h_1 be the maximum clique transversal number for G_1 and h_2 the maximum clique transversal number for G_2. If the clique number of G_i is larger than the clique number for G_{3-i} then $h = h(G) = h_i$. If both clique numbers are the same: $h = h_1 + h_2$.

Corollary 1. *The maximum clique transversal set problem can be solved in linear time for cographs.*

Graphs with Few P_4s. Recall that Jamison and Olariu extended the concept of cographs to those of graphs with few P_4s. They call a graph p-connected if for every partition of the vertex set into two non empty sets V_1 and V_2 some induced P_4 has elements on both sides of the partition. A p-connected graph is called *separable* if there is a partition of its vertex set such that every crossing P_4 has its midpoints in V_1 and its endpoints in V_2. Jamison and Olariu obtained an unique tree structure theorem.

Definition 3. *A graph is called a $(q, q - 4)$-graph if no set of at most q vertices induces more than $q - 4$ distinct P_4s.*

Lemma 8. *A p-connected component of a $(q, q - 4)$-graph is either a spider or it contains less than q vertices.*

We consider $(q, q - 4)$-graphs for fixed parameter q. A lot of classes that have been examined before can be expressed as $(q, q - 4)$-graphs. (For details and other examples and results see the various papers of Jamison and Olariu on related classes [2].)

The decomposition tree for $(q, q-4)$-graphs can be found in linear time by an algorithm of Bauman [2]. Briefly we show how to extend the result for cographs to $(q, q - 4)$-graphs. This class of graphs has again a decomposition tree, similar to the one for cographs with a additional node type, and with some restricted type of leaves which they typed (thin or thick) spiders.

It is easy to see that the maximum clique transversal set number of a thin spider with clique K and stable set S is 1 and the hitting set number for a thick spider is 2.

A decomposition tree for graphs with few $P4$s can be obtained in linear time (see, e.g., [2]). For the tree nodes that are different from the complete sum or union, we obtain easy algorithms.

Theorem 8. *There exist linear time algorithm to compute the maximum clique transversal set number for $(q, q - 4)$-graphs for each q.*

7   Comparability Graphs

In this section, we consider the maximum clique transversal problem for *comparability graphs*.

For a given comparability graph, its transitive orientation can be found in linear time [6]. In [5] it was shown that there exists an $O(M(n)+m\sqrt{n})$ algorithm to compute the clique transversal of a comparability graph. Here $M(n)$ is the time complexity needed to multiply two $n \times n$ matrices. In the following we show how to use transitively–oriented graphs to solve the maximum clique transversal set problem for comparability graphs (and hence for permutation graphs) in $O(m\sqrt{n})$ time.

Given a comparability graph $G = (V, E)$ and its transitively–oriented digraph $D = (V, A)$. According to Redei [16] every directed path in $D = (V, A)$ corresponds to a clique of G. We add a new vertex s to D and add arcs from s to every source vertex u in D. Correspondingly we add another new vertex t and arcs from every sink vertex w to t. We label s and t with numbers 0 and $|V|+1$, respectively. Let $D' = (V', A')$ denote this digraph. Clearly D' is cognizant of the incidence relationships in D.

For solving the maximum clique transversal set problem, we remove some vertices and arcs from D^* such that every s, t-path in the new digraph is of the same length in $D' = (V', A')$.

We use the following notation:

- $\ell(v)$ is the length of a longest path from v to t in D'.
- $d^-(v)$ and $d^+(v)$ represent the in-degree and out-degree respectively of v.
- $N^+(v) = \{w \mid (v, w) \in A'\}$ is the *out-neighbourhood* of a vertex v.

We omit the proof of the following lemma to keep the size within limits.

Lemma 9. *For $1 \leq i \leq |V'| + 1$, let $L = (s' = v_0, v_1, \ldots, v_i = t)$ be a s', t-path. If L is a longest s', t-path (an oriented geodesic), then each arc (v_{j-1}, v_j) in L has $\ell(v_{j-1}) = \ell(v_j) + 1$, where $0 < j \leq i$.*

By Lemma 9, we know that an arc (u, v) with $\ell(u) - \ell(v) \neq 1$ does not exist in any longest s, t-path in D'. We remove arcs that have the same property as (u, v). If this removal leads to vertices having in-degree zero, we know that these vertices except s cannot appear in any longest s, t-path.

We construct a digraph $G^* = (V^*, E^*)$ in $O(|V'| + |D'|)$ time as follows. By using a depth-first search, we compute $\ell(u)$ for each vertex u in D^* time (see, e.g., [7]). Next, we follow the ordering of vertices to check the relationships between u and $N^+(u)$ in the following:

1. If $\ell(u) - \ell(v) \geq 2$, then remove the arc (u, v) and decrease $d^+(u)$ and $d^-(v)$.
2. If $d^-(u) = 0$ and $u \neq s$, remove the vertex u and all its neighbours in D'. Of course we decrease $d^+(u)$ and $d^-(v)$.

Lemma 10. *All maximal paths in G^* are s, t-paths of the same length.*

Proof. By algorithm of constructing G^*, G^* only has arcs (u, v) with $\ell(u) - \ell(v) = 1$ and G^* has the same source s and sink t as D'. Hence, all maximal paths in G^* are s, t-paths. Let $L = (s, v_1, \ldots, t)$ be an s, t-path in G^*. The sequence $\ell(s), \ell(v_1), \ldots, \ell(t)$ of L is an ordered sequence with $\ell(v_{i-1}) = \ell(v_i) - 1$. By Lemma 9, L is an s, t-path of maximum length. Therefore all maximal paths in G^* are maximum s, t-paths. □

Lemma 11. *All longest s, t-paths of D' are also longest s, t-paths in G'.*

We have to omit the proof.

Lemma 12. *The maximum clique transversal problem of a comparability graph G can be reduced to the minimum s, t-vertex separating problem in G^*.*

We have to omit the proof. By [5], we have:

1. It takes $O(m\sqrt{n})$ time to compute the number of maximum vertex disjoint s, t-paths of G^* (see, e.g., [12]).
2. The number of maximum vertex disjoint s, t-paths of G^* is equal to the number of minimum s, t-vertex separators in G^* by Menger's theorem.

We obtain the following result.

Theorem 9. *Given a comparability graph $G = (V, E)$ with $|V| = n$ and $|E| = m$, the maximum clique transversal number of G can be computed in $O(m\sqrt{n})$ time.*

8   Distance Hereditary Graphs

Recall the definition of a distance hereditary graph:

A graph $G = (V, E)$ is called *distance hereditary* if each pair of vertices are equidistant in every connected induced subgraph containing them.

The following characterization, using so-called *twin sets*, appeared in [9].

Theorem 10. *Distance hereditary graphs can be defined recursively as follows:*

1. *A graph consisting of one vertex is distance hereditary. The twin set is the vertex itself. We call this a* **start operation**.
2. *If G_1 and G_2 are distance hereditary, then the union G of them is again distance hereditary. The twin set is the union of the twin sets of G_1 and G_2. We call this the* **false twin** *operation.*
3. *If G_1 and G_2 are distance hereditary, then the "join" obtained by connecting every vertex of the twin set of G_1 with the twin set of G_2 is again distance hereditary. The twin set is again the union of the twin sets of G_1 and G_2. We call this the* **true twin** *operation.*
4. *If G_1 and G_2 are distance hereditary, then G obtained by connecting every vertex of the twin set of G_1 with the twin set of G_2 is again distance hereditary. The twin set is the twin set of G_1. In this case we say that G_2 is* **attached** *to G_1.*

For each of the operations in Theorem 10 we show how to compute an optimal maximum clique transversal set.

We maintain for each distance hereditary graph $G = (V, E)$ with twin set T five parameters; $(\omega, h, \omega_t, h_t, h')$. We call this set of parameters the *character set of G*. These parameters represent the following:

ω and h are the clique number and maximum clique transversal number of G.

ω_t and h_t are the clique number and maximum clique transversal number of $G[T]$, i.e., the subgraph induced by the twin set T.

h' is the cardinality of a smallest set of vertices such that this set can hit the maximum cliques of $G[T]$ and G.

Lemma 13. *If G consists of one vertex, then $(\omega, h, \omega_t, h_t, h') = (1, 1, 1, 1, 1)$*

Lemma 14. *Assume G is obtained by a* **false twin** *operation of G_1 and G_2. Let T_1 and T_2 be the respective twin sets of G_1 and G_2. Hence $T = T_1 \cup T_2$. Let $(\omega_i, h_i, \omega_{t_i}, h_{t_i}, h'_i)$ be the parameter set for G_i $(i = 1, 2)$. Since G is the (disjoint) union of G_1 and G_2 in this case, we can use the formulas for the union and join to obtain ω and h. Also the twin set T is the disjoint union of T_1 and T_2. Hence ω_t and h_t can be also be computed using the formulas for the union and join.*

For the parameter h' we have to consider 4 cases.

1. *$\omega_{t_1} = \omega_{t_2}$ and $\omega_1 = \omega_2$. Then $h' = h'_1 + h'_2$.*
2. *$\omega_{t_1} = \omega_{t_2}$ and $\omega_i > \omega_{3-i}$ $(i = 1, 2)$. Then $h' = h'_i + h_{t_{3-i}}$.*
3. *$\omega_{t_i} > \omega_{t_{3-i}}$ and $\omega_1 = \omega_2$ $(i = 1, 2)$. Then $h' = h'_i + h_{3-i}$.*

4. $\omega_{t_i} > \omega_{t_{3-i}}$ and $\omega_i > \omega_{3-i}$ $(i = 1, 2)$. Then $h' = h'_i$.
5. $\omega_{t_i} > \omega_{t_{3-i}}$ and $\omega_i < \omega_{3-i}$ $(i = 1, 2)$. Then $h' = h_{t_i} + h_{3-i}$.

Proof. We show that the 4th case is true. According to the definition of h', h' is the smallest set of vertices such that this set can hit maximum cliques in $G[T]$ and G. Since $\omega_{t_i} > \omega_{t_{3-i}}$ and $\omega_i > \omega_{3-i}$, maximum cliques in $G[T]$ and G occur in $G[T_i]$ and G_i, respectively. Hence h' is equal to h'_i in this case.

$\square$

Lemma 15. *Assume G is obtained by* **attaching** *G_2 to G_1. Since $T = T_1$; $\omega_t = \omega_{t_1}$ and $h_t = h_{t_1}$. For the parameters ω and h we have to consider 5 cases.*

1. $\omega_{t_1} + \omega_{t_2} > \max(\omega_1, \omega_2)$. Then $\omega = \omega_{t_1} + \omega_{t_2}$, and $h = \min(h_{t_1}, h_{t_2})$.
2. $\omega_{t_1} + \omega_{t_2} = \omega_{3-i} > \omega_i$. Then $\omega = \omega_{3-i}$ and $h = \min(h'_{3-i}, h_{t_2} + h_1)$.
   $\omega_{t_1} + \omega_{t_2} = \omega_1 = \omega_2$. Then $\omega = \omega_1 = \omega_2$ and $h = \min(h'_1 + h_2, h'_2 + h_1)$.
3. $\omega_{t_1} + \omega_{t_2} < \omega_i > \omega_{3-i}$. Then $\omega = \omega_i$ and $h = h_i$.
4. $\omega_{t_1} + \omega_{t_2} < \omega_2 = \omega_1$. Then $\omega = \omega_1 = \omega_2$ and $h = h_1 + h_2$.

   For the parameter h' we consider 4 cases.

1. $\omega_{t_1} + \omega_{t_2} > \max(\omega_1, \omega_2)$. Then $h' = h_{t_1}$.
2. $\omega_{t_1} + \omega_{t_2} \leq \omega_1 > \omega_2$. Then $h' = h'_1$.
3. $\omega_{t_1} + \omega_{t_2} \leq \omega_2 > \omega_1$. Then $h' = h_{t_1} + h_2$.
4. $\omega_{t_1} + \omega_{t_2} \leq \omega_1 = \omega_2$. Then $h' = h'_1 + h_2$.

We omit the proof of this lemma.

Lemma 16. *Assume G is obtained by a* **true twin** *operation of G_1 and G_2. Since $T = T_1 \cup T_2$, we have $\omega_t = \omega_{t_1} + \omega_{t_2}$ and $h_t = \min(h_{t_1}, h_{t_2})$. For the parameters ω and h we have exactly the same cases and formulas as in Lemma 15.*

   For the parameter h' we have to consider 3 cases.

1. $\omega_{t_1} + \omega_{t_2} > \max(\omega_1, \omega_2)$. Then $h' = \min(h_{t_1}, h_{t_2})$.
2. $\omega_{t_1} + \omega_{t_2} \leq \omega_i = \max(\omega_i, \omega_{3-i})$ $(i = 1, 2)$. Then $h' = \min(h'_i, h_i + h_{t_{3-i}})$.
3. $\omega_{t_1} + \omega_{t_2} \leq \omega_1 = \omega_2$. Then $h' = \min(h'_1 + h_2, h'_2 + h_1)$.

The proof of Lemma 16 is similar to the proof of Lemma 15. We finally obtain the following result:

Theorem 11. *If $G = (V, E)$ is a distance hereditary graph, then the maximum clique transversal number can be obtained in linear time.*

9   Conclusion

Courcelle, Engelfriet and Rozenberg introduced the concept of *cliquewidth* in 1993. The family of graphs of cliquewidth at most k can be defined by k-expressions based on graph operations which use k vertex labels. As mentioned before, both cographs and distance hereditary graphs are contained in the graph family of cliquewidth at most 3. In Section 8 we have solved the maximum clique transversal problem for distance hereditary graphs in time $O(n + m)$. However until now it remains open to compute the maximum clique transversal problem for graphs with bounded clique-width at most k. For further research, it is worth studying.

References

1. Alber, J., H. Bodlaender, H. Fernau, and R. Niedermeier, Fixed parameter algorithms for planar dominating set and related problems. *Lecture Notes in Computer Science* **1851**, (2000), pp. 97–110. Proceedings SWAT'00.

2. Babel, L. and S. Olariu, On the structure of graphs with few P_4's, *Discrete Applied Mathematics* **84**, (1998), pp. 1–13.

3. Babel, L., T. Kloks, J. Kratochvíl, H. Müller, and S. Olariu, Algorithms for graphs with few P_4's. To appear.

4. Baker, B. S., Approximation algorithms for NP-complete problems on planar graphs, *Journal of the Association for Computing Machinery* **41**, (1994), pp. 153–180.

5. Balachandran, V., P. Nagavamsi, and C. P. Rangan, Clique transversal and clique independence on comparability graphs, *Information processing letters* **58**, (1996), pp. 181–184.

6. Brandstädt, A., V. B. Le, and J. P. Spinrad, *Graph classes–A Survey*, SIAM Monographs on Discrete Mathematics and Applications, Philadelphia, 1999.

7. Cai, Y. and M. C. Kong, Generating all maximal cliques and related problems for certain perfect graphs. *Proceedings of the Twenty-third Southeastern International Conference on Combinatorics, Graph Theory, and Computing* (Boca Raton, FL, 1992). Congr. Numer. **90** (1992), pp. 33–55.

8. M.-S. Chang, Y.-H. Chen, G. J. Chang, and J. H. Yan, Algorithmic aspects of the generalised clique transversal problem on chordal graphs, *Discrete Applied Mathematics* **66**, (1996), pp. 189–203.

9. M.-S. Chang, S.-Y. Hsieh, and G.-H. Chen, Dynamic programming on distance hereditary graphs. *Lecture Notes in Computer Science* **1350**, (1997), pp 344–353.

10. Chrobak, M. and D. Eppstein, Planar orientations with low out-degree and Compaction of adjacency Matrices. *Theoretical Computer Science* **86**, (1991), pp 243–266.

11. Corneil, D. G. and J. M. Keil, A dynamic programming approach to the domination set problem on k-trees, *SIAM J. Algebraic and Discrete Methods* **8**, (1987), pp. 535–543.

12. Even, S., *Graph Algorithms*, Computer Science Press, Rockville, MD, 1979.

13. Garey, M. R., and D. S. Johnson, *Computers and intractability–A guide to the theory of NP-completeness*, W. H. Freeman and company, New York, 1999.

14. Guruswami, V. and C. P. Rangan, Algorithmic aspects of clique transversal and clique-independent sets, *Discrete Applied Mathematics* **100**, (2000), pp. 183–202.

15. Itai, A. and M. Rodeh, Finding a minimum circuit in a graph, *SIAM Journal on Computing* **7**, (1978), pp. 413–423.

16. Redei, L., *Ein Konbinatorischer*, Acta Litt. Szeged, **7**, pp. 39–43, 1934.

17. Uehara, R., NP-complete problems on a 3-connected cubic planar graph and their applications, Technical Report TWCU-M-0004, Tokyo Woman's Christian University, 1996
   http://www.komazawa-u.ac.jp/~uehara/ps/triangle.ps.gz

18. Papadimitriou, C. and M. Yannakakis, The clique problem for planar graphs, *Information Processing Letters*, **13**, (1981), pp. 131–133.

On the Tree-Degree of Graphs

Maw-Shang Chang[1,*] and Haiko Müller[2]

[1] Department of Computer Science and Information Engineering
Chung Cheng University
Ming-Shiun, Chiayi 621, Taiwan
mschang@cs.ccu.edu.tw
[2] School of Computing
University of Leeds
Leeds LS2 9JT, United Kingdom
hm@comp.leeds.ac.uk

Abstract. *Every* graph is the *edge* intersection graph of subtrees of a tree. The *tree-degree* of a graph is the minimum maximal degree of the underlying tree for which there exists a subtree intersection model. Computing the tree-degree is NP-complete even for planar graphs, but polynomial time algorithms exist for outer-planar graphs, diamond-free graphs and chordal graphs. The number of minimal separators of graphs with bounded tree-degree is polynomial. This implies that the treewidth of graphs with bounded tree-degree can be computed efficiently, even without the model given in advance.

1 Introduction

An intersection model of a graph can be very useful to solve many problems that seem to be hard when the graph is given without the model. Examples of these problems are the maximum clique, treewidth, minimum fill-in, clique covering numbers, rankings, clique transversal, independent set, (vertex-)separators, (see, e.g., [6]). It is well-known that for every graph there exists an intersection model, which is the intersection of subsets of a set [8]. Unfortunately, this kind of model is too general to be useful to solve algorithmic problems. More appropriate are *geometric* intersection models. Well-known classes of graphs for which geometric intersection models exist are planar graphs, interval graphs, permutation graphs, cocomparability graphs, circle graphs, circular arc graphs, unit disk graphs, string graphs, etc. For graph classes not defined here explicitly we refer to the overview [6] and the classical work [10].

Given a graph, finding the intersection model is often not an easy task, because this implies a recognition algorithm for that class. Well-known examples are unit disks graphs [7], string graphs [16] and EPT-graphs [12]. For each of these classes the recognition is NP-hard.

One of the oldest examples of graphs with a nice intersection model is the class of *chordal graphs*. Chordal graphs are defined as those graphs without any

---

* corresponding author

A. Brandstädt and V.B. Le (Eds.): WG 2001, LNCS 2204, pp. 44–54, 2001.
© Springer-Verlag Berlin Heidelberg 2001

chordless cycle of length at least four. *One* characterization of chordal graph is the following [9]: A graph is chordal if and only if it is the intersection graph of subtrees of a tree. In this intersection tree model of chordal graphs, we only require that the the intersection of two subtrees is at least one common vertex. In this paper we investigate intersection graphs of subtrees of trees where vertices are adjacent if they share at least one edge (or line) of the underlying tree. For clarity, we shall refer to vertices and edges of the underlying tree as *points* and *lines*, respectively, in the rest of the paper.

2   Preliminaries

Let $G = (V, E)$ be a graph. As usual, we use $n = |V|$ and $m = |E|$. For a vertex $x \in V$ we denote by $G - x$ the graph $G[V \setminus \{x\}]$. For a subset $S \subset V$ we use $G - S$ to denote the induced subgraph $G[V \setminus S]$. A *clique* is a complete subgraph of G. We say that a clique covers an edge if the edge belongs to the clique.

Definition 1. *A* clique covering *of G is a family of cliques such that each edge of G is in at least one clique of the family. The minimum cardinality of such a family is the* clique covering number, *denoted by* $\mathrm{cc}(G)$.

Definition 2. *Given a tree T and a collection $\{T_x : x \in V\}$ of subtrees we construct a graph (V, E) such that $\{x, y\} \in E$ if and only if T_x and T_y have at least one line in common. We call a graph which can be represented in this way an* edge-intersection graph of subtrees in a tree *and the pair $(T, \{T_x : x \in V\})$ is called an* edge-intersection model.

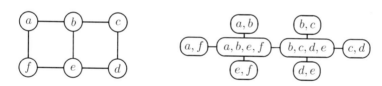

Fig. 1. The domino graph and an edge-intersection representation.

Lemma 1. *Every graph is an edge-intersection graph of subtrees of a star $K_{1,k}$ with maximum degree $\mathrm{cc}(G)$.*

Proof. Let $G = (V, E)$ be any graph and let the edges of G be covered by $k = \mathrm{cc}(G)$ cliques. Let T be a star $K_{1,k}$ with central point c. We use a bijection from the cliques in the cover to the points of $T - c$. For each vertex $x \in V$ let T_x be the subtree of T induced by c and the points corresponding with cliques containing x. Clearly, two vertices x and y are adjacent in G if and only if T_x and T_y share a line of T.   □

Clearly, not every graph is chordal, but, as shown in Lemma 1, every graph is an edge-intersection graph of some tree. In this paper we are interested in the minimum possible value of the maximal degree of such an underlying tree.

Definition 3. *Let $G = (V, E)$ be a graph. The minimum value ε for which there exists a tree T with maximal degree ε such that G can be represented as the edge-intersection of subtrees of T is called the* tree-degree *of G denoted by $\epsilon(G)$.*

Remark 1. It is easy to see that $\epsilon(G) = 0$ if and only if $E = \varnothing$. For connected graphs $G = (V, E)$ we have $\epsilon(G) = 1$ if and only if G is complete and $E \neq \varnothing$. On the other hand $\epsilon(G) = \max\{2, \epsilon(G_i) : 1 \leq i \leq k\}$ holds for graphs G with k connected components $G_i = (V_i, E_i)$ such that $E_i \neq \varnothing$ for $1 \leq i \leq k$, $k \geq 2$.

If x is an isolated vertex of G then $\epsilon(G) = \epsilon(G - x)$ because we can extend an arbitrary representation of $G - x$ as edge-intersection graph to a model of G by choosing any single point as subtree representing x. If $x \in V$ is adjacent to all vertices in $V \setminus \{x\}$ then $\epsilon(G) = \epsilon(G - x)$ because we can extend every model of $G - x$ by subtrees of T to a representation of G by choosing the whole tree T as a subtree representing x. Finally we have $\epsilon(G - x) \leq \epsilon(G)$ for all $x \in V$.

If G' is the graph obtained from G by shrinking the edge $\{x, y\}$ to a single vertex z then $\epsilon(G') \leq \epsilon(G)$ because we can define T_z to be the union of T_x and T_y.

The tree-degree of a graph can be the number of edges, e.g. for chordless cycles of length at least four or complete bipartite graphs, see Corollary 1.

Lemma 2. *G is an interval graph if and only if $\epsilon(G) \leq 2$.*

Proof. Let $(C_1, C_2, \ldots, C_k)$ be a consecutive clique arrangement of an interval graph $G = (V, E)$ [3]. We consider a path T with points p_i, $0 \leq i \leq k$ and lines $\{p_{i-1}, p_i\}$, $1 \leq i \leq k$. For each vertex $x \in V$ let T_x be the subpath of T induced by the set $\{p_{i-1}, p_i : x \in C_i\}$. Then G is exactly the edge-intersection graph of the subpath T_x of the path T.

Now let G be a graph with $\epsilon(G) \leq 2$. We consider a corresponding representation of G as edge-intersection graph of subpaths T_x of a path T. Obviously $\{x : p, q \in V(T_x)\}$ is a clique of G for each line $\{p, q\}$ of T. Since we may assume that G does not contain isolated vertices each subpath T_x contains at least one line. Hence the lines of T define a consecutive clique arrangement of G. $\square$

Lemma 3. *G is a chordal graph if and only if $\epsilon(G) \leq 3$.*

Proof. Chordal graphs are exactly the vertex-intersection graphs of subtrees of a tree [9]. For an arbitrary chordal graph $G = (V, E)$ we extend any vertex-intersection model $(T, \{T_x : x \in V\})$ to an edge-intersection model as follows:

1. To each point p of T we stick an additional pendant leaf p' and extend all subtrees T_x containing p to p'.
   This way we obtain an edge-intersection model for G.

2. Each inner point p of degree $d(p) \geq 4$ we split it into $d(p) - 2$ points $p_1, \ldots, p_{d(p)-2}$ connected in a path-like fashion by $d(p) - 3$ additional lines. Two neighbors of p become adjacent to p_1 and $p_{d(p)-2}$, respectively, and the remaining neighbors are matched to $p_2, \ldots, p_{d(p)-3}$. All subtrees containing p are modified in the same way.

The latter operation preserves the modeling property while decreasing the maximal degree of a point to at most 3.

Now let G be a graph that is not chordal. Hence G contains a chordless cycle of length $k \geq 4$. By Corollary 1 the tree-degree of G is at least k. □

One of the basic results that we need is a bound on the number of points in the underlying tree.

Lemma 4. *For each graph there is an edge-intersection representation on an optimal tree with at most $\frac{3}{2}m$ points.*

Proof. A tree is *optimal* for the graph G if its maximal degree is $\epsilon(G)$ and there is a collection of subtrees representing G. Let us call a line *high* if it is incident with two *branch points* (these are points of degree at least three) and *low* otherwise. If two subtrees share a line, we say the line *contains* the edge connecting the two vertices corresponding to the two subtrees.

We consider a low edge of an optimal tree T with minimum number of points. If all edges contained in this line are also contained in some other lines we can contract it without changing the maximal degree. Moreover the new tree still represents the same graph. Hence every low line of T contains an edge contained in no other line.

This means that the number of low edges in T is at most m, especially the number of leaves is at most m. Since the number of leaves in a tree is at least twice the number of branch points the latter number is bounded by $\frac{1}{2}m$. This implies that the number of high lines is at most $\frac{1}{2}m - 1$. Since the total number of lines is bounded by $\frac{3}{2}m - 1$ the tree T contains at most $\frac{3}{2}m$ points. □

3 Complete Separators in Edge-Intersection Graphs

Let $(T, \{T_x : x \in V\})$ be an optimal edge-intersection model of $G = (V, E)$ such that the number of points in T is minimal. We consider a line $\ell = \{a, b\}$ of T such that neither a nor b is a leaf. Removing ℓ from T results two disjoint subtrees T^a and T^b. Both T^a and T^b contain at least one subtree T_x for a suitable vertex x. Otherwise we can remove either $T^a - a$ or $T^b - b$ from T resulting a smaller model. Consequently, $\{x : T_x \text{ contains } \ell\}$ is a complete separator of G. Therefore every internal line of T corresponds to a complete separator. This leads to the following lemma:

Lemma 5. *If a graph has no complete separator, then it has an optimal edge-intersection tree which is a star.*

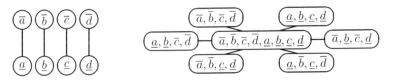

Fig. 2. The graph $4K_2$ and a representation of its complement.

Theorem 1. *For all graphs G holds $\epsilon(G) \leq \mathrm{cc}(G)$ with equality if G does not contain a complete separator.*

Proof. By Lemma 1 we have $\epsilon(G) \leq \mathrm{cc}(G)$. To prove equality, assume that G has no complete separator. By Lemma 5 there is an optimal representation of G as edge-intersection graph of a star $K_{1,k}$. Clearly each line of this star corresponds with a clique of G, and these cliques cover all edges of G. It follows $\mathrm{cc}(G) \leq k = \epsilon(G)$. □

Corollary 1. $\epsilon(K_{i,j}) = i \cdot j$ *and* $\epsilon(C_k) = k$ *for* $k > 3$.

Proof. It is easy to see that neither $K_{i,j}$ nor C_k contain a complete separator, and $\mathrm{cc}(K_{i,j}) = i \cdot j$ and $\mathrm{cc}(C_k) = k$ for $k > 3$. □

Lemma 6. *Let S be a complete separator of G. Then $M \leq \epsilon(G) \leq M+1$ holds where $M = \max\{\epsilon(G[C \cup S]) : G[C] \text{ is a connected component of } G - S\}$.*

Proof. Obviously the lemma is true for chordal graphs G. Moreover, $M \leq \epsilon(G)$ holds because $G[C \cup S]$ is an induced subgraph of G.

On the other hand, let $C^1, \ldots, C^d$ be the vertex sets of the connected components of $G - S$. We consider optimal models $(T^i, \{T^i_x : x \in C^i \cup S\})$ of $G[C^i \cup S]$ for $i = 1, \ldots, d$. Each tree T^i contains a point p^i that belongs to all subtrees T^i_s, $s \in S$. We create an additional point q^i adjacent to p^i for each i and construct a tree T by linking the trees T^i by a path $(q^1, q^2, \ldots, q^d)$. The subtrees T^i_s for $s \in S$ are linked in the same way to create T_s. All other subtrees T_x, $x \notin S$ are defined to be T^i_x if $x \in C^i$.

It is easy to see that $(T, \{T_x : x \in V\})$ is an edge-intersection representation of G and $\Delta(T) \leq M + 1$. □

Remark 2. An interesting special case with respect to Lemma 7 is that each subtree T^i contains a line ℓ^i that belongs to all subtrees T^i_s, $s \in S$. Under this circumstances we can subdivide ℓ^i by an additional point p^i. Proceeding as in the above proof we obtain $\epsilon(G) = M$.

However, in general such optimal models $(T^i, \{T^i_x : x \in C^i \cup S\})$ do not exist. In Figure 3 we give a graph O' obtained by adding two simplicial vertices to opposite faces of the octahedron O. Since no minimum edge clique cover of O contains opposite triangles we have $\epsilon(O) = 4$ but $\epsilon(O') = 5$.

Lemma 7. *Let S be a complete separator of $G = (V, E)$. Then there is an optimal edge-intersection model $(T, \{T_x : x \in V\})$ of G such that T has a line ℓ with $S = \{x : T_x \text{ contains } \ell\}$.*

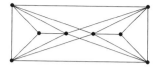

Fig. 3. The octahedron graph extended by two simplicial vertices.

Proof. Let $(T, \{T_x : x \in V\})$ be any optimal edge-intersection model of G and let C be the vertex set of a connected component of $G - S$. Since S is complete there is a point p contained in all subtrees T_s, $s \in S$. We split p into two points p_1 and p_2 joined by the line ℓ. A neighbor q of p becomes adjacent to p_1 if there is a vertex $c \in C$ such that T_c contains q, and q becomes adjacent to p_2 otherwise. All subtrees T_x, $x \in V$ are modified accordingly. Especially for $c \in C$ the new subtree T'_c does not contain p_2 since c has neighbors in $C \cup S$ only. Similar, for $d \in V \setminus (C \cup S)$ the new subtree T'_d does not contain p_1 since d has no neighbors in C. Finally, for $s \in S$ the new subtree T'_s contains both p_1 and p_2. The latter property ensures $S = \{x : T_x \text{ contains } \ell\}$.

Obviously splitting p into p_1 and p_2 does not increase the maximum degree of the host tree. Moreover the new model $(T', \{T'_x : x \in V\})$ represents the same graph G. □

4 Minimal Separators in Edge-Intersection Graphs

In this section we locate the minimal separators in an edge-intersection graph. A *minimal (vertex) separator* is minimal for two non-adjacent vertices, see [10] (page 82) for a precise definition. S is a minimal separator of G if and only if $G - S$ has two connected components such that each vertex in S has a neighbor in both of them.

Definition 4. *Let $G = (V, E)$ be a graph and let S be a separator of G. Then $C \subset V$ is a* full *component of S in G if every vertex in S has a neighbor in C.*

Remark 3. Obviously a set $S \subset V$ is a minimal a, b-separator of $G = (V, E)$ if and only if a and b are vertices in different full components of $G - S$.

We proceed with a lemma similar to Lemma 6. For $W \subseteq V$ let $G \mid W$ be the graph obtained from $G = (V, E)$ by shrinking W to a single vertex.

Lemma 8. *Let S be a complete minimal separator of $G = (V, E)$. Then $\epsilon(G) = \max\{\epsilon(G \mid (V \setminus (S \cup C))) : G[C] \text{ is a connected component of } G - S\}$.*

Proof. Let $G(C)$ be shorthand for $G \mid (V \setminus (S \cup C))$. Since S is a minimal separator there is a full component $D \neq C$ of $G - S$. Hence $G(C)$ is the graph $G[C \cup S]$ augmented by a vertex d with $N(d) = S$. This implies $\epsilon(G) \geq \epsilon(G(C))$ because we obtain $G(C)$ from G in two steps: first we remove all components of $G - S$ except C and D and then we shrink D to the single vertex d.

Let $C^1, \ldots, C^d$ be the vertex sets of the connected components of $G - S$. To show $\epsilon(G) \leq \max\{\epsilon(G(C^i)) : 1 \leq i \leq d\}$ we consider optimal models $(T^i, \{T_x^i : x \in C^i \cup S\})$ of $G(C^i)$ for $i = 1, \ldots, d$. Since S is a complete separator of $G(C^i)$ Lemma 7 ensures the existence of a line ℓ^i in T^i that belongs to all subtrees T_s for $s \in S$. Now we can follow the lines of Remark 2 to complete the proof. □

Remark 4. The assertion of the lemma remains true if S is a complete separator that is not minimal because every complete separator contains a complete minimal separator.

Next we consider the number of minimal separators in a graph of bounded tree-degree. Let S be *any* minimal a, b-separator, and let C_a and C_b be the full components of S containing a and b, respectively. Consider the two subtrees T_a' and T_b' formed by the union of subtrees corresponding to vertices of C_a and C_b, respectively. Hence T_a' and T_b' are two subtrees of T having at most one point in common, otherwise some vertex of C_a would be adjacent to some vertex in C_b. There is a unique shortest path starting from a point p of T_a' to a point q of T_b' such that the path does not pass any line of T_a' nor T_b'. Notice that p maybe equals q. Let L_a (resp. L_b) be the set of lines of T_a' (resp. T_b') incident with p (resp. q). By Remark 3, S corresponds to the set of subtrees containing a line of L_a and a line of L_b. Therefore every subtree T_s with $s \in S$ contains a line in L_a and another line incident with p but not in L_a. Any subtree containing a line of L_a and another line incident with p but not in L_a must be in S. Otherwise it should be in C_a, a contradition to the definition of L_a. Similarly, any subtree containing a line of L_b and another line incident with q but not in L_b must be in S.

Consider an edge-intersection graph $G = (V, E)$ and the intersection model of subtrees of a tree T with maximum degree $\varepsilon = \epsilon(G)$. Consider two subtrees T_a and T_b that do not share any line of T, i.e., a and b are non-adjacent vertices of G. Following the obsevation above the number of subsets of lines incident with a point of T is an upper bound of the number of minimal separators. For each internal point of T, the number of subsets of lines incident with it is at most 2^ε. There are at most $\frac{3}{2}m$ internal points. Since each separator correspondes with two complementary subsets of lines incident to a point the number of minimal separators is bounded by $3m \cdot 2^{\varepsilon-2}$.

Theorem 2. *There are at most $3m \cdot 2^{\varepsilon-2}$ minimal separators in a graph G with m edges and $\varepsilon = \epsilon(G)$.*

5   Treewidth

In this section we show that we do not need the intersection model to compute the treewidth of a graph with bounded tree-degree.

Theorem 3. *There exists a polynomial time algorithm to find the treewidth of any edge-intersection graph with bounded tree-degree.*

Proof. First we use the algorithm of [2] to list all minimal separators in a graph. In the previous section we have shown that the number of minimal separators in a graph with bounded tree-degree is linear, hence the listing algorithm runs in polynomial time.

Having the complete list of all minimal separators, we make use of the result of [5] which states that TREEWIDTH can be computed in polynomial time for any graph with a bounded number of minimal separators.   □

A more or less trivial upper bound for the treewidth of a graph with bounded tree-degree is the following.

Lemma 9. *If $G = (V, E)$ is an edge-intersection graph, then the treewidth is bounded by $\epsilon(G) \cdot \omega(G)$, where $\omega(G)$ is the cardinality of a maximum clique of graph G.*

Proof. Consider an underlying tree T which models G with maximum degree $\epsilon(G)$. Take the chordal embedding H of G by making two vertices of H adjacent if they have at least a point of T in common. The maximal cliques of H correspond with the subtrees of T that share a point in T. Consider a point p in T. Every line incident with p is contained in at most $\omega(G)$ subtrees. There are at most $\epsilon(G)$ lines in T incident with p.   □

6   Complexity of Computing the Tree-Degree of a Graph

In this section we show that computing the tree-degree is NP-complete even for planar graphs. We first show that computing $cc(G)$ for planar graphs G is NP-complete and then reduce this problem to the problem of computing tree-degrees for planar graphs.

It is well known that the problem CLIQUE COVER

INSTANCE: Graph $G = (V, E)$, positive integer k.
QUESTION: Is $cc(G) \leq k$?

is NP-complete [15,19,13] for general graphs. This problem was posed by Erdős, Goodman and Pósa [8] and somewhat later also by Lovász [17]. The parameter $cc(G)$ is computable in polynomial time for chordal graphs G [18], for graphs with maximum degree less than 6 [14], and for line graphs [19,20]. It cannot be approximated within a factor less than two for graphs in general [15] unless P=NP.

In the following we first show that the problem CLIQUE COVER is NP-complete for planar graphs. A *triangulation of the plane* is a planar graph where every face is a triangle. It is easy to see that every triangulation of the plane is 3-connected. If G is a triangulation of the plane without any triangle separator (i.e. every triangle of G is a face) then G is K_4-free except $G = K_4$. Therefore such a triangualtion does not contain any complete separator.

In [22] Uehara considers the problem INDEPENDENT TRIANGLE SET defined as follows:

INSTANCE: Graph $G = (V, E)$ and integer k.
QUESTION: Does G contains k triangles such that no two of them have an edge in common?

His proof shows that the problem INDEPENDENT TRIANGLE SET is NP-complete, even if restricted to triangulation of the plane without any triangle separator. In a triangulation of the plane that is not K_4 and contains no triangle separators, every edge is contained in two triangles and we can let all cliques in any clique cover be triangles. Suppose G is such a triangulation of the plane. Let $\mathcal{T}$ be the set of all triangles of G. It is easy to see that G contains k triangles such that no two of them have an edge in common if and only if G has a clique cover of $|\mathcal{T}| - k$ triangles. Therefore the problem CLIQUE COVER remains NP-complete even if restricted to triangulation of the plane.

Let the problem TREE-DEGREE be defined as follows:

INSTANCE: Graph $G = (V, E)$ and integer k.
QUESTION: Is $\epsilon(G) \leq k$?

Theorem 4. *The problem* TREE-DEGREE *is NP-complete, even if restricted to triangulations of the plane.*

Proof. By the NP-completeness of CLIQUE COVER and Theorem 1.   □

7 An Approach to Computing the Tree-Degrees for Some Classes of Graphs

We restrict our attention to chordal graphs first. By Remark 1 and Lemmas 2 and 3 we can compute the tree-degree of a chordal graph in linear time because both interval graphs and chordal graphs can be recognized in linear time. Moreover, the remark and the two lemmas characterize all graphs with tree-degree at most three. This allows us to concentrate on graphs with tree-degree at least four in the following algorithms.

Our next algorithm is based on Remark 2. Assume we are given a graph G without complee separators. By Theorem 1 we know $\epsilon(G) = \text{cc}(G)$, so we have to ensure that we are able to compute $\text{cc}(G)$. Now assume G has a complete separaor S. To apply Remark 2 we have to construct for each component $G[C]$ of $G - S$ a model of $G[C \cup S]$ such that all vertices in S are present at a common line. Both are possible for diamond-free graphs. (A *diamond* is the graph obtained from K_4 by removing one edge.) We say that G is *c-connected* if G does not contain a complete separator.

Algorithm TF
Input A diamond-free graph $G = (V, E)$.
Output $\epsilon(G)$.

1. Compute the set $\mathcal{S}$ of all inclusion-maximal complete separators of G.

2. Compute the set $\mathcal{B}$ of all maximal c-connected subgraphs of G.
3. Construct a tree $\mathcal{T}$ with nodes $\mathcal{S} \cup \mathcal{B}$ and arcs $\{S, G'\}$ whenever $S \subset V'$ for $G' = (V', E')$.
4. For each $G' \in \mathcal{B}$ compute the set $\mathcal{C}$ of all maximal cliques and construct an edge intersection representation on a star $K_{1,|\mathcal{C}|}$.
5. For each separator $S \in \mathcal{S}$ do
   (a) Create a path on $d_{\mathcal{T}}(S)$ points.
   (b) For each G' adjacent to S in $\mathcal{T}$ choose a line ℓ of the tree representing G' that is a line of T_s for all $s \in S$. Subdivide ℓ by an additional point.
   (c) Link these subdivision point to the point of the path in a 1–1 manner as shown in Figure 4.
6. Output $\epsilon(G) = \max\{|\mathcal{C}| : G' \in \mathcal{C}\}$ and stop.

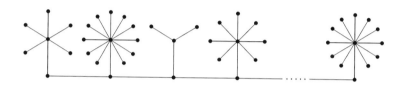

Fig. 4. Linking stars.

The correctness of this algorithm relays on Theorem 1, Remark 2 and the following simple fact.

Lemma 10. *A graph G is diamond-free if and only if every edge of G belongs to a unique maximal clique of G.*

This lemma implies that the set of maximal cliques is the unique edge-clique cover of a diamond-free graph. This set can be computed in linear time. Moreover, every complete separator is contained in a maximal clique.

Theorem 5. *The tree-degree of diamond-free graphs can be computed in time $O(nm)$.*

Proof. What remains to prove is the time bound. We apply Tarjan's algorithm [21] to find all complete separators in time $O(nm)$. Steps 2 to 6 can be executed within the same time bound.   □

Remark 5. The same algorithms works for outerplanar graphs too.

Acknowledgment

We thank Hans Bodlaender and Ton Kloks for fruitful discussions and anonymous referees for their helpful comments.

References

1. Arnborg, S., D. G. Corneil, and A. Proskurowski, Complexity of finding embeddings in a *k*-tree, *SIAM J. Alg. Discrete methods* **8**, (1987), pp. 277–284.

2. Berry, A., J.-P. Bordat, O. Cogis, Generating all the minimal separators of a graph, *Internat. J. Found. Comput. Sci.* **11** (2000), pp. 397–403.

3. Booth, K. S., G. S. Lueker, Testing for the consecutive ones property, interval graphs, and graph planarity using PQ-tree algorithms, *J. Comput. Syst. Sci.* **13** (1976), pp. 335–379.

4. Bodlaender, H. L., D. M. Thilikos, Tree width of graphs with small chordality, *Discrete Applied Mathematics* **79**, (1997), pp. 45–61.

5. Bouchitté, V., I. Todinca, Minimal triangulations for graphs with "few" minimal separators, Proceedings ESA'98, LNCS 1461, Springer, 1998, pp. 344–355.

6. Brandstädt, A., V. B. Le, and J. P. Spinrad, *Graph-classes–A Survey*, SIAM monographs on discrete mathematics and application, Philadelphia, (1999).

7. Breu, H., D. G. Kirkpatrick, Unit disk graph recognition is NP-hard, *Comput. Geom.* **9**, (1998), pp. 3–24.

8. Erdős, P., A. Goodman, L. Pósa, The representation of graphs by set intersection, *Cand. J. Math.* **18**, (1966), pp. 106–112.

9. Gavril, F., The intersection graphs of subtrees in trees are exactly the chordal graphs, *J. Combin. Theory* **16**, (1974), pp. 47–56.

10. Golumbic, M. C., *Algorithmic graph theory and perfect graphs*, Academic press, New York, (1980).

11. Golumbic, M. C., R. E. Jamison, Edge and vertex intersections of paths in trees, *Discrete Math.* **55**, (1995), pp. 151–159.

12. Golumbic, M. C., R. E. Jamison, The edge-intersection graphs of paths in a tree, *J. Comb. Theory B* **38**, (1985), pp. 8–22.

13. Holyer, I., The NP-completeness of some edge-partition problems, *SIAM J. Comput.* **4**, 1981, pp. 713–717.

14. Hoover, D. N., Complexity of graph covering problems for graphs of low degree, *JCMCC* **11**, 1992, pp. 187–208.

15. Kou, L. T., L. J. Stockmeyer and C. K. Wong, Covering edges by cliques with regard to keyword conflicts and intersection graphs, *Comm. ACM* **21**, 1978, pp. 135–139.

16. Kratochvíl, J., String graphs II. Recognising string graphs in NP-hard, *J. Comb. Theory B*, **52**, (1991), pp. 67–78.

17. Lovász, L., On coverings of graphs, in: P. Erdős and G. Katona eds., Proceedings of the Colloquium held at Tihany, Hungary, 1966, Academic Press, New York, 1968, pp. 231–236.

18. Ma, S., W. D. Wallis and J. Wu, Clique covering of chordal graphs, *Utilitas Mathematica* **36**, (1989), pp. 151–152.

19. Orlin, J., Contentment in graph theory, *Proc. of the Nederlandse Academie van Wetenschappen, Amsterdam, Series A*, **80**, 1977, pp. 406–424.

20. Pullman, N. J., Clique covering of graphs IV. Algorithms, *SIAM J. Comput.* **13**, (1984), pp. 57–75.

21. Tarjan, R. E., Decomposition by clique separators, *Discrete Mathematics* **55**, (1985), pp. 221–223.

22. Uehara, R., NP-complete problems on a 3-connected cubic planar graph and their application, Technical report TWCU-M-0004, Tokyo 1996.
http://www.komazawa-u.ac.jp/~uehara/ps/triangle.ps.gz

On Constrained Minimum Vertex Covers of Bipartite Graphs: Improved Algorithms[*]

Jianer Chen and Iyad A. Kanj

Department of Computer Science, Texas A&M University,
College Station, TX 77843-3112, USA
{chen,iakanj}@cs.tamu.edu

Abstract. The constrained minimum vertex cover problem on bipartite graphs arises from the extensively studied fault coverage problem for reconfigurable arrays. In this paper, we develop a new algorithm for the problem, in which classical results in matching theory and recently developed techniques in parameterized computation are nicely combined and extended. The algorithm is practically efficient with running time bounded by $O(1.26^k + kn)$, where k is the size of the constrained minimum vertex cover in the input graph. The algorithm is a significant improvement over the previous algorithms for the problem.

1 Introduction

As the density of VLSI chips increases, the probability of introducing defects on the chips during the fabrication process also increases. A number of reconfiguration strategies have been proposed. In particular, algorithms for repair of reconfigurable arrays using spare rows and columns have received considerable attention [7,9,16,17]. A typical reconfigurable array consists of a rectangular array plus a set of spare rows and spare columns. A defective element is repaired by replacing the row or the column containing the element with a spare row or a spare column. In general, the number of spare rows and columns is much smaller compared with the total number of rows and columns in the reconfigurable array, and the cost of reconfigurating an array is proportional to the number of replacement rows and columns. Therefore, it is desirable to use the minimum number of spare rows and columns to repair the defective elements, under the constraint that the numbers of replacement spare rows and columns do not exceed the total numbers of the available spare rows and columns, respectively.

Formally, let A be an $n \times m$ rectangular reconfigurable array. A *fault covering* of A consists of a set S_r of rows and a set S_c of columns in A such that every defective element in A is either in a row in S_r or in a column in S_c (or both). The fault covering is *minimum* if $|S_r| + |S_c|$ is the minimum. Suppose the array A is also provided with k_r spare rows and k_c spare columns. Then the problem is to either find a minimum fault covering (S_r, S_c) of A with $|S_r| \leq k_r$ and $|S_c| \leq k_c$, or report no such fault covering exists.

---

[*] This work is supported in part by NSF under Grant CCR-0000206.

A. Brandstädt and V.B. Le (Eds.): WG 2001, LNCS 2204, pp. 55–65, 2001.
© Springer-Verlag Berlin Heidelberg 2001

This problem can be easily converted to the *constrained minimum vertex cover problem on bipartite graphs*, (shortly MIN-CVCB), as follows. Let $G_A = (U \cup L, E)$ be a bipartite graph with $U = \{u_1, \ldots, u_n\}$ and $L = \{v_1, \ldots, v_m\}$, such that there is an edge between u_i and v_j if and only if the (i, j) element in the array A is defective. It is fairly easy to see that a vertex set K of k_1 vertices in U and k_2 vertices in V is a vertex cover of the graph G_A if and only if the corresponding k_1 rows and k_2 columns make a fault covering of the array A. Therefore, the problem can be formally stated as follows:

MIN-CVCB
Given a bipartite graph $G = (U \cup L, E)$ and two integers k_u and k_l, either find a minimum vertex cover of G with at most k_u vertices in U and at most k_l vertices in L, or report no such a vertex cover exists.

An extensive list of algorithms has been published in the literature for the MIN-CVCB problem and its variations (e.g., [7,9,16,17]). Most of these algorithms are heuristic and have no guaranteed performance. An algorithm proposed in [7] introduced the concept of *critical set* and used branch-and-bound technique based on the A^* algorithm in Artificial Intelligence [14]. No explicit analysis was given in [7] for this algorithm, but it is not hard to see that in the worst case the algorithm running time is at least $\Theta(2^{k_u+k_l} + m\sqrt{n})$. Experimental results show that this algorithm is much more favourable than previous algorithms and the running time is moderate for practical instances for the MIN-CVCB problem.

In the current paper, we develop simpler and more efficient algorithms for the MIN-CVCB problem. Classical results in matching theory and recently developed techniques in parameterized computation are nicely combined and extended. In particular, we observe that the classical Gallai-Edmonds Structure Theorem [8] applied to bipartite graphs will allow us to concentrate on the MIN-CVCB problem on bipartite graphs with perfect matching. In fact, Gallai-Edmonds Structure Theorem is a stronger (and more systematical) version of the critical set theory developed in [7]. Gallai-Edmonds Structure Theorem also provides a solid basis so that the recently developed technique in parameterized computation, *reduction to problem kernel*, can be nicely applied. We then use the classical Dulmage-Mendelsohn Decomposition [8] to decompose a bipartite graph with perfect matching into elementary bipartite subgraphs. This decomposition makes the other technique, *bounded search tree*, in parameterized computation become much more effective. Combining all these enables us to derive an $O(1.26^k + kn)$ time algorithm for the MIN-CVCB problem, where k is the size of a minimum vertex cover in the input graph. To see how significant this improvement is over the previous best algorithm [7], take a typical reconfigurable array of size 1024×1024 with 20 spare rows and 20 spare columns. A simple estimation shows that our algorithm is more than 10^8 times faster!

We point out that our results are also of theoretical interests in the current study of "efficient" exponential time algorithms for NP-hard problems [2,4,15], in particular they make a nice contribution to the study of parameterized algorithms for the minimum vertex cover problem, which has drawn much attention recently [1,3,12]. A recently developed algorithm by Fernau and Niedermeier [6] has running time $O(1.40^k + kn)$ and constructs a constraint (not necessarily minimum) vertex cover for bipartite graphs. This algorithm could also be used to solve the MIN-CVCB problem, but our algorithm is significantly faster and much simpler using elegant results in graph theory (instead of lengthy case-by-case combinatorial enumerations).

2   Reduction to Kernel by GE-Theorem

A vertex set K in a graph G is a *vertex cover* if every edge in G has at least one end in K. A graph $G = (V, E)$ is *bipartite* if its vertex set V can be bipartitioned into two sets U (the "upper part") and L (the "lower part"), such that every edge in G has one end in U and the other end in L. A bipartite graph is written as $G = (U \cup L, E)$ to indicate the vertex bipartition. The vertex sets U and L are called the U-*part* and the L-*part* of the graph G. A vertex is a U-*vertex* (resp. an L-*vertex*) if it is in the U-part (resp. the L-part) of G. A *matching* M in a graph G is a set of edges in G such that no two edges in M share a common end. A vertex v is *matched* if it is an end of an edge in M and *unmatched* otherwise. The matching M is *perfect* if all vertices in G are matched. It is well-known that for a bipartite graph, the number of edges in a maximum matching is equal to the number of vertices in a minimum vertex cover [8]. Given a matching M in a graph G, an *alternating path* (with respect to M) is a simple path $\{v_1, v_2, \ldots, v_r\}$ such that v_1 is unmatched and the edges $[v_{2k}, v_{2k+1}]$ are in M for all k. An alternating path is an *augmenting path* if it has odd length and both endpoints are unmatched. A matching M in a graph G is maximum, if and only if there is no augmenting path in G with respect to M [8]. A maximum matching of a graph with n vertices and m edges can be constructed in time $O(m\sqrt{n})$ [10], and for a bipartite graph G, a minimum vertex cover of G can be constructed from a maximum matching in G in linear time [8].

We start with a theorem, which answers an open problem posed in [7].

Theorem 1. *The* MIN-CVCB *problem is NP-complete.*

We now show the classical Gallai-Edmonds Structure Theorem (shortly, GE-Theorem) [8] can be applied to significantly reduce the size of the MIN-CVCB problem. Let $G = (V, E)$ be a bipartite graph. For a subset S of vertices in G, let $G(S)$ be the subgraph induced by S. Let D be the set of vertices in G that are unmatched in at least one maximum matching, let A be the set of vertices in $V - D$ that are adjacent to a vertex in D, and let $C = V - D - A$.

Proposition 1. (GE-Theorem) *Let G be a bipartite graph with the sets D, A, and C defined above. Then* (1) *D is an independent set in which no vertex is contained in any minimum vertex cover for G;* (2) *A is the intersection of all minimum vertex covers for G;* (3) *the subgraph $G(C)$ has a perfect mathing.*

The sets D, A, and C can be constructed via a maximum matching M in the bipartite graph G, as follows. Let W be the set of unmatched vertices under

M. Define $\overline{W}$ to be the set of vertices such that a vertex v is in $\overline{W}$ if and only if v is reachable from a vertex in W via an alternating path of even length. By the definition, $W \subseteq \overline{W}$. Since exchanging the edges in M and $E - M$ along any alternating path from a vertex v' in W to a vertex v in $\overline{W}$ will result in a maximum matching that does not contain v, we have $\overline{W} \subseteq D$. Now for any vertex w not in $\overline{W}$, we know $[w, w']$ is an edge in M for some vertex w'. Since w is not in $\overline{W}$, w' is not connected from any vertex in W by an alternating path of odd length. Therefore, after removing w from the graph G, there is no augmenting path in $G - \{w\}$ with respect to the matching $M - [w, w']$. Thus, a maximum matching in the graph $G - \{w\}$ contains $|M| - 1$ edges. This shows that the vertex w must be in every maximum matching in the graph G so w is not in D. In consequence, the set $\overline{W}$ is exactly the set D in Proposition 1. With the set D known, the sets A and C can be constructed easily.

Theorem 2. *The time for solving an instance* $\langle G; k_u, k_l \rangle$ *of* MIN-CVCB, *where G is a bipartite graph of n vertices and m edges, is* $O(m\sqrt{n} + t(k_u + k_l))$, *where $t(k_u + k_l)$ is the time for solving an instance* $\langle G'; k_u', k_l' \rangle$ *for* MIN-CVCB, *where $k_u' \le k_u$, $k_l' \le k_l$, and G' has a perfect matching.*

Proof. Let $G = (U \cup L, E)$, and the instance $\langle G; k_u, k_l \rangle$ asks for a minimum vertex cover for G with at most k_u U-vertices and at most k_l L-vertices. We apply the algorithm described above to construct the sets D, A, and C in GE-Theorem. The algorithm has its time complexity dominated by that of constructing a maximum matching in G, which is bounded by $O(m\sqrt{n})$. Now let $G' = G(C)$. Since the set A is contained in *every* minimum vertex cover for G and no vertex in the set D is in *any* minimum vertex cover, if the set A contains h_u U-vertices and h_l L-vertices, then the graph G has a minimum vertex cover with at most k_u U-vertices and at most k_l L-vertices if and only if the graph G' has a minimum vertex cover with at most $k_u' = k_u - h_u$ U-vertices and at most $k_l' = k_l - h_l$ L-vertices. By GE-Theorem, the graph G' has a perfect matching. $\qquad\square$

The technique of *reduction to problem kernel* has been widely used in the study of parameterized computation and in the research in fault coverage of reconfigurable arrays. Theorem 2 provides a much stronger version for this technique than previously proposed versions in both areas.

Hasan and Liu [7] proposed the concept of *critical sets*. The critical set of a bipartite graph G is the intersection of all minimum vertex covers for G, which is, equivalently, the set A in GE-Theorem. However, Hasan and Liu were unable to characterize the structure of the remaining graph when the critical set is removed from the graph G. By lack of this observation, they proposed to interlace the process of branch-and-bound searching and the process of constructing the critical set [7]. On the other hand, Theorem 2 indicates explicitly that the remaining graph G' after removing the critical set A and the independent set D has a perfect matching. This fact immediately limits the size of the graph G'. Moreover, since every minimum vertex cover for the graph G' contains exactly one end from every edge in a perfect matching in G', a careful study can show that the construction of critical sets during the branch-and-bound process proposed in [7] is totally unnecessary. This observation can significantly speed up the searching process. Finally, having a perfect matching in the graph G' will enable us to derive further powerful structure techniques in the searching process, which will be illustrated in detail in the next section.

The reduction to problem kernel technique has been very powerful in designing efficient parameterized algorithms [1,3,6,12]. Fernau and Niedermeier have considered a generalized version of the MIN-CVCB problem (in which the minimization of the vertex cover is not required), and described how to reduce the graph size to $2k_u k_l$. Theorem 2 enables us to improve this bound to $2(k_u + k_l)$: The facts that the graph G' has a perfect matching and that a minimum vertex cover of G' has at most k'_u U-vertices and at most k'_l L-vertices necessarily bound the number of vertices in G' by $2(k'_u + k'_l) \leq 2(k_u + k_l)$. In the development of parameterized algorithms for minimum vertex cover for general graphs, Chen, Kanj, and Jia [3] proposed to apply the classical Nemhauser-Trotter Theorem [11] that can reduce the problem size to $2k$, where k is the size of a minimum vertex cover. Similar to GE-Theorem, Nemhauser-Trotter Theorem decomposes the vertices of a graph into three parts D', A', and C', where D' is an independent set, A' is a subset of *some* minimum vertex cover, and the induced subgraph $G(C')$ has its minimum vertex cover of size at least half of C'. When applied to bipartite graphs, GE-Theorem, which decomposes the graph into three parts D, A, and C, is much stronger than Nemhauser-Trotter Theorem in all aspects: the set A is contained in *every* minimum vertex cover for G while the set A' is only contained in *some* vertex cover for G. In particular, for the application of the MIN-CVCB problem, the set A' cannot be automatically included in the minimum vertex cover being searched. Moreover, since the subgraph $G(C)$ has a perfect matching, it follows directly that the minimum vertex cover for $G(C)$ contains at least half of the vertices in C. By lack of having a perfect matching, the subgraph $G(C')$ also cannot take the advantages we will use in the searching process, which will be described in the next section.

3   Efficient Search by DM-Decomposition

According to Theorem 2, we can concentrate on instances $\langle G; k_u, k_l \rangle$ of the MIN-CVCB problem, where G is a bipartite graph that has a perfect matching and has at most $2(k_u + k_l)$ vertices.

A connected bipartite graph $B = (U \cup L, E)$ is *elementary* if every edge of B is contained in a perfect matching in B. An elementary bipartite graph B has exactly two minimum vertex covers, namely U and L [8]. This property makes our searching process very efficient on the elementary bipartite graph B: we should exclusively include either U or L in the minimum vertex cover.

The classical Dulmage-Mendelsohn Decomposition Theorem (shortly DM-Theorem) [8] provides a nice structure for bipartite graphs with perfect matching so that the above suggested searching process can be applied effectively.

Proposition 2. (DM-Theorem).   *A bipartite graph $G = (U \cup L, E)$ with perfect matching can be decomposed into elementary bipartite subgraphs $B_i = (U_i \cup$*

L_i, E_i), $i = 1, \ldots, r$, such that every edge in G between two subgraphs B_i and B_j with $i < j$ must have its U-vertex in B_i and its L-vertex in B_j.

The subgraphs B_i will be called *blocks*. A block B_i is a *d-block* if $|U_i| = |L_i| = d$. Edges connecting two different blocks will be called "inter-block edges".

Lemma 1. *Let $G = (U \cup L, E)$ be a bipartite graph with perfect matching. Then the decomposition of G given in Proposition 2 can be constructed in time $O(|E|^2)$.*

Proof. An edge e in G is an inter-block edge if and only if e is not contained in any perfect matching in G [8]. Consider the algorithm given in Figure 1.

DM-Decomposition
Input: a bipartite graph $G = (U \cup L, E)$ with perfect matching
Output: the blocks $B_1, \ldots, B_r$ in DM-Theorem
1. construct a perfect matching M in G;
2. **for** each edge $e = [u, v]$ in G **do**
      if the graph $G^- = G - \{u, v\}$ has no perfect matching
      then mark e as an "inter-block edge";
3. remove all inter-block edges;
4. sort the remaining connected components: $B_1, \ldots, B_r$, so that every
      inter-block edge connects a U-vertex in a lower indexed block to an
      L-vertex in a higher indexed block.

Fig. 1. Dulmage-Mendelsohn Decomposition

Note that an edge $e = [u, v]$ is in a perfect matching in the graph G if and only if the graph $G - \{u, v\}$ has a perfect matching. Therefore, Step 2 of the algorithm **DM-Decomposition** correctly determines the inter-block edges. By DM-Theorem, each connected component after Step 3 is a block for the graph G. Now applying the topological sorting in a straightforward way will order the blocks so that each inter-block edge in the graph G connects a U-vertex in a lower indexed block to an L-vertex in a higher indexed block.

Constructing the perfect matching M takes time $O(|E|\sqrt{|E|})$ [10]. Consider an edge $e = [u, v]$ in G. If e is in M then obviously e is not an inter-block edge. If e is not in M, then after removing the vertices u and v and their incident edges, the matching M becomes a matching M^- of $|M| - 2$ edges in the remaining graph G^-. Thus, the remaining graph G^- has a perfect matching if and only if there is an augmenting path in G^- with respect to the matching M^-. Since testing the existence of an augmenting path can be done in linear time [8], we conclude that we can decide whether an edge in G is an inter-block edge in time $O(|E|)$. Therefore, Step 2 of the algorithm **DM-Decomposition** takes time $O(|E|^2)$, which dominates the complexity of the algorithm.   □

Now we are ready to describe the main body of our algorithm for the MIN-CVCB problem. Let $\langle G; k_u, k_l \rangle$ be an instance of the MIN-CVCB problem, where $G = (U \cup L, E)$ is a bipartite graph with perfect matching, and G has at most $2(k_u + k_l)$ vertices. We first apply DM-Theorem that decomposes the graph G into blocks $B_i = (U_i \cup L_i, E_i)$, such that every inter-block edge connects a U-vertex in a block of lower index to an L-vertex in a block of higher index. It is easy to see that a minimum vertex cover of the graph G must be the union of minimum vertex covers of the blocks. Since the only two minimum vertex covers of the block B_i are U_i and L_i, the constrained minimum vertex cover of the graph G with at most k_u U-vertices and at most k_l L-vertices must be the union of properly selected U-parts and L-parts from the blocks.

The execution of our algorithm is depicted by a search tree whose leaves correspond to the potential constrained minimum vertex covers K of the graph G with at most k_u U-vertices and at most k_l L-vertices. Each internal node of the search tree corresponds to a branch in the searching process. Let $F(k_u + k_l)$ be the number of leaves in the search tree for finding a minimum vertex cover of at most k_u U-vertices and at most k_l L-vertices in the bipartite graph G. We list the possible situations in which we branch in our search process.

Case 1. A block B_i is a d-block with $d \geq 3$.

Let $B_i = (U_i \cup L_i, E_i)$. Since the constrained minimum vertex cover K of the graph G either contains the entire U_i and is disjoint from L_i, or contains the entire L_i and is disjoint from U_i, we branch in this case by either including the entire U_i in K (and removing L_i from the graph) or including the entire L_i in K (and removing U_i from the graph). In each case, we add at least 3 vertices in the constrained minimum vertex cover K. Thus, this branch satisfies the recurrence relation:

$$F(k_u + k_l) \leq 2F(k_u + k_l - 3) \tag{1}$$

A block sequence $[B'_1, \ldots, B'_h]$ is a *block chain* if there is an edge from the U-part of B'_i to the L-part of B'_{i+1} for all $1 \leq i \leq h - 1$. Observe that the *chain implication* can speed up the searching process significantly. For example, if we include the L-part of the block B'_1 in the minimum vertex cover K, then the U-part of B'_1 must be excluded. Since there is an edge in G from the U-part of B'_1 to the L-part of B'_2, we must also include the entire L-part of B'_2 in K, which, in consequence, will imply that the L-part of B'_3 must also be included, and so on. Thus, the L-part of B'_1 in the minimum vertex cover K implies that the L-parts of all blocks in the chain must be in K. Similarly, the U-part of B'_h in K implies that the U-parts of all blocks in the chain must be in K. This observation enables us to handle many cases very efficiently.

Case 2. A 2-block is an interior block in a block chain.

Let $[B', B, B'']$ be a block chain, where B is a 2-block. Then including the L-part of B in the minimum vertex cover K forces at least 3 vertices in K (the L-parts of B and B''), and excluding the L-part of B also forces at least 3 vertices in K (the U-parts of B and B'). Thus, in this case, the branch satisfies the recurrence relation (1).

Case 3. The U-parts of three different 2-blocks are connected to the L-part of the same 1-block, or the U-part of a 1-block is connected to the L-parts of three different 2-blocks.

Suppose that the U-parts of the 2-blocks B_1', B_2', and B_3' are connected to the L-part of the 1-block B_0'. Then including the U-part of B_0' in the minimum vertex cover K forces at least 7 vertices (the U-parts of all the four blocks) in K, and excluding the U-part of B_0' forces at least 1 vertex (the L-part of B_0') in K. Therefore, the branch satisfies the recurrence relation

$$F(k_u + k_l) \leq F(k_u + k_l - 7) + F(k_u + k_l - 1) \tag{2}$$

Similar analysis shows that (2) is also applicable to the other subcase.

Case 4. All Cases 1-3 are excluded.

Because Case 1 is excluded, there are only 1-blocks and 2-blocks. Since Case 2 is excluded, either the U-part or the L-part of each 2-block is not incident on any inter-block edges. Therefore, we can re-order the blocks as:

$$B_1', \ldots, B_x', B_{x+1}', \ldots, B_y', B_{y+1}', \ldots, B_z' \tag{3}$$

where B_1', ..., B_x' are 2-blocks whose L-parts are not incident on inter-block edges, B_{x+1}', ..., B_y' are 1-blocks, and B_{y+1}', ..., B_z' are 2-blocks whose U-parts are not incident on inter-block edges, such that every inter-block edge goes from the U-part of a lower indexed block to the L-part of a higher indexed block. Note that for any index i, $1 \leq i \leq z$, the U-parts of B_1', ..., B_i' plus the L-parts of B_{i+1}', ..., B_z' make a minimum vertex cover for the bipartite graph G. Now to construct a constrained minimum vertex cover K of the bipartite graph G with at most k_u U-vertices and at most k_l L-vertices, we have the following situations.

If $k_u + k_l > k$, where k is the size of a minimum vertex cover of the graph G, then we can pick an index i such that the U-parts of the blocks B_1', ..., B_i' consist of either $k_u - 1$ or k_u vertices (such an index always exists since all blocks are 1-blocks and 2-blocks). Now the U-parts of the blocks B_1', ..., B_i' together with the L-parts of the blocks B_{i+1}', ..., B_z' make a minimum vertex cover with at most k_u U-vertices and at most $k - (k_u - 1) \leq k_l$ L-vertices.

Thus, we can assume that $k_u + k_l = k$.

If $2x \leq k_u \leq 2x + (y - x)$, then let $i = k_u - x$. The minimum vertex cover K consisting of the U-parts of B_1', ..., B_i' and the L-parts of B_{i+1}', ..., B_z' contain $(i - x) + 2x = k_u$ U-vertices and $k - k_u = k_l$ L-vertices. Thus, K is a desired minimum vertex cover satisfying the constraint.

If $k_u < 2x$ and k_u is even, then the U-parts of the first $k_u/2$ 2-blocks plus the L-parts of the rest of the blocks make a desired minimum vertex cover.

Now consider the case $k_u < 2x$ and k_u is odd. If the graph G has no 1-blocks, then obviously there is no minimum vertex cover with k_u U-vertices and k_l L-vertices. Thus we assume that the graph G has at least one 1-block.

It is easy to see that there is a minimum vertex cover of exactly 1 U-vertices if and only if there is a 1-block whose L-vertex has degree 1, and there is a minimum vertex cover of exactly 3 U-vertices if and only if there is a 1-block whose L-vertex is adjacent to the U-part of at most one 2-block. Thus, we can further assume $k_u \geq 5$.

Pick the 1-block B of the minimum index i in sequence (3). Since Case 3 is excluded, there are at most two 2-blocks B' and B'' with their U-parts adjacent to the L-vertex of B. Note that since Case 2 is excluded, no inter-block edges can be incident on the L-parts of the blocks B' and B''. Now re-order the sequence (3) by removing B, B', and B'' from the sequence then re-inserting B', B'', B in the front of the sequence. Since the L-parts of B' and B'' are not adjacent to inter-block edges and inter-block edges from the L-part of B only go to the U-parts of B' and B'', the new sequence still has the property that every inter-block edge links the U-part of a lower indexed block to the L-part of a higher indexed block. Now the U-parts of the blocks B', B'', and B plus the U-parts of the next $(k_u - 5)/2$ 2-blocks in the new sequence and the L-parts of the rest of the blocks give a minimum vertex cover for G with k_u U-vertices and k_l L-vertices. In case the L-vertex of B is adjacent to the U-part of fewer than two 2-blocks, the desired minimum vertex cover K can be constructed similarly.

Finally, the case $k_u > 2x + (y - x)$ can be reduced to the previous case by reversing the sequence (3), exchanging the U-part and L-part of the bipartite graph G, and exchanging the numbers k_u and k_l.

Thus, in Case 4 the MIN-CVCB problem can be solved in linear time.

4   Putting All Together

We summarize all the discussions in the algorithm given in Figure 2.

CVCB-Solver
Input: a bipartite graph $G = (U \cup L, E)$ and two integers k_u and k_l
Output: a minimum vertex cover K of G with at most k_u U-vertices and
        at most k_l L-vertices, or report no such a vertex cover exists
1. **for** each U-vertex u of degree $> k_l$ **do**
      include u in K and remove u from G; $k_u = k_u - 1$;
2. **for** each L-vertex v of degree $> k_u$ **do**
      include v in K and remove v from G; $k_l = k_l - 1$;
3. apply GE-Theorem to reduce the instance so that G is a bipartite
    graph with perfect matching and has at most $2(k_u + k_l)$ vertices;
4. apply DM-Decomposition to decompose G into blocks $B_1, \ldots, B_r$;
5. while one of the Cases 1-3 in section 3 is applicable, branch accordingly;
6. solve the instance for Case 4 in section 3.

Fig. 2. The main algorithm

Steps 1-2, taking time $O(kn)$, make immediate decisions on high degree vertices. If a U-vertex u of degree larger than k_l is not in K, then all neighbors of u should be in K, which would exceed the bound k_l. Thus, such U-vertices should be automatically included. Similar justification applies to L-vertices of degree larger than k_u. After Steps 1-2, the degree of the vertices in the graph is bounded by $k' = \max\{k_u, k_l\}$. Since each vertex can cover at most k' edges, the number of edges in the resulting graph must be bounded by $k'(k_u + k_l) \leq (k_u + k_l)^2$, otherwise the graph cannot have a desired minimum vertex cover.

Theorem 2 allows us to apply GE-Theorem in Step 3 to further reduce the bipartite graph G so that G has a perfect matching. The running time of this step is bounded by $O((k_u + k_l)^3)$. The number of vertices in the graph G now is bounded by $2(k_u + k_l)$.

Step 4 applies DM-Theorem to decompose the graph G into elementary bipartite subgraphs. By Lemma 1, this takes time $O((k_u + k_l)^4)$.

As we have discussed in Section 3, if any of Cases 1-3 is applicable, we can branch in our searching process with the recurrence relations (1) and (2). Eventually, we reach Case 4 in Section 3, which can be solved in linear time.

It is easy to verify that $F(k_u + k_l) = 1.26^{k_u+k_l}$ satisfies the recurrence relations (1) and (2). Thus, the running time of Step 5 in our algorithm is bounded by $O(1.26^{k_u+k_l}(k_u + k_l)^2)$, where the factor $(k_u + k_l)^2$ is the size of the graph resulted from Step 4. Using the general technique given in [13], we can further reduce the time complexity of Step 5 to $O(1.26^{k_u+k_l})$. Combining all these, we conclude with the following theorem.

Theorem 3. *The* MIN-CVCB *problem is solvable in time* $O(1.26^{k_u+k_l} + kn)$.

5   Concluding Remarks

Parameterized computation theory has proved very useful in designing practical algorithms in industrial applications, in particular for many NP-hard optimization problems. In this paper, we have developed a simple and significantly improved algorithm for the MIN-CVCB problem, which arises from the extensively studied fault coverage problem for reconfigurable arrays in the area of VLSI manufacturing. Our algorithm is conceptually simple, easy to implement, and nicely combines classical graph structural results with the recently developed techniques in parameterized computation. For typical applications, our algorithm is faster than the previous known algorithms by several orders of magnitude.

Our improvement is on the reduction of the base in the exponential running time, which is of great theoretical and practical importance in the study of NP-hard optimization problems. For example, reducing the base by 0.01 in the parameterized vertex cover algorithm has resulted in upto 60% improvement in the running time for the application program used in biochemistry [5]. Our algorithm reduces the base by more than 0.14 (from 1.4^k to 1.26^k) from the previous known algorithms for the MIN-CVCB problem.

We point out that our investigation emphasizes on the practicality of the algorithms, which includes simplicity (both conceptual and algorithmic) and effectiveness. The running time of our algorithm could be further improved slightly if we examine in more depth the combinatorial structures in a case-by-case exhaustive manner, which is something that we tried to avoid since it would deprive our algorithm from its simplicity and practicality. We might also apply the dynamic programming techniques, as illustrated in [3], which will improve the algorithm running time to $O(1.247^{k_u+k_l} + kn)$. However, this improvement is at the cost of exponential space.

References

1. R. BALASUBRAMANIAN, M, R. FELLOWS, AND V. RAMAN, An improved fixed parameter algorithm for vertex cover, *Information Processing Letters 65*, (1998), pp. 163-168.

2. R. BEIGEL AND D. EPPSTEIN, 3-coloring in time $O(1.3446^n)$: a no-MIS algorithm, *Proc. 36th IEEE Symp. on Foundations of Computer Science*, (1995), pp. 444-452.

3. J. CHEN, I. A. KANJ, AND W. JIA, Vertex cover: further observations and further improvement, *Lecture Notes in Computer Science 1665* (WG'99), (1999), pp. 313-324.

4. *DIMACS Workshop on Faster Exact Solutions for NP-Hard Problems*, Princeton, February 23-24, 2000.

5. R. G. DOWNEY, M. R. FELLOWS, AND U. STEGE, Parameterized complexity: A framework for systematically confronting computational intractability, *AMS-DIMACS Proceedings Series 49*, F. Roberts, J. Kratochvil, and J. Nesetril, eds., (1999), pp. 49-99.

6. H. FERNAU AND R. NIEDERMEIER, An efficient exact algorithm for constraint bipartite vertex cover, *Lecture Notes in Computer Science 1672* (MFCS'99), (1999), pp. 387-397.

7. N. HASAN AND C. L. LIU, Minimum fault coverage in reconfigurable arrays, *Proc. 18th Int. Symp. on Fault-Tolerant Computing* (FTCS'88), (1988), pp. 348-353.

8. L. LOVÁSZ AND M. D. PLUMMER, *Matching Theory*, Annals of Discrete Mathematics 29, North-Holland, 1986.

9. C. P. LOW AND H. W. LEONG, A new class of efficient algorithms for reconfiguration of memory arrays, *IEEE Trans. Comput. 45*, (1996), pp. 614-618.

10. S. MICALI AND V. VAZIRANI, An $O(\sqrt{|V|} \cdot |E|)$ algorithm for finding maximum matching in general graphs, *Proc. 21st IEEE Symp. on the Foundation of Computer Science*, (1980), pp. 17-27.

11. G. L. NEMHAUSER AND L. E. TROTTER, Vertex packing: structural properties and algorithms, *Mathematical Programming 8*, (1975), pp. 232-248.

12. R. NIEDERMEIER AND P. ROSSMANITH, Upper bounds for vertex cover further improved, *Lecture Notes in Computer Science 1563* (STACS'99), (1999), pp. 561-570.

13. R. NIEDERMEIER AND P. ROSSMANITH, A general method to speed up fixed-parameter-tractable algorithms, *Information Processing Letters 73*, (2000), pp. 125-129.

14. N. J. NILSSON, *Principles of Artificial Intelligence*, Tioga Publishing Co., 1980.

15. R. PATURI, P. PUDLAK, M. E. SAKS, AND F. ZANE, An improved exponential-time algorithm for k-SAT, *Proc. 39th IEEE Symp. on Foundations of Computer Science*, (1998), pp. 628-637.

16. W. SHI AND W. K. FUCHS, Probabilistic analysis and algorithms for reconfiguration of memory arrays, *IEEE Trans. Computer-Aided Design 11*, (1992), pp. 1153-1160.

17. M. D. SMITH AND P. MAZUMDER, Generation of minimal vertex cover for row/column allocation in self-repairable arrays, *IEEE Trans. Comput. 45*, (1996), pp. 109-115.

$(k, +)$–Distance-Hereditary Graphs[*]
(Extended Abstract)

Serafino Cicerone, Gianluca D'Ermiliis, and Gabriele Di Stefano

Dipartimento di Ingegneria Elettrica, Università dell'Aquila
I-67040 Monteluco di Roio - L'Aquila, Italy
{cicerone,dermiliis,gabriele}@ing.univaq.it

Abstract. In this work we introduce, characterize, and provide algorithmic results for $(k, +)$–distance-hereditary graphs. These graphs can be used to model interconnection networks with desirable connectivity properties; a network modeled as a $(k, +)$–distance-hereditary graph can be characterized as follows: *if some nodes have failed, as long as two nodes remain connected, the distance between these nodes in the faulty graph is bounded by k plus the distance in the non-faulty graph.* The class of all these graphs is denoted by $\mathrm{DH}(k, +)$ By varying the parameter k, classes $\mathrm{DH}(k, +)$ form a hierarchy that represents a parametric extension of the well-known class of distance-hereditary graphs, and include all graphs.

1 Introduction

A fundamental problem in any parallel or distributed system is the efficient communication of data between processors. Such efficiency depends on the *routing scheme* defined over the system, that is the set of paths for each possible pair of processors. The efficiency of a routing scheme is meanly measured in terms of its *stretch factor* and *dilation*. The stretch factor (dilation) is the maximum ratio (difference) between the length of a path defined by the scheme and the shortest path between the same pair of processors.

In this work we are interested in networks in which routing schemes coincide with shortest paths and node failures may occur. Distances are always computed by means of shortest paths in the subnetwork that is induced by the non-faulty components. In this context, the decrease of the efficiency of the communication depends only on the topology of the networks.

To measure this efficiency degradation, some parameters about the topology can be defined. In [6] the authors defined the notion of *stretch number*, while in this paper we introduce the *dilation number* $\partial(G)$ of a graph G. It is defined as the smallest k such that $G \in \mathrm{DH}(k, +)$, where a network modeled as a graph belonging to the class $\mathrm{DH}(k, +)$ can be characterized as follows: *if some nodes have failed, as long as two nodes remain connected, the distance between these*

---

[*] Work partially supported by the Italian MURST Project "Teoria dei Grafi ed Applicazioni".

A. Brandstädt and V.B. Le (Eds.): WG 2001, LNCS 2204, pp. 66–77, 2001.
© Springer-Verlag Berlin Heidelberg 2001

nodes in the faulty graph is at most k plus *the distance in the non-faulty graph.* Elements of $DH(k, +)$ are called $(k, +)$–distance-hereditary graphs. By varying the parameter k, classes $DH(k, +)$: (i) form a hierarchy that represents a parametric extension of the well-known class of distance-hereditary graphs [15], (ii) include all the graphs.

Given the relevance of $(k, +)$–distance-hereditary graphs in the area of communication networks, our purpose is to provide characterization and algorithmic results about the introduced graphs.

Related works. In literature there are several papers devoted to fault-tolerant network design, mainly starting from a given desired topology and introducing fault-tolerance to it (e.g., see [3,14,16]).

Papers [6,7] present several results about $(k, *)$–distance-hereditary graphs, that is graphs whose induced distance is bounded by a multiplicative factor k. In [12], a study about similar concepts is performed: they give characterizations for graphs in which *no delay* occurs in the case that a *single* node fails. These graphs are called *self-repairing*. In [8], authors introduce and characterize new classes of graphs that guarantee constant stretch factors k even when a multiple number of *edges* have failed. In a first step, they do not limit the number of edge faults at all, allowing for *unlimited* edge faults. Secondly, they examine the more realistic case where the number of edge faults is *bounded* by a value ℓ. The corresponding graphs are called k–self-spanners and (k, ℓ)–self-spanners, respectively. In both cases, the names are motivated by strong relationships to the concept of k–*spanners* [17]. Related works are also those concerning distance-hereditary graphs. Distance-hereditary graphs have been investigated to design interconnection network topologies [5,10,11], and several papers have been devoted to them (see [2] and references therein).

Results. First, we formally introduce $(k, +)$–distance-hereditary graphs and provide some preliminary results. An initial characterization is given in terms of the dilation number. Then, we remark differences about structural properties between $(k, *)$–distance-hereditary graphs and $(k, +)$–distance-hereditary graphs. Starting from these observations, we introduce the notion of "twin graph" G^* of an arbitrary graph G. This graph has the remarkable property that $G \in DH(k, +)$ if and only if $G^* \in DH(k, +)$. Thanks to this notion, we are able to provide a characterization of graphs G in $DH(k, +)$ based on cycle-chord conditions of its twin graph G^*.

Since we show that the recognition problem for the new graph classes is Co-NP-Complete (for k not fixed), then we investigate in more detail the smallest class among the new ones, i.e., class $DH(1, +)$. In this context, our main result consists of listing all the forbidden induced subgraphs of every $G \in DH(1, +)$. A theoretical consequence of this characterization is that it allows us to show that the recognition problem of class $DH(1, +)$ can be solved in polynomial time.

This extended abstract is organized as follows. Notation and basic concepts used in this work are given in Section 2, while Section 3 formally introduces $(k, +)$–distance-hereditary graphs and provide some preliminary results. Section 4 states the Co-NP-Completeness result. Section 5 characterizes graphs in $DH(1, +)$ and provide a polynomial time recognition for them. Finally, Section 6 lists some open problems.

2   Notation

In this work we consider finite, simple, loopless, undirected and unweighted graphs $G = (V, E)$ with node set V and edge set E. We use standard terminology, some of which are briefly reviewed here.

A *subgraph* of G is a graph having all its nodes and edges in G. Given a subset S of V, the *induced subgraph* $\langle S \rangle$ of G is the maximal subgraph of G with node set S. $|G|$ denotes the cardinality of V. If x is a node of G, by $N_G(x)$ we denote the *neighbors* of x in G, that is, the set of nodes in G that are adjacent to x, and by $N_G[x]$ we denote the *closed neighborhood* of x, that is $N_G(x) \cup \{x\}$. $G - S$ is the subgraph of G induced by $V \setminus S$.

A sequence of pairwise distinct nodes $(x_0, x_1, \ldots, x_n)$ is a *path* in G if $(x_i, x_{i+1}) \in E$ for $0 \leq i < n$, and is an *induced path* if $\langle \{x_0, \ldots, x_n\} \rangle$ has n edges. A graph G is *connected* if for each pair of nodes x and y of G there is a path from x to y in G.

A *cycle* C_n in G is a path $(x_0, \ldots, x_{n-1})$ where also $(x_0, x_{n-1}) \in E$. Two nodes x_i and x_j are *consecutive* in C_n if $j = (i+1) \bmod n$ or $i = (j+1) \bmod n$. A *chord* of a cycle is an edge joining two non-consecutive nodes in the cycle. H_n denotes an *hole*, i.e., a cycle with n nodes and without chords. The *chord distance* of a cycle C_n is denoted by $cd(C_n)$, and it is defined as the minimum number of consecutive nodes in C_n such that every chord of C_n is incident to some of such nodes. We assume $cd(H_n) = 0$.

The length of a shortest path between two nodes x and y in a graph G is called *distance* and is denoted by $d_G(x, y)$. Moreover, the length of a longest induced path between them is denoted by $D_G(x, y)$. We use the symbols $P_G(x, y)$ and $p_G(x, y)$ to denote a longest and a shortest induced path between x and y, respectively. Sometimes, when no ambiguity occurs, we use $P_G(x, y)$ and $p_G(x, y)$ to denote the sets of nodes belonging to the corresponding paths.

If x and y are two nodes of G such that $d_G(x, y) \geq 2$, then $\{x, y\}$ is a *cycle-pair* if there exist a path $p_G(x, y)$ and a path $P_G(x, y)$ such that $p_G(x, y) \cap P_G(x, y) = \{x, y\}$. In other words, if $\{x, y\}$ is a cycle-pair, then the set $p_G(x, y) \cup P_G(x, y)$ induces a cycle in G.

3   Preliminary Results

In this section we first formally define $(k, +)$–distance-hereditary graphs and then provide some preliminary results.

Definition 1. *Let k be a real number. A graph $G = (V, E)$ is a $(k, +)$–distance-hereditary graph if for each connected induced subgraph G' of G:*

$$d_{G'}(x, y) \leq d_G(x, y) + k, \quad \text{for each } x, y \in G'.$$

The class of all the $(k, +)$–distance-hereditary graphs is denoted by $\mathrm{DH}(k, +)$.

Notice that the above definition holds for both connected and disconnected graphs.

Definition 2. *Let G be a graph, and $\{x, y\}$ be a pair of connected nodes in G. Then:*

1. *the dilation number $\partial_G(x, y)$ of the pair $\{x, y\}$ is given by $\partial_G(x, y) = D_G(x, y) - d_G(x, y)$;*
2. *the dilation number $\partial(G)$ of G is the maximum dilation number over all possible pairs of connected nodes, that is, $\partial(G) = \max_{\{x,y\}} \partial_G(x, y)$;*
3. *$\mathcal{D}(G)$ is the set of all the pairs of nodes inducing the dilation number of G, that is, $\mathcal{D}(G) = \{\{x, y\} \mid \partial_G(x, y) = \partial(G)\}$.*

The dilation number can be used to provide a first characterization of graphs in $\mathrm{DH}(k, +)$.

Theorem 1. *Let G be a graph. $G \in \mathrm{DH}(k, +)$ if and only if $\partial(G) \le k$.*

Proof. To prove this theorem we show that $\partial(G) = \min\{t : G \in \mathrm{DH}(t, +)\}$.

By Definition 2, $\partial(G) = \max_{\{x,y\}} D_G(x, y) - d_G(x, y)$, and then $\partial(G) \ge D_G(x, y) - d_G(x, y)$ for each pair of connected nodes $x, y \in V$. If $G' = (V', E')$ is a connected induced subgraph of G, then $\partial(G) \ge d_{G'}(x, y) - d_G(x, y)$ for each $x, y \in V'$. Hence $d_{G'}(x, y) \le \partial(G) + d_G(x, y)$ for each $x, y \in V'$. By the generality of G', it follows that $G \in \mathrm{DH}(\partial(G), +)$.

By contradiction, let us suppose that there exists an integer $t < \partial(G)$ such that $G \in \mathrm{DH}(t, +)$. Let $\{x, y\} \in \mathcal{D}(G)$, and $G' = \langle P_G(x, y) \rangle$. In this case we have that $d_{G'}(x, y) = D_G(x, y)$, and hence the relation $D_G(x, y) - d_G(x, y) = \partial(G) > t$ implies that

$$d_{G'}(x, y) = D_G(x, y) > t + d_G(x, y).$$

Then $G \notin \mathrm{DH}(t, +)$, a contradiction. □

The following two lemmas list some basic properties of $(k, +)$–distance-hereditary graphs.

Lemma 1. *The following facts hold:*

1. *$\mathrm{DH}(0, +)$ coincides with the class of distance-hereditary graphs;*
2. *$\mathrm{DH}(k, +) = \mathrm{DH}(\lfloor k \rfloor, +)$;*
3. *$\mathrm{DH}(k_1, +) \subseteq \mathrm{DH}(k_2, +)$ for each $k_1 \le k_2$;*
4. *If $(x, y) \in E$ then $\partial_G(x, y) = 0$. As a consequence:*
   - *if $\partial(G) = 0$ then $\mathcal{D}(G)$ contains every pairs of connected nodes of G;*
   - *if $\partial(G) > 0$ then $d_G(x, y) \ge 2$ for each pair $\{x, y\} \in \mathcal{D}(G)$;*
5. *if G contains n nodes, then $\partial(G) \le \max\{0, n - 4\}$, and for each $n \in \mathbb{N}$ there exists a graph G' such that $\partial(G') = n$;*
6. *$\mathrm{DH}(k, +) \subset \mathrm{DH}(1 + \frac{k}{2}, *)$ for each $k \ge 1$.*

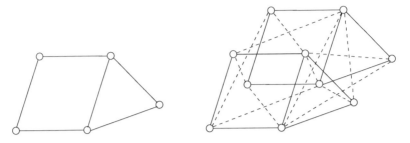

Fig. 1. *A graph G and its twin graph G^*. Dashed lines represent edges in E^3 (see Definition 3).*

Lemma 2. *The class* $\mathrm{DH}(k, +)$ *is closed under taking induced subgraphs.*

Proof. Let G be a graph in $\mathrm{DH}(k, +)$, G' be a connected induced subgraph of G, and $x, y \in G'$. Relationships $D_{G'}(x, y) \leq D_G(x, y)$ and $d_{G'}(x, y) \geq d_G(x, y)$ are straightforward, whereas $D_G(x, y) \leq k + d_G(x, y)$ holds because G belongs to $\mathrm{DH}(k, +)$. Hence

$$D_{G'}(x, y) \leq D_G(x, y) \leq k + d_G(x, y) \leq k + d_{G'}(x, y)$$

and the theorem follows. If G' is not connected the same argument can be applied to every connected component of G'.   □

In the remainder of this section, we provide a characterization of $(k, +)$–distance-hereditary graphs based on cycle-chord conditions. To this purpose, we first introduce the notion of *twin* of a given graph.

Definition 3. *Let* $G = (V, E)$ *be a graph. The* twin graph *of* G *is a graph* $G^* = (V^*, G^*)$ *such that* $V^* = V^1 \cup V^2$ *and* $E^* = E^1 \cup E^2 \cup E^3$, *where:*

- $V^1 = \{v^1 \mid v \in V\}$;
- $V^2 = \{v^2 \mid v \in V\}$;
- $E^1 = \{(u^1, v^1) \mid (u, v) \in E\}$;
- $E^2 = \{(u^2, v^2) \mid (u, v) \in E\}$;
- $E^3 = \{(u^1, v^2), (u^2, v^1), \mid (u, v) \in E\}$.

Then $|V^*| = 2 \cdot |V|$ and $|E^*| = 4 \cdot |E|$. Informally, G^* is composed by two "copies" of G (corresponding to the subgraphs $G^1 = (V^1, E^1)$ and $G^2 = (V^2, E^2)$, both isomorphic to G) which are joined by edges in E^3 (see Fig. 1, where edges in E^3 are represented by dashed lines).

The name "twin graph" is due to the fact that, for each pair (v^1, v^2) of nodes in G^* such that $v_1 \in V^1$ and $v_2 \in V^2$, v^2 is a *false twin* of v^1 in G^*. Two vertices v and v' are *twins* if they have the same neighborhood; we distinguish between *false twins* when $N(v) = N(v')$ and *true twins* when $N[v] = N[v']$. Let $G = (V, E)$ be a graph and $u \in V$. Operation $\gamma(G, u)$ (see [2]) extends G by creating a false twin of u. The resulting graph is $G' = (V \cup \{u'\}, E \cup \{(u', v) \mid v \in N_G(u)\})$. Hence, the twin graph G^* can be obtained from G by applying operation $\gamma(G, v)$ to each node v of G.

Lemma 3. *Let $G = (V, E)$ be a graph and let $S = (V_S, E_S)$ be an induced subgraph of G^*. There exists an induced subgraph T of G isomorphic to S and with no false twins if and only if S has no false twins.*

Proof. Omitted. ☐

Lemma 4. *$G \in \mathrm{DH}(k,+)$ if and only if $G^* \in \mathrm{DH}(k,+)$.*

Proof. $\Longrightarrow$: If we assume $G^* \notin \mathrm{DH}(k,+)$, then there exist a pair of nodes $\{u, v\} \in V^*$ such that $D_{G^*}(u,v) - d_{G^*}(u,v) > k$. Let $P^*(u,v) = (u = v_0^{i_0}, v_1^{i_1}, \ldots, v_n^{i_n} = v)$ and $p^*(u,v) = (u = u_0^{\ell_0}, u_1^{\ell_1}, \ldots, u_m^{\ell_m} = v)$, $i_j, \ell_j \in \{1,2\}$ and $0 \leq j \leq m$, be the longest and the shortest induced paths connecting u and v respectively. Then, by Lemma 3, $P^1 = (v_0^1, v_1^1, \ldots, v_n^1)$ and $p^1 = (u_0^1, u_1^1, \ldots, u_m^1)$ are also induced paths in G^1 connecting $u_0^1 = v_0^1$ to $u_m^1 = v_n^1$. Since $D_{G^1}(u_0^1, u_m^1) - d_{G^1}(u_0^1, u_m^1) \geq |P^1| - |p^1| = |P^*| - |p^*| > k$, then $G^1 \notin \mathrm{DH}(k,+)$. Finally, since G^1 is isomorphic to G, also $G \notin \mathrm{DH}(k,+)$.

$\Longleftarrow$: By hypothesis $G^* \in \mathrm{DH}(k,+)$ and since G is an induced subgraph of G^*, then, by Lemma 2, $G \in \mathrm{DH}(k,+)$. ☐

Lemma 5. *Let G be a graph such that $\partial(G) > 0$. Then, $\mathcal{D}(G^*)$ contains a cycle-pair of G^*.*

Proof. Omitted. ☐

Theorem 2. *Let G be a graph and $k \geq 1$ an integer. Then, $G \in \mathrm{DH}(k,+)$ if and only if $cd(C_n) \geq \frac{n-k}{2} - 1$ for each cycle C_n, $n > k+4$, of G^*.*

Proof. $\Longrightarrow$: By hypothesis, $G \in \mathrm{DH}(k,+)$. By contradiction, let us suppose that a cycle $C_n, n > k + 4$, exists in G^* such that $cd(C_n) < \frac{n-k}{2} - 1$. Let $(x, v_1, v_2, \ldots, v_q, y, u_1, u_2, \ldots u_p)$, $p+q+2=n$, be the cycle C_n, and $\{v_1, v_2, \ldots, v_q\}$ the set of nodes giving the chord distance of C_n; then, $d_{G^*}(x,y) \leq cd(C_n) + 1 < \frac{n-k}{2}$. It follows that, since

$$D_{G^*}(x,y) \geq n - d_{G^*}(x,y) > n - \frac{n-k}{2},$$

then

$$D_{G^*}(x,y) - d_{G^*}(x,y) > n - \frac{n-k}{2} - \frac{n-k}{2} = k.$$

This is a contradiction because, by Lemma 4, $G^* \in \mathrm{DH}(k,+)$.

$\Longleftarrow$: By contradiction, let us suppose $G \notin \mathrm{DH}(k,+)$. By Lemma 4, $G^* \notin \mathrm{DH}(k,+)$. Since $k \geq 1$, then $\partial(G^*) > 0$. Hence, by Lemma 5 and Fact 4 of Lemma 1 there exists a cycle-pair $\{x, y\} \in \mathcal{D}(G^*)$ such that $d_{G^*}(x,y) \geq 2$ and inducing a cycle C_n with n nodes. By contradiction hypothesis, we have that $D_{G^*}(x,y) > k + d_{G^*}(x,y)$. The inequality

$$n = D_{G^*}(x,y) + d_{G^*}(x,y) > k + d_{G^*}(x,y) + d_{G^*}(x,y) = k + 2 \cdot d_{G^*}(x,y)$$

implies that $d_{G^*}(x, y) < \frac{n-k}{2}$. Moreover, since $cd(C_n) = d_{G^*}(x, y) - 1$, then

$$cd(C_n) < \frac{n-k}{2} - 1.$$

Since this is a contradiction, then $G \in DH(k, +)$. $\qquad\square$

4   Recognition Problem for DH$(k, +)$

Although Theorems 1 and 2 provide characterizations for $(k, +)$–distance-hereditary graphs, they cannot be used to devise an efficient algorithm to solve the recognition problem for the class DH$(k, +)$. In this section we study this problem when k is not fixed.

Definition 4. *Dilation Number Problem:*
INSTANCE: *A graph* $G = (V, E)$, *an integer* $q \geq 0$.
QUESTION: $\partial(G) > q$?

The NP-completeness of this problem can be shown by providing a polynomial transformation from the NP-complete problem *Induced Path* (cf. [13], GT23).

Theorem 3. *Dilation Number is NP-complete.*

Proof. Omitted. $\qquad\square$

If we fix $k = 0$, then the recognition problem for the class DH$(k, +)$ can be solved in linear time [1]. If we consider k not fixed, then the recognition problem for the class DH$(k, +)$ is exactly the complementary problem of *Dilation Number*. As a consequence, the following complexity result can be stated.

Corollary 1. *If k not fixed, the recognition problem for the class* DH$(k, +)$ *is Co-NP-complete.*

5   Studying Graphs in DH$(1, +)$

In the following we investigate in more detail the smallest class among the new ones, i.e., class DH$(1, +)$.

Lemma 6. *Let G be a graph not containing the following graphs as induced subgraphs:*

- H_n, *for each* $n \geq 6$;
- *cycles* C_6 *with* $cd(C_6) = 1$;
- *cycles* C_7 *with* $cd(C_7) = 1$;
- *cycles* C_8 *with* $cd(C_8) = 1$.

Then, G does not contain, as induced subgraphs, cycles C_n with $n \geq 6$ and $cd(C_n) \leq 1$.

Proof. Omitted.

Lemma 7. *Let G be a graph and let G^* its twin graph. $G \in \mathrm{DH}(1,+)$ if and only if the following graphs are not induced subgraphs of G^*:*

   i. H_n, for each $n \geq 6$;
   ii. cycles C_6 with $cd(C_6) = 1$;
   ii. cycles C_7 with $cd(C_7) = 1$;
   iv. cycles C_8 with $cd(C_8) = 1$;
   v. cycles C_{2i+4} with $cd(C_{2i+4}) = i$, for each $i \geq 2$.

Proof. $\Longrightarrow$: Holes H_n, $n \geq 6$, have dilation number at least 2. Cycles with 6, 7, or 8 nodes and chord distance 1 have dilation number equal to 2, 3, and 4, respectively. Cycles C_{2i+4} with chord distance equal to i have dilation number at least $2i+4-2 \cdot (cd(C_{2i+4})+1) = 2$. Then they are forbidden induced subgraphs for every graph belonging to $\mathrm{DH}(1,+)$, and, in particular they are forbidden both for G and, by Lemma 4, for G^*.

$\Longleftarrow$: We prove that if $G \notin \mathrm{DH}(1,+)$ then G^* contains one of the forbidden subgraphs.

If $G \notin \mathrm{DH}(1,+)$ then, by Theorem 2, G^* contains a cycle C_n, $n \geq 6$, as induced subgraph such that $0 \leq cd(C_n) < \frac{n-3}{2}$. In what follows, let us denote $cd(C_n)$ as q.

If $q = 0$ then we obtain the holes H_n, $n \geq 6$. If $1 \leq q < \frac{n-3}{2}$ and $n = 6,7,8$, then we obtain the other forbidden subgraphs.

Now, we have to show that each cycle C_n with $n \geq 9$ and chord distance q such that $1 \leq q < \frac{n-3}{2}$ contains a forbidden subgraph. In this case, let us assume the cycle C_n be induced by the nodes of the two node-disjoint paths $P_G(x,y) = (x, u_1, u_2, \ldots, u_p, y)$ and $p_G(x,y) = (x, v_1, v_2, \ldots, v_q, y)$ such that $p+q+2 = n$. In this cycle we denote by r_j the largest index j' such that v_j and $u_{j'}$ are connected by a chord of C_n, i.e. $r_j = \max\{j' \mid (v_j, u_{j'}) \text{ is a chord of } C_n\}$; we assume r_j undefined when v_j is not incident to a chord of C_n. Informally, r_j gives the *rightmost* chord incident to v_j. Notice that, since $q \geq 1$, r_1 is defined.

If $r_1 > 3$ then the subgraph of C_n induced by the nodes $v_1, x, u_1, \ldots, u_{r_1}$ is a cycle with at least 6 nodes and chord distance at most 1. According to Lemma 6, this subgraph contains one subgraph among those listed in items from (i.) to (iv.) of the statement.

In the remainder of this proof we assume $r_1 \leq 3$. Since r_1 is defined, let $C_{n'}$ be the subgraph of C_n induced by the nodes $v_1, v_2, \ldots, v_q, y, u_p, u_{p-1}, \ldots, u_{r_1}$. Informally, $C_{n'}$ is one of the two cycles obtained by "cutting" C_n by means of chord (v_1, u_{r_1}). Cycle $C_{n'}$ has $n' \geq n - 3$ nodes (because $r_1 \leq 3$) and chord distance at most $q - 1$.

Now we show that either C_n coincides with the graph listed in item (v.) of the statement, or $C_{n'}$ is forbidden. In the latter case we can recursively apply to $C_{n'}$ this proof.

According to Theorem 2, $C_{n'}$ is forbidden when $cd(C_{n'}) < \frac{n'-3}{2}$. Since $cd(C_{n'}) \leq q - 1$, this corresponds to show that $\frac{n'-3}{2} > q - 1$ holds; and, since

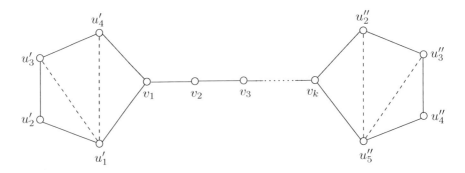

Fig. 2. *The clepsydra graph $cl(k)$ (see Definition 5). The dashed edges may or may not exist.*

$n' \geq n - 3$, in turns it corresponds to show that $\frac{(n-3)-3}{2} > q - 1$ holds. This relation is equivalent to the following inequality:

$$n \geq 2q + 5. \tag{1}$$

Since we are considering a chord distance q such that $1 \leq q < \frac{n-3}{2}$, then $n \geq 2q + 4$. This implies that Eqn. 1 trivially holds when $n \geq 2q + 5$, and it remains to be analyzed the case $n = 2q + 4$. In this case the cycle C_n under consideration has size $n = 2q + 4$ and chord distance $cd(C_n) = q$: this means that C_n is a forbidden graph (see item (v.) of the statement). This concludes the proof.   □

The following definition introduces the notion of *clepsydra* graph (see Fig. 2).

Definition 5. *Let $C'_5 = (u'_1, u'_2, u'_3, u'_4, u'_5)$ be a cycle such that $cd(C'_5) \leq 1$ and $deg(u'_5) = 2$, $P_k = (v_1, v_2, \ldots, v_k)$ be a path with $k \geq 1$, and $C''_5 = (u''_1, u''_2, u''_3, u''_4, u''_5)$ be a cycle such that $cd(C''_5) \leq 1$ and $deg(u''_1) = 2$. A clepsydra of order k is a graph $cl(k) = (V, E)$ such that:*

- $V = \{u'_1, u'_2, u'_3, u'_4, v_1, v_2, \ldots, v_k, u''_2, u''_3, u''_4, u''_5\}$;
- *if $u, v \in V$ then $(u, v) \in E$ if and only if one of the following condition holds:*
  1. $(u, v) \in \{(u'_1, v_1), (u'_4, v_1), (v_k, u''_2), (v_k, u''_5)\}$
  2. *u and v are adjacent in C'_5, P_k, or C''_5*

Notice that, since there are 3 possible cycles C_5 having chord distance at most 1, then for a fixed k there exist 9 possible clepsydrae $cl(k)$.

Lemma 8. *Let C_n, $n \geq 4 + 2i$ and $i \geq 1$, be a cycle having chord distance $cd(C_n) \leq i$. If $C_n \in DH(\frac{3}{2}, *)$ then C_n contains a clepsydra as induced subgraph.*

Proof. Let $(x, v_1, v_2, \ldots, v_q, y, u_p, u_{p-1}, \ldots, u_1)$ be the cycle C_n, and let $(v_1, v_2, \ldots, v_q)$ be the nodes giving the chord distance of C_n. From hypothesis, it follows that $p + q + 2 \geq 4 + 2i$ and $q \leq i$. From these two inequalities, we get a lower bound for p:

$$p \geq 2i + 2 - q \geq 2i + 2 - i = i + 2.$$

According to the definition of chord distance, it follows that:

$$s_{C_n}(x, y) \geq \frac{p+1}{i+1} \geq \frac{i+3}{i+1} = 1 + \frac{2}{i+1}.$$

This observation implies that, if $i = 1$ or $i = 2$ then $s_{C_n}(x, y) > \frac{3}{2}$ and hence $C_n \notin DH(\frac{3}{2}, *)$. We prove the theorem for $i \geq 3$ by showing that either C_n contains a clepsydra or C_n contains a cycle $C_{n'}$ such that $n' \geq 4 + 2(i-1)$ and $cd(C_{n'}) = i - 1$ (and hence we can recursively apply to $C_{n'}$ this proof).

In the remainder of the proof, if $1 \leq j \leq q$ then we denote by r_j the largest index j' such that v_j and $u_{j'}$ are connected by a chord of C_n, i.e. $r_j = \max\{j' \mid (v_j, u_{j'}) \text{ is a chord of } C_n\}$; we assume r_j undefined when v_j is not incident to a chord of C_n. Informally, r_j gives the *rightmost* chord incident to v_j. In a similar way, $l_j = \min\{j' \mid (v_j, u_{j'}) \text{ is a chord of } C_n\}$ gives the *leftmost* chord incident to v_j. By definition of chord distance, l_1 and l_q are both defined; l_j, $1 < j < q$, may or may not be defined.

Now, assuming l_j, $1 < j \leq q$, defined, let $C_{n'}$ and $C_{n''}$ be the subgraphs of C_n induced by the nodes $x, v_1, v_2, \ldots, v_j, u_{l_j}, u_{l_j - 1}, \ldots, u_1$ and $v_j, v_{j+1}, \ldots, v_q, y$, $u_p, u_{p-1}, \ldots, u_{l_j}$, respectively. Informally, $C_{n'}$ and $C_{n''}$ are the cycles obtained by "cutting" C_n by means of chord (v_j, u_{l_j}).

Cycle $C_{n'}$ has $n' = l_j + j + 1$ nodes; as a consequence, cycle $C_{n''}$ has

$$n'' = n - n' + 2 \geq 4 + 2i - (l_j + j + 1) + 2 = 2i + 5 - l_j - j$$

nodes. Moreover, it is easy to see that $cd(C_{n'}) \leq j - 1$ and $cd(C_{n''}) \leq q - j + 1$.

We may assume that $n' < 4 + 2(j-1)$, otherwise $n' \geq 4 + 2(j-1)$ and $cd(C_{n'}) \leq j - 1$ imply that we can recursively apply this proof to $C_{n'}$. Similarly, we may assume that $n'' < 4 + 2(q - j + 1)$, otherwise $n'' \geq 4 + 2(q - j + 1)$ and $cd(C_{n''}) \leq q - j + 1$ imply that we can recursively apply this proof to $C_{n''}$. Inequality $n' < 4 + 2(j-1)$ can be rewritten as $l_j + j + 1 < 4 + 2(j-1)$, from which $l_j < j + 1$ and hence $l_j \leq j$ follows. Similarly, from $n'' \geq 2i + 5 - l_j - j$ and $n'' < 4 + 2(q - j + 1)$ we get $l_j \geq 2(i - q) + j$. Relations $l_j \leq j$ and $l_j \geq 2(i - q) + j$ hold only if $l_j = j$.

Now, if v_2 is not incident to a chord, let $k = \min\{j' \mid l_{j'} \text{ is defined}\}$. Since l_q is defined, it follows $3 \leq k \leq q$. Now, let $C_{n'''}$ be the subgraphs of C_n induced by the nodes $x, v_1, v_2, \ldots, v_k, u_{l_k}, u_{l_k - 1}, \ldots, u_1$. Since $l_k = k \geq 3$, cycle $C_{n'''}$ has $n''' = 2k + 1 \geq 7$ nodes and chord distance equal to 1. This contradicts $C_n \in DH(\frac{3}{2}, *)$ since $s(C_{n'''}) > \frac{3}{2}$.

As a consequence, l_2 must be defined and, since $l_j = j$, then $l_2 = 2$. Chord (v_2, u_2) creates the cycle $C_5' = (x, v_1, v_2, u_2, u_1)$ having chord distance at most 1 (possibly, chords incident to v_1).

Symmetrically, the arguments above can be applied to show that r_{q-1} must be defined and that $r_{q-1} = p - 1$. Hence, chord $(v_{q-1}, u_{r_{q-1}})$ creates the cycle $C_5'' = (v_{q-1}, v_q, y, u_p, u_{p-1})$ having chord distance at most one (chord possibly incident to v_q). Finally, cycles C_5' and C_5'' along with path $(v_2, v_3, \ldots, v_q - 1)$ form the requested clepsydra. □

In [7] the authors proved that $G \in DH(\frac{3}{2}, *)$ if and only if G does not contain graphs listed in items from (i.) to (iv.) of Theorem 7 as induced subgraphs.

As a consequence, we can characterize any graph G in $DH(1,+)$ in terms of the following forbidden subgraphs of G^*: clepsydra and forbidden subgraphs of $DH(\frac{3}{2},*)$. Since none of these subgraphs have false twins, by Lemma 3, we can characterize graphs in $DH(1,+)$ without referring to twin graphs:

Theorem 4. *Let G be a graph. $G \in DH(1,+)$ if and only if the following graphs are not induced subgraphs of G:*

   i. H_n, for each $n \geq 6$;
  ii. cycles C_6 with $cd(C_6) = 1$;
 iii. cycles C_7 with $cd(C_7) = 1$;
 iv. cycles C_8 with $cd(C_8) = 1$ or $cd(C_8) = 2$;
  v. clepsydrae.

This theorem give us a basis to device a polynomial algorithm for the recognition of graphs in $DH(1,+)$. By Lemma 1, we know that $DH(1,+) \subset DH(\frac{3}{2},*)$, and comparing the characterization of graphs in $DH(\frac{3}{2},*)$ recalled above and 4, it follows that a graph G belongs to $DH(1,+)$ if it belongs to $DH(\frac{3}{2},*)$ and does not contain a clepsydra as induced subgraph. In [7], the authors give a polynomial algorithm for the recognizing whether a graph G belongs to $DH(\frac{3}{2},*)$; as a consequence, for our purposes it is sufficient to devise a polynomial time algorithm to check whether G contains a clepsydra as induced subgraph. Here is a brute force algorithm.

For each connected component of G, find all the cycles C_5 with $cd(C_5) \leq 1$. Then, consider a pair of cycles (C_5', C_5'') with at most one node in common, and two nodes $v_1 \in C_5'$ and $v_k \in C_5''$ both with degree 2 in C_5' and C_5'' respectively. Consider the induced subgraph G' of G obtained by removing all the nodes in C_5' and C_5'', but v_1 and v_k. If v_1 and v_k are connected in G', then G contains a clepsydra as induced subgraph. If necessary, repeat the above steps for all the possible pairs of cycles and nodes v_1 and v_k.

As regards the time complexity of the algorithm, it is obviously polynomial. The correctness, not given here, is obtained by proving that either C_5', C_5'', and the path connecting v_1 and v_k in G' induce a clepsydra or, otherwise, there exists another clepsydra in G.

Theorem 5. *The recognition problem for the class $DH(1,+)$ can be solved in polynomial time.*

6   Conclusions

In this paper we have introduced the $(k,+)$–distance-hereditary graphs, a parametric extension of the class of distance-hereditary graphs. These graphs can model communication networks having desirable connectivity properties.

In spite of the results provided in this work, many problems are left open. Among the most relevant, what is the largest constant k such that the recognition problem for $DH(k,+)$ can be solved in polynomial time? Moreover, several algorithmic problems are solvable in polynomial time for $DH(0,+)$. Can some of these results be extended to $DH(k,+)$, $k > 0$?

References

1. H. J. Bandelt and M. Mulder. Distance-hereditary graphs. *Journal of Combinatorial Theory, Series B*, 41(2):182–208, 1986.
2. A. Brandstädt, V. B. Le and J. P. Spinrad. *Graph classes - a survey*. SIAM Monographs on Discrete Mathematics and Applications, Philadelphia, 1999.
3. J. Bruck, R. Cypher, and C.-T. Ho. Fault-tolerant meshes with small degree. *SIAM J. on Computing*, 26(6):1764–1784, 1997.
4. S. Cicerone and G. Di Stefano. Graph classes between parity and distance-hereditary graphs. *Discrete Applied Mathematics*, 95(1-3): 197–216, August 1999.
5. S. Cicerone, G. Di Stefano, and M. Flammini. Compact-Port routing models and applications to distance-hereditary graphs. In *6th Int. Colloquium on Structural Information and Communication Complexity (SIROCCO'99)*, pages 62–77, Carleton Scientific, 1999.
6. S. Cicerone and G. Di Stefano. Graphs with bounded induced distance. *Discrete Applied Mathematics*, 108(1-2): 3–21, January 2001.
7. S. Cicerone and G. Di Stefano. Networks with small stretch number. In *Proc. 26th International Workshop on Graph-Theoretic Concepts in Computer Science, WG2000*, pages 95–106. Lecture Notes in Computer Science, vol. 1928, Springer-Verlag, 2000.
8. S. Cicerone, G. Di Stefano, and D. Handke. Survivable networks with bounded delay: The edge failure case (Extended Abstract). In *Proc. 10th Annual International Symp. Algorithms and Computation, ISAAC'99*, pages 205–214. Lecture Notes in Computer Science, vol. 1741, Springer-Verlag, 1999.
9. G. D'Ermiliis. Topologie di reti con particolari caratteristiche metriche. Master thesis, Faculty of Engineering, University of L'Aquila, 2000.
10. G. Di Stefano. A routing algorithm for networks based on distance-hereditary topologies. In *3rd Int. Colloquium on Structural Information and Communication Complexity (SIROCCO'96)*, 1996.
11. A. H. Esfahanian and O. R. Oellermann. Distance-hereditary graphs and multidestination message-routing in multicomputers. *Journal of Comb. Math. and Comb. Computing*, 13:213–222, 1993.
12. A. M. Farley and A. Proskurowski. Self-repairing networks. *Parallel Processing Letters*, 3(4):381–391, 1993.
13. M.R. Garey and D.S. Johnson. *Computers and Intractability. A Guide to the Theory of NP-completeness*. W.H. Freeman, 1979.
14. J. P. Hayes. A graph model for fault-tolerant computing systems. *IEEE Transactions on Computers*, C-25(9):875–884, 1976.
15. E. Howorka. Distance hereditary graphs. *Quart. J. Math. Oxford*, 2(28):417–420, 1977.
16. F. T. Leighton, B. M. Maggs, and R. K. Sitaraman. On the fault tolerance of some popular bounded-degree networks. *SIAM J. on Computing*, 27(6):1303–1333, 1998.
17. D. Peleg and A. Schaffer. Graph spanners. *Journal of Graph Theory*, 13:99–116, 1989.

On the Relationship
between Clique-Width and Treewidth
(Extended Abstract)

Derek G. Corneil[1] and Udi Rotics[2]

[1] Department of Computer Science, University of Toronto, Toronto, Ontario, Canada
[2] School of Mathematics and Computer Science, Netanya Academic College,
Netanya, Israel

Abstract. Treewidth is generally regarded as one of the most useful parameterizations of a graph's construction. Clique-width is a similar parameterizations that shares one of the powerful properties of treewidth, namely: if a graph is of bounded treewidth (or clique-width), then there is a polynomial time algorithm for any graph problem expressible in Monadic Second Order Logic, using quantifiers on vertices (in the case of clique-width you must assume a clique-width parse expression is given). In studying the relationship between treewidth and clique-width, Courcelle and Olariu showed that any graph of bounded treewidth is also of bounded clique-width; in particular, for any graph G with treewidth k, the clique-width of $G \leq 4 * 2^{k-1} + 1$.

In this paper, we improve this result to the clique-width of $G \leq 3 * 2^{k-1}$ and more importantly show that there is an exponential lower bound on this relationship. In particular, for any k, there is a graph G with treewidth $= k$ where the clique-width of $G \geq 2^{\lfloor k/2 \rfloor - 1}$.

1 Introduction

One of the most fruitful graph theoretical developments of the last few decades has been the concept of treewidth, pioneered by Robertson and Seymour. Loosely speaking, the treewidth of a graph captures a way of constructing the graph in a "tree like" fashion. The lower the treewidth of a graph, the closer it is to being a tree (connected treewidth 1 graphs are precisely trees). One of the major results in this area is that any problem expressible in Monadic Second Order Logic (which includes many NP-complete graph problems) when restricted to graphs of bounded treewidth k, has a linear time algorithm (albeit with a constant that grows exponentially with k). Although this result is very far reaching, it is still somewhat dissatisfying since many classes of "tame" graphs, for example cliques, have arbitrarily high treewidth, yet have simple linear time algorithms for most of the problems mentioned above.

The clique-width of a graph is another attempt to parameterize the construction of a graph so that sweeping claims can be made about the graph's tractability for polynomial time solutions to difficult problems.

A. Brandstädt and V.B. Le (Eds.): WG 2001, LNCS 2204, pp. 78–90, 2001.
© Springer-Verlag Berlin Heidelberg 2001

The clique-width of a graph G, denoted by $cwd(G)$, is defined as the minimum number of labels needed to construct G, using the four graph operations: creation of a new vertex v with label i (denoted $i(v)$), disjoint union ($\oplus$), connecting vertices with specified labels (η) and renaming labels (ρ). The construction of a graph G using the above four operation is represented by an algebraic expression called a k-*expression* where k is the number of labels used in the expression. More details are given in section 2. This notion was first introduced by Courcelle, Engelfriet and Rozenberg in [6] and has been studied extensively in recent years.

For example, cographs (graphs with no induced P_4s) are exactly the graphs of clique-width at most 2, and trees have clique-width at most 3, [9]. P_4-sparse (every 5 vertices have at most one P_4) and P_4-tidy graphs (no induced P_4 has more than one partner, where a partner is a vertex whose inclusion in the P_4 results in at least two distinct P_4s) have clique-width at most 4, [7]. The $(q, q-4)$ graphs for $q \geq 4$ and $(q, q-3)$ graphs for $q \geq 7$ have clique-width at most q [7,14]. A (q, t) graph is a graph in which every subgraph induced by q vertices contains at most t induced P_4's.

As mentioned above, the motivation for studying clique-width is analogous to that of treewidth. In particular, problems defined by Monadic Second Order Logic formulas, using quantifiers on vertices but not on edges, can be solved in polynomial time on any class of graphs $\mathcal{C}$ of clique-width at most k, for some fixed k, assuming that the k-expression defining the input graph is given. For details, cf. [7,8]. In addition, polynomial time algorithms can be obtained for other problems such as chromatic number, edge dominating set [13] and ID_q-partition problems [10], on any class of graphs $\mathcal{C}$ of clique-width at most k, for some fixed k, assuming that the k-expression defining the input graph is given.

This raises the obvious question about the relationship between treewidth and clique-width. Courcelle and Olariu [9] proved the following theorem.

Theorem 1 ([9]). *If the treewidth of G is k, then $cwd(G) \leq 2^{k+1} + 1 (= 4 * 2^{k-1} + 1)$.*

This theorem guarantees that any class of graphs of bounded treewidth is also of bounded clique-width and thus, from the perspective of algorithmic tractability, clique-width is a more powerful concept. Note that since the clique-width of a clique on n vertices is at most 2 and its treewidth is $n - 1$, the gap between a graph's treewidth and clique-width can be arbitrarily high. The above theorem also raises the questions of whether the bound can be improved (note that for trees, $k = 1$ and the theorem guarantees that the clique-width of a tree is ≤ 5, whereas it is known that the clique width of a tree is ≤ 3) and more importantly, whether the exponential bound represented in the theorem can be replaced by a polynomial bound. To make this more precise, we let $\mathcal{G}_k = \{G : twd(G) = k\}$ where $twd(G)$ denotes the treewidth of G, and want to determine whether $f(\mathcal{G}_k) = max\{cwd(G) : G \in \mathcal{G}_k\}$ can grow polynomially with k.

In answer to these two questions, our paper proves the following two theorems:

Theorem 2. $f(\mathcal{G}_k) \geq 2^{\lfloor k/2 \rfloor - 1}$. *(i.e., for any k, there is a graph G where $twd(G) = k$ and $cwd(G) \geq 2^{\lfloor k/2 \rfloor - 1}$.)*

Theorem 3. *If the treewidth of G is k, then $cwd(G) \leq 3 * 2^{k-1}$.*

A notion similar to clique-width is the notion of NLC-width. The NLC-width of a graph G, denoted as $NLCwidth(G)$, is the minimum number of labels needed to construct G using two graph operations called union ($\times_S$) and relabeling ($\circ_R$), see [15] for details. For every graph G, $NLCwidth(G) \leq cwd(G)$ and $cwd(G) \leq 2 \times NLCwidth(G)$ [12]. If G is of treewidth at most k then $NLCwidth(G) \leq 4 * 2^{k-1} - 1$, [15]. Using these results, Johansson [12] showed that if G is of treewidth at most k then $cwd(G) \leq 4 * 2^{k-1}$, thereby slightly tightening the Courcelle-Olariu bound.

By the relationship between clique-width and NLC-width mentioned above we obtain from Theorem 2 that for every k, there is a graph G, where $twd(G) = k$ and $NLCwidth(G) \geq 2^{\lfloor k/2 \rfloor - 2}$. Similarly, we obtain from Theorem 3 that if the treewidth of G is k, then $NLCwidth(G) \leq 3 * 2^{k-1}$.

Overview of the paper:

The second section of the paper introduces the notation and definitions used throughout the paper. In particular, we present the concept of k-trees and partial k-trees. (Note that partial k-trees encompass an alternative definition of graphs having treewidth k.) We also describe how to construct the k-trees that will be used in the proofs of the two theorems. Sections 3 and 4 present overviews of the proofs of Theorem 2 and 3 respectively. The paper ends with concluding remarks and directions for further research.

2   Notation and Definitions

The graphs we consider in this paper are undirected and loop-free. For a graph G and a vertex u of G, we denote by $N_G(u)$ the *neighborhood* of u in G, which is the set of all vertices in G which are adjacent to u. We denote by $N_G[u]$ the *closed neighborhood* of u in G which is equal to $N_G(u) \cup \{u\}$.

We first give more details on the definition of clique-width presented above. The *clique-width* of a graph G, denoted by $cwd(G)$, is defined as the minimum number of labels needed to construct G, using the four graph operations: creation of a new vertex v with label i (denoted $i(v)$), disjoint union ($\oplus$), connecting vertices with specified labels (η) and renaming labels (ρ). The operation $\eta_{i,j}$ ($i \neq j$) adds all edges (that are not already present) between every vertex of label i and every vertex of label j. The operation $\rho_{i \to j}$ renames all vertices of label i with label j. An expression built from the above four operations using k labels is called a k-*expression*. Each k-expression t uniquely defines a labeled graph $val(t)$ where the labels are integers $1, \ldots, k$ associated with the vertices and each vertex has exactly one label. We say that a k-expression t defines a graph G if G is equal to the graph obtained from the labeled graph $val(t)$ after removing its labels. The clique-width of a graph G is equal to the minimum k such that there exists a k-expression defining G.

For a k-expression t defining a graph G, we denote by $tree(t)$ the parse tree constructed from t in the usual way. The leaves of this tree are the vertices of G with their initial labels, and the internal nodes correspond to the operations of t and can be either binary corresponding to $\oplus$, or unary corresponding to η or ρ. If $k = cwd(G)$ we say that t is a *clique-width expression* of G and we say that $tree(t)$ is a *clique-width parse tree* for G. For a vertex u of G and an internal node a of $tree(t)$, such that u occurs (as a leaf) in the subtree of $tree(t)$ rooted at a, the *label of u at a* is defined as the label that u has immediately before the operation a is applied on the subtree of $tree(t)$ rooted at a.

As mentioned above, G being a partial k-tree is equivalent to G having treewidth at most k. We now define k-trees and present notation for the k-tree that will be used to establish the lower bound presented in Section 3.

We denote by K_k the clique with k vertices. For a vertex v and a set of vertices S we say that v is *universal* to S if v is adjacent to all the vertices in S. A graph G is a k-*tree* if G is either a K_k or is formed from a k-tree G', by adding a new vertex that is universal to a K_k in G'. Note that 1-trees are precisely trees. G is a *partial k-tree* (i.e., $twd(G) \le k$) if G is a partial subgraph of a k-tree. This recursive definition of k-trees immediately suggests that the construction of a given k-tree G can be represented by a labeled tree T_G. To illustrate this concept, we now describe F, the k-tree that will be used to establish the lower bound and the construction tree T_F corresponding to F. In this and in the next section we denote T_F by T.

We let the initial K_k be on vertices $\{1, 2, \cdots, k\}$. When a vertex x is added to an existing k-tree, we let $S_x \cup \{x'\}$ denote the K_k to which x is universal. The vertex x' is called the *parent of x*, represented by $p(x)$. S_x, the remaining vertices in the K_k, are called the *sources of x*. In the construction tree T, x becomes a child of the vertex x', and S_x is the label associated with x. To "start" the tree T, we let k be the root of T. v, the first vertex added to F is a child of k and $S_v = \{1, 2, \cdots, k-1\}$. To continue the construction of F (as represented by T), we let v have k children $v_1, v_2, \cdots, v_k$ where $S_{v_i} = (S_v \cup \{k\}) - \{i\}$. Thus v_i is adjacent to the K_k formed on $\{1, \cdots, i-1, i+1, \cdots, k, v\}$. Often we will denote v_i by "$-i$", thereby indicating that the vertex i in $S_v \cup p(v)$ has been "replaced" by v, the parent of v_i. In this case we say that v_i is the $-i$ child of its parent.

Note that this process of giving any node in T k children can continue indefinitely. (In particular, node x has children $x_1, x_2, \cdots, x_k$ where $S_{x_i} = (S_x \cup p(x)) -$ the ith element of $S_x \cup p(x)$.) For x a node of T, we let T_x denote the subtree of T rooted at x and let L_x^h (the *hth level of T_x*) denote the nodes of T_x at distance h from x.

F, the graph used in Section 3, consists of S_v, k and v as described above as well as T_v consisting of the complete k-ary tree up to level $\alpha = (2k+1)2^{\lfloor k/2 \rfloor + 1}$.

Thus all leaves of T are in L_v^α. Note that each node of T corresponds to a vertex of F.

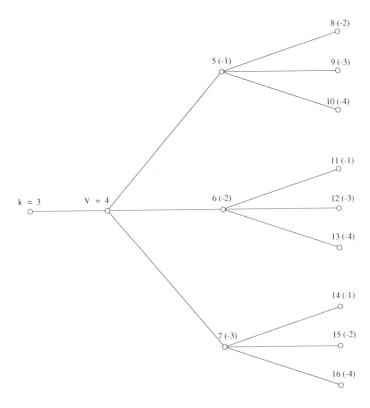

Fig. 1. The beginning of the construction tree T for the graph F when $k = 3$.

Example 1. Figure 1 illustrates the beginning of the construction tree T for the graph F when $k = 3$. The root of this tree is k. v, the first vertex added to F is vertex 4 and is the child of the root. The first level of the tree T_v is $L_v^1 = \{5, 6, 7\}$ and the second level of T_v is $L_v^2 = \{8, 9, \cdots, 16\}$. We omit from the figure the labels S_u of the nodes of the tree. Instead, we employ a (-i) label for a node u indicating that u is the $-i$ child of its parent. For example, node 12 is the -3 child of node 6 which is the -2 child of node 4. The set of sources of vertex 12 is $S_{12} = \{1, 4\}$. The parent of vertex 12 is $p(12) = 6$. Figure 2 illustrates the graph corresponding to the construction tree T of Figure 1. Note that the graph of Figure 2 is the subgraph of the graph F for $k = 3$, induced by the vertices $\{1, 2, \cdots, 16\}$.

For x a vertex of F, such that $x \notin \{1, \cdots, k\}$ we say that T_x is a *complete tree* if L_x^{2k+1} is not empty (i.e., the distance in T between x and v is at most

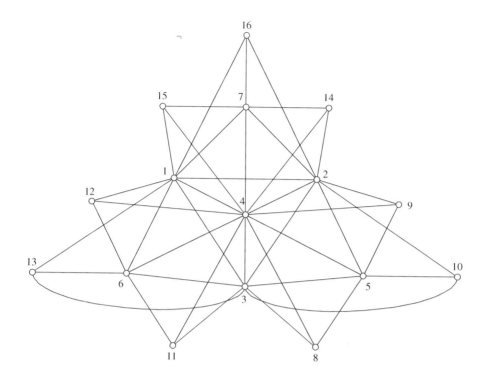

Fig. 2. The graph corresponding to the construction tree of Figure 1.

$\alpha - (2k + 1)$.) We now state, without proofs, some properties of complete trees that will be used in subsequent arguments.

Claim 1 *Let T_u be a complete tree in T. Then for every $R \subseteq S_u$, there exists vertex $r_R \in N_F[u] \cap T_u$ such that $S_{r_R} \cap S_u = R$. Furthermore we can choose these vertices such that the set $\mathcal{R} = \{r_R : R \subseteq S_u\} - \{u\}$ is an independent set.*

Corollary 1. *Let T_u be a complete tree in T. Then for any $Q \subset S_u$ with $q_1, q_2 \in Q$, there exist vertices $u_1, u_2 \in T_u$, at level at most $2k + 1$ of T_u, such that u_1 is neither an ancestor nor a descendent of u_2, $S_{u_i} = Q - \{q_i\} \cup S'_{u_i}$ where $S'_{u_i} \cap Q = \emptyset, i = 1, 2$, all vertices of S'_{u_i} are in T_u and $S'_{u_1} \cap S'_{u_2} = \emptyset$.*

For $x \in T$ we let P_x denote a path $(x = x_0, x_1, \cdots, x_\beta)$ of T_x where x_β is a leaf and :

$$\begin{aligned} &- S_{x_i} = S_{x_{i-1}}, \ i > 0 \\ &- p(x_i) = x_{i-1}, \ i > 0. \end{aligned} \qquad (\star)$$

For any vertex $x \in T$ where T_x is a complete tree and $S \subseteq S_x$ we let: $X_x^S = \{y : y \in L_x^{2k+1} \wedge S_y \cap S_x = S\}$ (i.e., X_x^S denotes the set of vertices at level $2k + 1$ that have S as the only S_x vertices in their sources), and $P_x^S = \{P_y : y \in X_x^S\}$ (i.e., P_x^S denotes the paths rooted at $y \in X_x^S$ that satisfy $(\star)$ above).

Claim 2 $|X_x^S| \geq (k - |S|)^{k+1}$.

In order to argue about the clique-width of F, we let $\mathcal{T}$ denote a clique-width expression tree for F. For a an internal node of $\mathcal{T}$, we let $\mathcal{T}_a$ denote the subtree of $\mathcal{T}$ rooted at a and let V_a represent the leaves (i.e., vertices of F) of $\mathcal{T}_a$. As the following claim shows, having a set of vertices outside V_a together with certain neighbours inside V_a establishes a lower bound on the number of distinct labels required at a (i.e., to label the vertices of V_a at node a of $\mathcal{T}$).

Claim 3 *Let X be a set of l vertices in $V - V_a$ where for each subset of these vertices, there is a vertex inside V_a adjacent to all vertices of this subset of X and to no other vertices in X. Then at least 2^l labels are required to label the vertices of V_a.*

For $\mathcal{T}_1$ a subtree of $\mathcal{T}$ and P_y a path of T satisfying $(\star)$, we say that P_y is *full with respect to* $\mathcal{T}_1$ if all vertices of P_y are in $\mathcal{T}_1$. Given a P_x^S we will often choose a lowest possible internal node of $\mathcal{T}$, say a, such that some path in P_x^S is full with respect to $\mathcal{T}_a$. (i.e., for some path $P \in P_x^S$ all of its vertices are in V_a; P is said to be *full with respect to* a.) If $T_u \cap V_a = \emptyset$, for some u, we say that T_u is *empty with respect to* a.

The following claim is the key to the lower bound result. First, it is the bottleneck for improving our lower bound; in particular, $2^{\lfloor k/2 \rfloor - 1}$ is the largest value for which we could find a proof and any improvement in this value would be a major step towards achieving a better lower bound. Second, the claim is the basis for the inductive proof of the lower bound. Roughly speaking, the proof first establishes a vertex a in $\mathcal{T}$ such that S^*, a small set of vertices of F all must have unique labels at a. These unique labels are required because of the full path with respect to a. Now, the existence of a subtree of T that is empty with respect to a allows other vertices of T to be identified so that the claim may be reused. By doing so, a new vertex is added to S^* and the process of incrementing S^* by one vertex at a time continues until the lower bound is surpassed.

Claim 4 *Assume $cwd(F) < 2^{\lfloor k/2 \rfloor - 1}$. Let x be a node in T where T_x is a complete tree, let S be a strict subset of S_x, and let a be a lowest node in $\mathcal{T}$ such that some path in P_x^S is full with respect to a. Then there is a $y \in X_x^S$ such that T_y is empty with respect to a.*

3 Lower Bound

The purpose of this section is to give an overview of the techniques used to establish the lower bound stated in Theorem 2.

Theorem 2 $f(\mathcal{G}_k) \geq 2^{\lfloor k/2 \rfloor - 1}$.

In particular we will examine the k-tree F defined in Section 2 and proceed by assuming $cwd(F) < 2^{\lfloor k/2 \rfloor - 1}$.

We let $\mathcal{T}$ be a clique-width expression tree for F. Initially we will examine the vertices in $X_v^{2,\cdots,k-1}$ and let a be a lowest node in $\mathcal{T}$ such that some path in $P_v^{2,\cdots,k-1}$ is full with respect to a. We will then show that at least $\lfloor k/2 \rfloor$ vertices of $\{2,\cdots,k-1\}$ are in V_a and have unique labels at a. We then proceed by induction by identifying other vertices in T and looking at b, a lowest node in $\mathcal{T}$ such that some path rooted at one of these vertices is full with respect to b. We show that $V_a \subseteq V_b$ and that the set of vertices requiring unique labels at a also requires unique labels at b and that this set must be augmented by a new vertex that requires a new label. This augmentation continues until eventually this set has cardinality equal to $2^{\lfloor k/2 \rfloor - 1}$, thereby proving the theorem. α, the depth of T is chosen so that all subtrees used to generate full paths, are complete and have depth at least $2^{\lfloor k/2 \rfloor} + 2k + 1$. The length of the full paths generated by these subtrees will be at least $2^{\lfloor k/2 \rfloor}$, since the root of each such path is at level $2k + 1$ of its corresponding subtree.

Following the above outline, we start by stating some claims about $\mathcal{T}_a$.

Claim 5 *At least $\lfloor k/2 \rfloor$ vertices of $S_v - \{1\}$ (i.e., $\{2,\cdots,k-1\}$) are in V_a and each must have a unique label at a.*

We now prepare for the general induction step, each execution of which will add one more vertex requiring a new unique label. Thus executing the induction step $2^{\lfloor k/2 \rfloor - 1} - \lfloor k/2 \rfloor$ times will complete the proof of the theorem. First we let S^* denote a subset of $S_v - \{1\}$ such that all elements of S^* are in V_a and $|S^*| = \lfloor k/2 \rfloor$. (Note that there may be more elements of $S_v - \{1\}$ that are in V_a but we only consider $\lfloor k/2 \rfloor$ of them.) By Claim 5 the set S^* exists and all the vertices of S^* have different labels at a.

Let u be a vertex in $X_v^{S_v - \{1\}}$ where T_u is empty with respect to a. (u is guaranteed by Claim 4.) Let s_1 and s_2 be arbitrary vertices in S^*. By Corollary 1, T_u contains vertices u_1 and u_2, at level at most $2k + 1$ of T_u, such that u_1 is neither an ancestor nor a descendant of u_2 and:

- $S_{u_i} = S_{u_i}^* \cup S_{u_i}'$ $i = 1, 2$ where $S_{u_i}^* = S^* - \{s_i\}$
- $S_{u_i}' \cap S^* = \emptyset$ $i = 1, 2$
- $S_{u_1}' \cap S_{u_2}' = \emptyset$.

Note that $|S_{u_i}^*| = \lfloor k/2 \rfloor - 1$. Let u_i' be an arbitrary element in S_{u_i}' for $i = 1, 2$ and set $U = \{u_1, u_2\}$. Clearly $S^* = \bigcup \{S_{u_i}^* : u_i \in U\}$. Let $P_U = \bigcup \{P_{u_i}^{S_{u_i} - \{u_i'\}} : u_i \in U\}$ and define b to be a lowest vertex in $\mathcal{T}$ such that some path in P_U is full with respect to b.

Claim 6 *b is an ancestor of a.*

Claim 7 *For all $u_i \in U$, there is a $y \in X_{u_i}^{S_{u_i} - \{u_i'\}}$ such that T_y is empty with respect to b.*

Claim 8 *All S^* vertices have different labels at b.*

Let $u_1 \in U$ satisfy the existence of a vertex $z \in X_{u_1}^{S_{u_1}-\{u_1'\}}$ such that z is the root of a full path with respect to b.

Claim 9 *There are at least $\lfloor k/2 \rfloor$ vertices of $S_{u_1} - \{u_1'\}$ that are in V_b.*

Since $|S_{u_1}^*| = \lfloor k/2 \rfloor - 1$, there is a vertex $\widetilde{u_1} \in S_{u_1}' - \{u_1'\}$ that is in V_b.

Claim 10 *At b, the label of $\widetilde{u_1}$ is different than the labels needed for the S^* vertices.*

Thus this step has shown that at least $\lfloor k/2 \rfloor + 1$ distinct labels are required at b. To continue the process, we will augment the set U and define c to be a lowest vertex in $\mathcal{T}$ such that some path in P_U is full with respect to c. It will then be shown that c is an ancestor of b and that all vertices in S^* must have different labels at c. A new vertex will be identified that must have a different label at c than all vertices in S^* and it will be added to S^*. This process continues by renaming c to be b and augmenting U. For augmenting U we shall use the two vertices $u_{1,1}$ and $u_{1,2}$ guaranteed by the following claim.

Claim 11 *Let $\widetilde{s}$ be an arbitrary vertex in $S_{u_1}^*$. Then there exists vertices $u_{1,1}$ and $u_{1,2}$ in T_{u_1} such that $u_{1,1}$ is neither an ancestor nor a descendant of $u_{1,2}$ and:*

- $S_{u_{1,1}} = S_{u_{1,1}}^* \cup S_{u_{1,1}}' : S_{u_{1,1}}^* = S_{u_1}^*; \ S_{u_{1,1}}' \cap S^* = \emptyset$
- $S_{u_{1,2}} = S_{u_{1,2}}^* \cup S_{u_{1,2}}' : S_{u_{1,2}}^* = S_{u_1}^* - \{\widetilde{s}\} \cup \{\widetilde{u_1}\}; \ S_{u_{1,2}}' \cap S^* = \emptyset, \ S_{u_{1,1}}' \cap S_{u_{1,2}}' = \emptyset$.

Augmenting U: u_1 (a vertex in U such that there is $z \in X_{u_1}^{S_{u_1}-\{u_1'\}}$ where z is the root of a full path with respect to b) will be replaced by the vertices $u_{1,1}$ and $u_{1,2}$ satisfying the conditions defined in Claim 11. As before, we let $u_{1,1}'$ and $u_{1,2}'$ be arbitrary elements in $S_{u_{1,1}}'$ and $S_{u_{1,2}}'$ respectively. After reindexing the elements of U to be $u_1, u_2, \cdots$, we define:

- $S^* = \bigcup \{S_{u_i}^* : u_i \in U\}$ and
- $P_U = \bigcup \{P_{u_i}^{S_{u_i}-\{u_i'\}} : u_i \in U\}$.

We now state various claims about U.

Claim 12 *The following hold for U:*

1. *No vertex in U is an ancestor (in T) of any other vertex in U.*
2. *$\forall u \in U, S_u = S_u^* \cup S_u'$ where $|S_u^*| = \lfloor k/2 \rfloor - 1$.*
3. *$\forall u, w$ distinct elements of $U, S_u' \cap S_w = \emptyset$.*
4. *$\forall u \in U$, there is a vertex $y \in P_u^{S_u-\{u'\}}$ such that T_y is empty with respect to b.*

5. *All vertices in S^* have different labels at b.*

We now define c to be a lowest vertex in $\mathcal{T}$ such that some path in P_U is full with respect to c. (Note that c could be b.) In order for the process to continue, it can be shown that analogues (where b is replaced by c) to Claims 6, 7, 8, 9, and 10 hold. Where appropriate, the proofs use the facts established in Claim 12.

Finally, we rename c to be b and again augment U by replacing u_1 with the vertices $u_{1,1}$ and $u_{1,2}$ guaranteed by Claim 11. This augmentation continues until $|S^*| > 2^{\lfloor k/2 \rfloor - 1}$ thereby completing the proof of Theorem 2. Note that for these augmentation steps to be valid, for every vertex u in U the depth of the tree T_u must be at least $2k + 1 + 2^{\lfloor k/2 \rfloor}$. This is assured by our initial choice of α, the depth of T_v having been set to $(2k + 1)2^{\lfloor k/2 \rfloor + 1}$.

4   Upper Bound

We now turn our attention to establishing the upper bound stated in Theorem 3.

Theorem 3 *If the treewidth of G is k, then $cwd(G) \leq 3 * 2^{k-1}$.*

To prove this theorem, we will first prove it for G a k-tree and then indicate the modifications for partial k-trees. We start by developing notation for G, consistent with that presented in Section 2. We then present an algorithm that is guaranteed to generate a clique-width expression tree that uses at most $3 * 2^{k-1}$ labels. It should be clear to the reader that this algorithm is an extension of the standard algorithm that shows that the clique-width of a tree (i.e., a 1-tree) is at most 3.

Given G, let T be a construction tree for it and let x be an arbitrary vertex in T where $S_x = \{1, 2, \cdots, k-1\}$ and $p(x) = k$. x can be the root of many subtrees of the construction tree of G. Each child of x will be a $-i$ node for some $i \in \{1, 2, \cdots, k\}$. (i.e., such a node will be universal to the k-clique consisting of $\{1, 2, \cdots, k\} - \{i\} \cup \{x\}$.) Note that the subtree rooted at a $-i$ child of x could contain vertices universal to any subset of $\{1, 2, \cdots, k\} - \{i\} \cup \{x\}$. To handle this situation, the recursive labeling of the subtree will assign up to 2^k labels to the vertices of the subtree, thereby allowing all possible connections into $\{1, 2, \cdots, k\} - \{i\} \cup \{x\}$.

The Algorithm:

The algorithm will proceed by bottom-up dynamic programming on T. At each stage it will assume that the subtrees whose roots are children of a particular x have been labeled using at most $3 * 2^{k-1}$ labels and that each vertex in the subtree rooted at a $-i$ child of x is labeled with one of 2^k labels indicating the subset of $\{1, 2, \cdots, k\} - \{i\} \cup \{x\}$ to which it is universal. Clearly this assumption holds for the leaves of T.

The algorithm, using at most $3 * 2^{k-1}$ labels, will label the vertices of T_x so that each vertex in T_x ends with one of 2^k labels indicating the subset of

$\{1, 2, \cdots, k\}$ to which it is adjacent. For $1 \leq i \leq k$, let y_i denote a $-i$ child of x. Let $\mathcal{T}_i = \{T_{y_i} : y_i \text{ is a } -i \text{ child of } x\}$. All vertices in $\mathcal{T}_i$ adjacent to the same subset of $\{1, 2, \cdots, k\} - \{i\} \cup \{x\}$ will be assumed to have the same label.

We now describe the steps of the algorithm for labeling the vertices in T_x.

1. Construct a new union node with the two children being the expression tree for $\mathcal{T}_1$ and x. x is given a new label, so we now have $2^k + 1$ labels being used. The labels of vertices that are going to be involved in subsequent η operations are considered to be "active". Thus at this stage we have $2^k + 1$ "active" labels consisting of the label of x and the labels corresponding to all subsets of $\{2, 3, \cdots, k, x\}$. Now we add all edges to x and COLLAPSE label class $X \cup \{x\}$ into X for each $X \subseteq \{2, 3, \cdots, k\}$. Thus we have $2^{k-1} + 1$ "active" labels. We let T' denote the current expression tree.

2. For i from 2 to k consider the subtrees $\mathcal{T}_i$ rooted at the $-i$ children of x: For each i, we augment T' by adding a new union node with the two children being the expression tree for $\mathcal{T}_i$ and T'. We make sure via ρ operations that the vertices in the expression tree for $\mathcal{T}_i$ that are adjacent to subsets of $\{i + 1, \cdots, k\}$ will have the same label as the vertices in T' which are adjacent to the same subsets of $\{i + 1, \cdots, k\}$. Any vertices of $\mathcal{T}_i$ adjacent to $\{1, \cdots, i - 1\}$ union a subset of $\{i + 1, \cdots k, x\}$ are given new "active" labels. Now we add all edges to x and COLLAPSE label class $X \cup \{x\}$ into X for each $X \subseteq \{1, \cdots, k\} - \{i\}$.

The algorithm ensures that vertices in different subtrees adjacent to the same subset of $\{1, 2, \cdots, k\}$ will have the same label. (For example, if $k = 4$, then step 1 ensures that x which is adjacent to $\{1, 2, 3, 4\}$ has an "active" label; furthermore, in this step all vertices adjacent to each $X \subseteq \{2, 3, 4\}$ have "active" labels. When $i = 2$ in step 2, vertices adjacent to any of the subsets: $\{1\}, \{1, 3\}, \{1, 4\}, \{1, 3, 4\}$ will be given "active" labels. When $i = 3$, the subsets $\{1, 2\}, \{1, 2, 4\}$ are added and when $i = 4$, the subset $\{1, 2, 3\}$ is added.) The COLLAPSE operation ensures that once an edge has been added from a vertex y to x, the label of y is reset to be that of X where $X \cup \{x\}$ is the subset of $\{1, 2, \cdots, k\} \cup \{x\}$ to which y is adjacent.

To prove the theorem, we show that the above algorithm constructs G and uses no more than $3 * 2^{k-1}$ labels. The fact that G is constructed follows by a straightforward induction argument. The proof that the number of labels is bounded by $3 * 2^{k-1}$ is contained in the journal version of the paper.

5   Concluding Remarks

The most important result in this paper is the fact that the clique-width of a graph can be exponentially higher than its treewidth. We fully expect that the bound expressed in Theorem 2 can be improved; the most interesting question, however, is to find a lower bound that confirms the bound achieved through some algorithm. In particular, we expect that the upper bound expressed in Theorem

3 is the best possible. Note that it is correct for $k = 1$ (trees) and for $k = 2$ (shown by a tedious exhaustive argument).

It is well appreciated that the most important open problems in the study of clique-width are the resolution of the general recognition problem (given G and integer k, is $cwd(G) \leq k$ - strongly believed to be NP-complete) and the fixed k recognition problem for $k \geq 4$. (The fixed k recognition problem can be solved in polynomial time for $k \leq 3$, in particular the case when $k = 1$ is trivial, for $k = 2$ these are the cographs [9] which can be recognized in linear time [5] and for $k = 3$ an $O(n^2m)$ algorithm is presented in [4]). It is interesting to note that the corresponding problems for treewidth are resolved. In particular, Arnborg, Corneil and Proskurowski [1] showed that the general treewidth recognition problem (given G and integer k, is $twd(G) \leq k$) is NP-complete whereas the fixed recognition problem is in P. Bodlaender [2] later presented a linear time algorithm for the fixed k treewidth recognition problem.

It seems as though the stumbling block for both problems is the difficulty in developing good arguments to provide strong lower bounds on the clique-width of a graph. Hopefully the techniques presented in Section 3 can assist.

Another avenue of promising research is the development of polynomial time algorithms to determine the clique-width of restricted families of graphs, especially those where the clique-width can be arbitrarily large. Such families include permutation graphs (even bipartite permutation graphs [3]), planar graphs and interval graphs (even unit interval graphs) [11]. Again progress in this area depends on lower bound arguments that show that the clique-width achieved by a particular algorithm is the best possible. It is interesting to note that it is an open and seemingly difficult problem to find a clique-width algorithm for k-trees, even for $k = 2$.

Acknowledgements

The authors wish to thank the Natural Science and Engineering Research Council of Canada for their financial support and the Fields Institute at the University of Toronto for hosting the authors' visit during this research.

References

1. S. Arnborg, D. G. Corneil and A. Proskurowski, " Complexity of finding embeddings in a k-tree" *SIAM J. Alg. Discrete Methods* 8 (1987) 277-284.
2. H. L. Bodlaender, "A linear time algorithm for finding tree-decompositions of small treewidth" *SIAM J. Comput.* 25 (1996) 1305-1317.
3. A. Brandstädt and V.V. Lozin, "On the linear structure and clique-width of bipartite permutation graphs" *Rutcor Research Report* 29-2001 (2001).
4. D. G. Corneil, M. Habib, J. M. Lanlignel, B. Reed and U. Rotics, "Polynomial time recognition of clique-width ≤ 3 graphs (Extended Abstract)" accepted to Latin American Theoretical INformatic, LATIN'2000.
5. D. G. Corneil, Y. Perl and L. Stewart, "A linear recognition algorithm for cographs" *SIAM J. Comput.* 14 (1985) 926-934.

6. B. Courcelle, J. Engelfriet and G. Rozenberg, "Handle-rewriting hypergraphs grammars" *J. Comput. System Sci.* 46 (1993) 218-270.

7. B. Courcelle, J.A. Makowsky and U. Rotics, "Linear time solvable optimization problems on graphs of bounded clique-width" *Theory of Computing Systems* 33 (2000) 125-150.

8. B. Courcelle, J.A. Makowsky, and U. Rotics, "On the fixed parameter complexity of graph enumeration problems definable in monadic second order logic" to appear in *Disc. Appl. Math.*

9. B. Courcelle and S. Olariu, "Upper bounds to the clique-width of graphs" *Disc. Appl. Math.* 101 (2000) 77–114.

10. M.U. Gerber and D. Kobler, "Algorithms for vertex partitioning problems on graphs with fixed clique-width" *submitted* (2000).

11. M.C. Golumbic and U. Rotics, "On the clique-width of some perfect graph classes" *Internat. J. Found. Comput. Sci* 11 (2000) 423–443.

12. O. Johansson, "Clique-decomposition, NLC-decomposition, and modular decomposition - relationships and results for random graphs" *Congressus Numerantium* 132 (1998) 39-60.

13. D. Kobler and U. Rotics, "Polynomial algorithms for partitioning problems on graphs with fixed clique-width (Extended Abstract)" Proceedings of the 12th Annual ACM-SIAM Symposium on Discrete Algorithms, (2001) 468-476.

14. J.A. Makowsky and U. Rotics, "On the classes of graphs with few P_4's" *International Journal of Foundations of Computer Science* 10 (1999) 329–348.

15. E. Wanke, "k-NLC graphs and polynomial algorithms" *Discrete Applied Math.* 54 (1994) 251-266.

Planarity of the 2-Level Cactus Model

Sabine Cornelsen[1], Yefim Dinitz[2], and Dorothea Wagner[1]

[1] Dept. of Computer and Information Science, University of Konstanz, Germany
{Sabine.Cornelsen, Dorothea.Wagner}@uni-konstanz.de
[2] Dept. of Computer Science, Ben-Gurion University, Israel
dinitz@cs.bgu.ac.il

Abstract. The 2-level cactus introduced by Dinitz and Nutov in [5] is a data structure that represents the minimum and minimum+1 edge-cuts of an undirected connected multi-graph G in a compact way. In this paper, we study planarity of the 2-level cactus, which can be used, e.g., in graph drawing. We give a new sufficient planarity criterion in terms of projection paths over a spanning subtree of a graph. Using this criterion, we show that the 2-level cactus of G is planar if the cardinality of a minimum edge-cut of G is not equal to 2, 3 or 5. On the other hand, we give examples for non-planar 2-level cacti of graphs with these connectivities.

1 Introduction

Edge connectivity is a fundamental structural property of a graph. In the last decade, not only the properties of minimum cuts but also the number [14,12,9] and structure [1] of near minimum cuts were examined. Galil and Italiano [8] and Dinitz and Westbrook [3,7] developed models for all 1 and 2 cuts and all 2 and 3 cuts, respectively. Based on these two models, Dinitz and Nutov introduced the so called 2-level cactus model – a data structure that represents the minimum and minimum+1 edge cuts of an undirected multi-graph with connectivity $\lambda \geq 3$ in a compact way [5]. There is no other so compact model, and no other compact graph model for these cuts known, for the best of our knowledge. The above models imply, in particular, fast incremental maintenance algorithms [8,7,5].

The 2-level cactus model (or "2-level cactus", for simplicity) generalizes the cactus model of all minimum cuts [4]. In case of odd connectivity $\lambda > 3$, the 2-level cactus is really a cactus, that is a connected graph in which every edge is contained in at most one simple cycle. Some cuts, however, are represented only implicitly in the graph of the model. In order to reduce this implicitness, we add some auxiliary edges. We call the resulting graph the extended 2-level cactus. The main question considered in this paper is whether the modeling graph is planar, for both odd and even cases.

The proof of planarity is based on properties of the set of projection paths of auxiliary edges, that is the set of (shortest) paths in the 2-level cactus between the end nodes of auxiliary edges. To obtain planarity, we give a new sufficient planarity criterion, generalizing a corollary to the criterion of MacLane [11].

A. Brandstädt and V.B. Le (Eds.): WG 2001, LNCS 2204, pp. 91–102, 2001.
© Springer-Verlag Berlin Heidelberg 2001

The question of planarity is not only of graph theoretical interest, it is also useful for algorithmic purposes. But the main reason, why we study planarity of the 2-level cactus, is an application in graph drawing. We are interested in drawing all small cuts ("bottlenecks") of a graph. Clearly, for a presentation on a screen, planarity is a key property. A first approach in visualizing the minimum cuts is done in [2], utilizing the cactus model for all minimum cuts [4]. The 2-level cactus might be a next step in this direction.

This paper is organized as follows. Sect. 2 first introduces MacLane's planarity criterion. Then, we generalize a corollary of MacLane's criterion. Sect. 3 introduces the 2-level cactus. In Sect. 4, we show, first, that it is planar for connectivity $\lambda = 1$ and for any even connectivity $\lambda \geq 4$. In case of odd connectivity, we show that the 2-level cactus together with the auxiliary edges is planar if $\lambda > 5$. We also give examples of non-planar (extended) 2-level cacti of graphs with connectivity $\lambda = 2$, $\lambda = 3$, and $\lambda = 5$. We conclude the paper by Sect. 5 with some remarks on how we would like to choose the faces of a planar embedding.

2   Planarity of Trees with Additional Edges

Let $E_1 \Delta E_2 = E_1 \setminus E_2 \cup E_2 \setminus E_1$ be the *ring sum* of two sets E_1 and E_2. Let $\mathfrak{E}_G$ be the vector space on the subsets of an edge-set E of a graph G over $\mathbb{F}_2$ under the ring sum operation Δ. The set $\mathfrak{Z}_G$ of all cycles and unions of edge-disjoint cycles is a subspace of the vector space $\mathfrak{E}_G$ and is called the *cycle space* of G. A *2-basis* of G is a basis of the cycle space of G, such that every edge occurs in at most two elements of this basis.

Theorem 1 (Planarity Criterion of MacLane [11]). *A graph is planar if and only if it has a 2-basis. Moreover, any 2-basis of a 2-connected graph consists of all but one facial cycle of some of its planar representations.*

A short proof of MacLane's planarity criterion can be found in [13].

A basis of the cycle space can be constructed from a spanning tree: Let T be a spanning tree of a connected graph G. For an edge $e = \{v, w\}$ in $G - T$, let p_e denote the set of edges on the path in T between v and w; called the *projection path* of e. Then, $\{\{e\} \cup p_e | e$ edge in $G - T\}$ is a basis of the cycle space $\mathfrak{Z}_G$. Thus, there is the following immediate corollary of MacLane's Planarity Criterion.

Corollary 1. *A graph G is planar if there is a spanning tree T of G such that every edge in T is contained in at most two projection paths.*

The following lemma gives a sufficient planarity criterion under somewhat weaker conditions.

Lemma 1. *A graph G is planar if there is a spanning tree T of G such that for any edge $e = \{v, w\}$ in T, the number of projection paths that contain e and one more edge in T incident to v is at most two.*

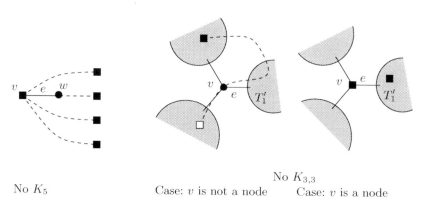

No K_5 Case: v is not a node   Case: v is a node

Fig. 1. Illustration of the proof of Lemma 1. Rectangularly shaped vertices are nodes.

Proof. Let $G = (V, B \dot\cup S)^1$ be a connected graph such that $T = (V, B)$ is a spanning tree of G that fulfills the condition of the lemma. We use Kuratowski's Theorem [10].

G does not contain a K_5: Suppose G contains a subdivision of a K_5 as a subgraph. We call the five vertices of this K_5 nodes. Let T' be the smallest subtree of T that contains all nodes of the K_5. Let v be a leaf of T'. Then, v is a node. Let $e = \{v, w\}$ be the edge incident to v in T'. At most one of the four subdivision paths of the K_5 that connect v to the other nodes can contain vertex w. This situation is illustrated in Fig. 1 (left). Thus, there are at least three subdivision paths that (i) connect the two connected components T_1 and T_2 of $T - \{e\}$ and that (ii) do not contain w. Such a subdivision path contains at least one edge $s \in S$ with one end vertex in T_1 and one end vertex in T_2. The projection path of s contains e and one more edge incident to w. Hence, there are at least three projection paths containing e and another edge incident to w. This contradicts the precondition.

G does not contain a $K_{3,3}$: Suppose G contains a subdivision of a $K_{3,3}$. We call the six vertices of this $K_{3,3}$ nodes. We distinguish the two parts of the $K_{3,3}$ as white nodes and black nodes. Let T' be the smallest subtree of T that contains all nodes of the $K_{3,3}$. Let v be a vertex that has maximum degree in T'.

Let us first consider the case that $\deg_{T'}(v) \geq 3$ and v is not a node as illustrated in Fig. 1 (middle). At most one subdivision path can contain v. So, there is at least one connected component T_1' of $T' - \{v\}$ that contains none of the end nodes of such a subdivision path. Let e be the edge that connects T_1' to v in T'. Each subdivision path incident to a node in T_1' and a node that is not in T_1' must contain an edge from S whose projection path contains e and another edge incident to v. If T_1' contains b black nodes and w white nodes, there are

---

[1] With $M_1 \dot\cup M_2$ we denote the disjoint union of two sets M_1 and M_2.

$b(3 - w) + w(3 - b)$ such paths. As we have $1 \leq b + w \leq 4$, there are at least 3 such paths. This contradicts the precondition.

Now consider the case that $\deg_{T'}(v) \geq 3$ and v is a node – say a black node as illustrated in Fig. 1 (right). Let T'_1 be a connected component of $T' - \{v\}$ that contains at least one black node. Let e be the edge that connects T'_1 to v in T'. If T'_1 contains b black nodes and w white nodes, there are $b(3 - w) + w(3 - b - 1)$ subdivision paths between nodes in T'_1 and nodes not in $T'_1 + \{v\}$. As we have $1 \leq b + w \leq 3$, there are at least 3 such paths and thus at least 3 projection paths containing e and another edge incident to v – a contradiction.

If $\deg_{T'}(v) = 2$, tree T' is a path. Without loss of generality, we can assume that v is the third node in the path. Then, we can use the same argumentation as in the previous case. $\square$

3   The 2-Level Cactus Model

Throughout the rest of this paper, let $G = (V, E)$ be an undirected connected multi-graph. Even though there might be several edges of G that are incident to the same two vertices v and w, we denote each of them by $\{v, w\}$. For two subsets $S, T \subset V$ let $E(S, T) := \{\{s, t\} \in E \mid s \in S, \ t \in T\}$ denote the set of edges in E that are incident to a vertex in S and to a vertex in T and let $c(S, T) := |E(S, T)|$ be the cardinality of this set. A non-empty proper subset S of V *induces* the *cut* $E(S, \overline{S})$ of G. A *2-cut* is a cut of cardinality 2. Let $\lambda = \min_{\emptyset \subsetneq S \subsetneq V} c(S, \overline{S})$ denote the minimum cardinality of a cut of G.

A cut that is induced by S *divides* a subset T of V if none of the two sets $S \cap T$ and $\overline{S} \cap T$ is empty. Two cuts $E(S, \overline{S})$ and $E(T, \overline{T})$ are *crossing*, if none of the four *corner sets* $S \cap T$, $S \cap \overline{T}$, $\overline{S} \cap T$, and $\overline{S} \cap \overline{T}$ is empty. If not, they are *parallel*. A cut that is induced by a corner set is called a *corner cut*. $G' = (V', E')$ results from G by *shrinking* a subset S of V, if $V' = (V \setminus S) \dot\cup \{v_S\}$ and $E' = (E \setminus E(V, S)) \cup \{\{v, v_S\}; \{v, s\} \in E(\overline{S}, S)\}$, that is every incidence of an edge in E to a vertex in S is replaced by an incidence to v_S, omitting loops. Two cuts C and C' *induce* the *quotient graph* that results from G by shrinking the four corner sets.

For $\lambda \geq 3$, Dinitz and Nutov developed in [5] a compact model for the λ and $(\lambda + 1)$-cuts of a graph G, called the 2-level cactus model. In the following, we briefly sketch this model and summarize those properties that we use to prove the planarity of the 2-level cactus.

Generally, a *model* for a family F of cuts of G is a triple $(\mathcal{G}, \varphi, \mathcal{F})$ such that the model graph $\mathcal{G} = (\mathcal{N}, \mathcal{E})$ is an undirected multi-graph, $\mathcal{F}$ is a set of cuts of $\mathcal{G}$, and $\varphi : V \longrightarrow \mathcal{N}$ is a mapping with $\varphi^{-1}(\mathcal{F}) = F$, where $\varphi^{-1}(E(S, \overline{S}))$ is defined to be $E(\varphi^{-1}(S), \overline{\varphi^{-1}(S)})$. We say that a cut $\mathcal{C}$ of $\mathcal{G}$ *models* the cut $\varphi^{-1}(\mathcal{C})$. The elements of $\mathcal{N}$ are called *nodes* and a *node* $\nu \in \mathcal{N}$ with $\varphi^{-1}(\nu) = \emptyset$ is called an *empty node*.

In [5], models for the set F of λ and $(\lambda + 1)$-cuts are built in the following way. Set F is divided into the set of those λ-cuts not crossing any other λ-cut in F, called the set of all *basic cuts* F^{bas}, the set of all remaining cuts in F that do not cross any cut in F^{bas}, called the set of all *local cuts* F^{loc}, and the set of all cuts that cross at least one of the cuts in F^{bas}, called the set of all *global cuts* F^{glb}. F^{bas} can be modeled by the tree $\mathcal{T}^{\text{bas}}$. See Fig. 2a,b for an illustration to the construction of such a tree.

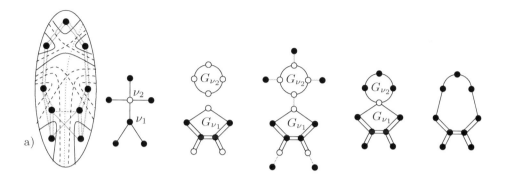

Fig. 2. a) A set of cuts of a graph (the graph edges are shown grey). Continuous curves indicate basic cuts and dashed curves indicate local cuts. The dotted curve represents a global cut. b) Tree $\mathcal{T}^{\text{bas}}$ and c) two local models $\mathcal{G}_{\nu_1}$ and $\mathcal{G}_{\nu_2}$ that d) are implanted instead of ν_1 and ν_2 into $\mathcal{T}^{\text{bas}}$. White nodes are empty nodes. The remainders of the edges in $\mathcal{T}^{\text{bas}}$ are grey. e) These remainders can be contracted at the end of the implantation process. f) The opening of the white halo node in (e).

For a node $\nu \in \mathcal{N}$, let $V_\nu^1, \ldots, V_\nu^k$ be the subsets of V that are mapped on the connected components of $\mathcal{T}^{\text{bas}} - \nu$, and let G_ν be the *quotient graph of ν* that is the graph resulting from G by shrinking the sets $V_\nu^1, \ldots, V_\nu^k$ into a single vertex each. Given any subset $\widetilde{F}^{\text{loc}}$ of the local cuts, it can be partitioned into F_ν^{loc}, $\nu \in \mathcal{N}$, where F_ν^{loc} is defined as the set of those cuts in $\widetilde{F}^{\text{loc}}$, that do not divide any of the sets V_ν^i. Assume, for all $\nu \in \mathcal{N}$, there is a model for the cut set F_ν^{loc} (they can be considered as cuts of the graph G_ν). Then, a model for $F^{\text{bas}} \cup \widetilde{F}^{\text{loc}}$ is built by "implanting" for every node $\nu \in \mathcal{N}$ with $F_\nu^{\text{loc}} \neq \emptyset$ a model $\mathcal{G}_\nu$ – called *local model* – for F_ν^{loc} into $\mathcal{T}^{\text{bas}}$. See Fig. 2c,d,e for illustration. The nodes μ of $\mathcal{G}_\nu$ with $\varphi^{-1}(\mu) = V_\nu^i$, for some i, are called *halo nodes*. We also call the vertices of G_ν that correspond to the shrunken sets V_ν^i *halo vertices*. Finally, depending on whether λ is even or odd, the global cuts and the local cuts in $F^{\text{loc}} \setminus \widetilde{F}^{\text{loc}}$ are added in a suitable way.

Odd connectivity. Every global cut is modeled by a 2-cut of $\mathcal{T}^{\text{bas}}$. Moreover, for a 2-cut $\{e_1, e_2\}$ in $\mathcal{T}^{\text{bas}}$, let $p_{e_1 e_2}$ be the set of edges on the path between e_1 and e_2 in $\mathcal{T}^{\text{bas}}$, with e_1 and e_2 included. If cut $\{e_1, e_2\}$ models a global cut, then, any 2-cut $\{e_1', e_2'\} \subset p_{e_1 e_2}$ does also model a cut in $F^{\text{loc}} \cup F^{\text{glb}}$ ([5], Lemma 5.1). The cuts that are modeled by a 2-cut of $\mathcal{T}^{\text{bas}}$ are called *degenerate* cuts. Let P be the set of inclusion-maximal sets $p_{e_1 e_2}$, such that $\{e_1, e_2\}$ models a $(\lambda + 1)$-cut of G. The elements of P are called *generating paths*.

Lemma 2 ([6] Lemma 5.4 and proof of Lemma 5.5).

1. *Any two generating paths have at most one edge in common.*
2. *Let $\lambda > 3$. Let $p \in P$ and $\{v, w\} \in p$. Except $\{v, w\}$, there are at most two edges incident to v in $\mathcal{T}^{\mathrm{bas}}$ such that there exists a generating path containing $\{v, w\}$ and such an edge.*

For the local models, we consider only $\lambda > 3$. Let $\widetilde{F}^{\mathrm{loc}}$ be the set of non-degenerate local cuts plus the set of corner cuts of non-degenerate local cuts. Let ν be a node of $\mathcal{T}^{\mathrm{bas}}$.

Lemma 3 ([5] Lemma 5.4). *Let C, C' be two crossing non-degenerate cuts. Then, the quotient graph induced by C and C' is a simple cycle with $\frac{\lambda+1}{2}$ edges between adjacent vertices.*

From this lemma and the fact that for $\lambda > 3$ no degenerate cut in $\widetilde{F}^{\mathrm{loc}}$ crosses another cut in $\widetilde{F}^{\mathrm{loc}}$, Dinitz and Nutov conclude in [5] that there exists a tree of cycles which is a suitable local model for each node ν with $F_\nu^{\mathrm{loc}} \neq \emptyset$. Implanting these local models into $\mathcal{T}^{\mathrm{bas}}$ results in a cactus tree type graph $\mathcal{G}$, which will be called *2-level cactus*.

To make the generating paths, and thus the 2-cuts of $\mathcal{G}$ modeling the global cuts, visible in a drawing of the 2-level cactus $\mathcal{G}$, let us extend $\mathcal{G}$ as follows. For each generating path p, consider the corresponding sequence of edges in $\mathcal{G}$ and add an auxiliary edge e_p connecting the first and last end node of this sequence to $\mathcal{E}$. We call the result *extended 2-level cactus* $\mathcal{G}^+$. The set of edges on the shortest path in $\mathcal{G}$ between the two end nodes of e_p is called the *projection path* of e_p. Note, that it follows from Lemma 6 Item 1 in Sect. 4 that the projection paths are unique up to multiple edges.

Even connectivity. For a node ν of $\mathcal{T}^{\mathrm{bas}}$, the local model $\mathcal{G}_\nu$ is either a simple cycle or it can be described as a tree $\mathcal{T}_\nu$ plus the halo nodes, where each halo node is connected by two additional edges to $\mathcal{T}_\nu$. In the latter case, the following property holds. For a halo node μ, let p_μ be the set of edges on the path in $\mathcal{T}_\nu$ between the two end nodes of the two edges incident to μ, and let P be the set of these paths. Lemma 5.6 in [5] gives the following properties of the paths in P.

Lemma 4.

1. *Two elements of P have at most one edge in common.*
2. *An edge of $\mathcal{T}_\nu$ is contained in at most two elements of P.*

Let C be a global cut. Then, there is exactly one non-empty node ν of $\mathcal{T}^{\mathrm{bas}}$ such that C divides $\varphi^{-1}(\nu)$. Moreover, C contains one or two sets of $\frac{\lambda}{2}$ edges corresponding to an edge of a cycle that was implanted instead a node μ in the neighborhood of ν in $\mathcal{T}^{\mathrm{bas}}$. To model these cuts, the halo node of $\mathcal{G}_\nu$ that was implanted into edge $\{\nu, \mu\}$ of $\mathcal{T}^{\mathrm{bas}}$ is "opened", which means the halo node is deleted and corresponding pairs of edges are merged. See Fig. 2f for illustration.

Suppose now that every edge in the tree $\mathcal{T}_\nu$ of a local model that is contained in k elements of P is replaced by $3 - k$ parallel edges and that every tree edge of $\mathcal{T}^{\text{bas}}$ that was not contracted after implanting is replaced by a pair of parallel edges. Then, the 2- and 3-cuts of the resulting 2-level cactus $\mathcal{G}$ model the λ- and $(\lambda + 1)$-cuts of G, respectively. This is the 2-level cactus tree model for the even case.

4   Planarity of the 2-Level Cactus

Odd connectivity. In Sect. 2, we have shown that a tree with additional edges, such that the projection paths fulfill the properties of Lemma 2 is planar. We cannot apply Item 2 of Lemma 2 for $\lambda = 3$ and, it turns out that in this case the 2-level cactus is not planar in general. For example (see Fig. 3a,b), if we take $G = K_4$, then the extended 2-level cactus is K_5.

In case of $\lambda \geq 5$, the tree $\mathcal{T}^{\text{bas}}$ extended by the above auxiliary edges is planar, by Lemma 2. However, when $\lambda = 5$, implanting the local model might destroy planarity. An example is shown in Fig. 3c,d. The 2-level cactus in Fig. 3d contains a subdivision of a K_5 with the 5 white nodes as nodes of the K_5.

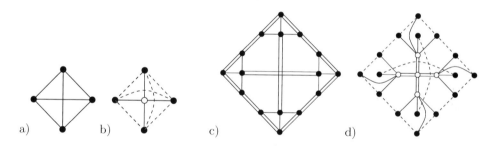

Fig. 3. Example of planar graphs of a) connectivity 3 and c) connectivity 5, and their non-planar extended 2-level cacti b) and d). φ is represented by the location of the vertices and nodes. In the extended 2-level cacti, white nodes are empty nodes, black edges are tree-edges, dashed edges represent the generating paths, and double edges are those of the implanted local model.

To show that for $\lambda \geq 7$, implanting the local models preserves planarity, we will use the following trick. First, we will consider all cycles of $\mathcal{G}$ and modify them in $\mathcal{G}^+$. Then, we show that Lemma 1 can be applied to the thus modified extended 2-level cactus $\mathcal{G}^+$. Second, we will restore the original extended 2-level cactus $\mathcal{G}^+$, and show that planarity is preserved. We start with the following observation (see also [6] Lemma 5.2).

Lemma 5. *Let e_1 and e_2 be two edges of a generating path. Let V_1, V_2 and V' be the set of vertices of G that are mapped on the connected components of $\mathcal{G} - \{e_1, e_2\}$ such that V' induces the $(\lambda + 1)$-cut that is modeled by $\{e_1, e_2\}$. Then, there are exactly $\frac{\lambda - 1}{2}$ edges connecting V_1 and V_2.*

Proof. Let ϵ, ϵ_1 and ϵ_2 be the number of edges between V_1 and V_2, V_1 and V', and V_2 and V', respectively. Then we have $\epsilon + \epsilon_1 = \lambda$, $\epsilon + \epsilon_2 = \lambda$ and $\epsilon_1 + \epsilon_2 = \lambda + 1$. Thus $\epsilon = \frac{\lambda - 1}{2}$. $\square$

An edge of an implanted local model is called a tree-edge if it is contained in a 2-cycle, and it is called a cycle-edge if it is contained in a cycle of length greater than 2. Applying Lemma 3 and Lemma 5, we can show

Lemma 6.

1. *A projection path contains at most one edge of each simple cycle.*
2. *Let $\lambda \geq 5$. Each cycle-edge is contained in at most one projection path.*
3. *Let $\lambda \geq 7$. Each tree-edge of an implanted local model is contained in at most two projection paths.*

Proof.

1. Let p be a projection path that contains the edges $\{v_1, v_2\}, \ldots, \{v_{l-1}, v_l\}$ of a simple cycle $c = v_1, \ldots, v_k$. Suppose p contains more than one edge of c. As p takes the shortest path on c, it follows that $k > l$. Let V_1, V_2, V_l, and V_k be the subsets of V that are mapped on the connected components of $\mathcal{G} - \{\{v_1, v_2\}, \{v_{l-1}, v_l\}, \{v_l, v_{l+1}\}, \{v_k, v_1\}\}$ such that a vertex of V_i is mapped on v_i. Then, the two cuts that are induced by $V_1 \cup V_2$ and by $V_2 \cup V_l$, respectively, are crossing $(\lambda + 1)$-cuts. By Lemma 3, there are no edges between V_1 and V_l. On the other hand, let us consider the generating path $e_1, e_2, \ldots, e_r$, corresponding to p. By Lemma 5 applied to the pair $\{e_1, e_r\}$, there are at least $\frac{\lambda - 1}{2}$ edges between V_1 and V_l, a contradiction.
2. Let c be the set of edges of a simple cycle in a local model $\mathcal{G}_\nu$ and let $\{v_1, v_2\}$ be an edge in c. For $i = 1, 2$, let $\mathcal{N}_i$ be the set of nodes in the connected component of $\mathcal{G} - c$ that contains v_i, and let V_i be the subset of V that is mapped on $\mathcal{N}_i$. Suppose edge $\{v_1, v_2\}$ is contained in at least two projection paths p_1 and p_2. Let e_i^j, $i, j = 1, 2$, be the edge on p_i with end nodes in $\mathcal{N}_j$ that was incident to ν in $\mathcal{T}^{\text{bas}}$ before implanting $\mathcal{G}_\nu$. By Lemma 2 Item 1, we know that $e_1^1 \neq e_2^1$ or $e_1^2 \neq e_2^2$, so we may assume $e_1^2 \neq e_2^2$. Let v_{p_1} and v_{p_2} be the end-nodes of p_1 and p_2 in $\mathcal{N}_2$. For $i = 1, 2$, let $e_i \in p_i$ be the edge incident to v_{p_i}, and let V_{p_1} and V_{p_2} be the subsets of V_2 that are mapped on the connected components of $\mathcal{G} - \{e_1, e_2\}$ that contain v_{p_i}. See Fig. 4a for illustration. Then, by Lemma 5 there are $\frac{\lambda - 1}{2}$ edges between V_1 and V_{p_1} and between V_1 and V_{p_2}, respectively, and thus, there are at least $\lambda - 1$ edges connecting V_1 and V_2. On the other hand, by Lemma 3, $c(V_1, V_2) = \frac{\lambda + 1}{2}$, contradicting the precondition that $\lambda \geq 5$.
3. Let e, e' be a 2-cycle of a local model, and let ϵ be the number of projection paths containing either e or e'. Let V_1 and V_2 be the sets that are mapped on the connected components of $\mathcal{G} - \{e, e'\}$. With the same argumentation as in Item 2, per projection path that contains e or e', there are at least $\frac{\lambda - 1}{2}$ distinct edges between V_1 and V_2. As $\{V_1, V_2\}$ induces a $(\lambda + 1)$-cut, ϵ must fulfill the inequality $\epsilon \cdot \frac{\lambda - 1}{2} \leq \lambda + 1$. For $\lambda \geq 7$, it follows $\epsilon < 3$. $\square$

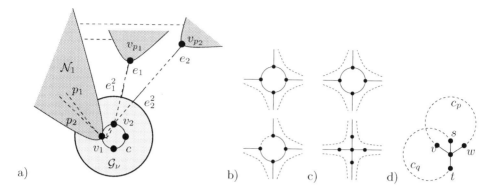

Fig. 4. a) Illustration of the proof of Lemma 6 Item 2. b), c) Modification of the cycles in the local model b) in case not all and c) in case all cycle edges are contained in a projection path, that is indicated by a dashed curve. d) Illustration of the final proof of planarity in the odd case.

Let us consider the following modifications on the cycles. Let c be a simple cycle. If c contains edges that are not contained in a projection path, we choose one of these edges and declare it to be an auxiliary edge. If each edge of c is contained in a projection path, we replace c by a star. That means, we delete all edges of c, add an additional vertex v_c to $\mathcal{G}^+$ and connect all vertices of c to v_c. See Fig. 4b,c for illustration. We call the so modified graph $\mathcal{G}'$ and we denote by $S' \supseteq S$ the set of all auxiliary edges of $\mathcal{G}'$ and by $\mathcal{T}'$ the spanning tree of $\mathcal{G}'$ that is induced by the edges that are not in S'.

The only case that e is contained in the projection path of an edge in $S' \setminus S$ is if e is a cycle-edge of a local model. Thus, by the above lemma and by Lemma 2, $\mathcal{G}'$ with its spanning tree $\mathcal{T}'$ fulfills the condition of Lemma 1. Hence, $\mathcal{G}'$ is planar.

It remains to show that we can restore the deleted cycle-edges into $\mathcal{G}'$ without producing edge crossings. Let c be the set of edges of a cycle of $\mathcal{G}$ that we have replaced by a star. Let $\{v, w\}$ be an edge of c. We will show, that in any embedding of $\mathcal{G}'$, edges $\{v, v_c\}$ and $\{w, v_c\}$ are neighboring in the cyclic order around v_c. Suppose not: then v and w divide the adjacency list of v_c into two non-empty parts S and T. As the graph induced by $c - \{v, w\}$ is connected, there is an edge $\{s, t\}$ in c with $s \in S$ and $t \in T$. Let p be the projection path that contains edge $\{v, w\}$. Let $e_p \in S$ be the auxiliary edge that represents p and let p' be the projection path of e_p in $\mathcal{G}'$. Similarly, let q be the projection path that contains edge $\{s, t\}$, let e_q be the corresponding edge in S and q' the corresponding path in $\mathcal{G}'$. Then, on one hand, the two cycles induced by the edge sets $c_p = p' \cup e_p$ and $c_q = q' \cup e_q$ are vertex disjoint except v_c: If not, the graph that is induced by $c_p \cup c_q$ has at least four faces and thus, the graph induced by $p' \cup q'$, which is a sub graph of tree $\mathcal{T}'$, would contain a cycle. But on the other hand, there is an edge of c_q in the inner part of c_p and an edge of c_q in the outer part of c_p, contradicting the planarity of $\mathcal{G}'$. See Fig. 4d for illustration.

Even connectivity. The corollary to MacLane's planarity criterion shows that a tree with additional edges, such that the projection paths fulfill the properties of Lemma 4 is planar. Thus, the local models are planar. Because the local models are stuck together in a tree-structure, proceeding from a leaf, we can open the halo nodes without losing planarity.

Connectivities One and Two. Finally, for completeness, we consider the prototypes of the 2-level cactus. The cut model of the inclusion minimal 1- and 2-cuts was introduced in [8]; it is a tree of edges and cycles and thus, is planar. In the case $\lambda = 2$, a cut model for all minimum and minimum+1 cuts is described in [3]; it is constructed in a similar way as the 2-level cactus in the even case. In general, it is not planar. An example is a K_4 with every edge broken into two by a new vertex. This graph and its non-planar 2-level cactus is shown on the right. We summarize this section in the following theorem.

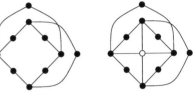

Theorem 2. *For a connected multi-graph with edge-connectivity $\lambda = 1$ or with even edge-connectivity $\lambda \geq 4$ the 2-level cactus is planar, for odd edge-connectivity $\lambda \geq 7$ the extended 2-level cactus is planar, and for all other connectivities there are examples of non-planar (extended) 2-level cacti.*

5   Final Remarks

For an extended planar 2-level cactus (the odd case), it would be nice if:

(a) the auxiliary edges are all on the same (outer) face and
(b) each cycle that is generated by an auxiliary edge and its projection path is the boundary of a face.

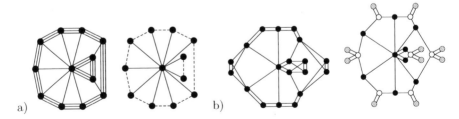

Fig. 5. a) A graph of odd connectivity $\lambda = 7$ and its 2-level cactus $\mathcal{G}$. There is no embedding of $\mathcal{G}$ such that all dashed auxiliary edges are on the same face. b) A graph of even connectivity $\lambda = 4$ and its 2-level cactus $\mathcal{G}$. All non-grey nodes belong to a single local model $\mathcal{G}_\nu$. There is no embedding of $\mathcal{G}_\nu$ such that its halo nodes (the white nodes) are on the same face. With an increasing number of vertices, both examples can be extended to arbitrary odd or even connectivity, respectively.

In this case, $\mathcal{G}$ is completely contained inside a planar drawing of $\mathcal{G}^+$, and the projection path that corresponds to an auxiliary edge can be discovered more easily. Moreover, the interior of each simple cycle of a local model will then be empty. In the even case we wish that:

(a) the halo nodes of any local model $\mathcal{G}_\nu$ are all on the same (outer) face of $\mathcal{G}_\nu$.

In this case, the tree structure of $\mathcal{T}^{bas}$ can be better recognized in a planar drawing of $\mathcal{G}$.

Unfortunately, there are examples (see Fig. 5 and Fig. 6a) that show that the above mentioned properties of the (extended) planar 2-level cactus are not true in general. However, it follows from Mac Lane's planarity criterion that in the even case, Property (a) is true for the 2-connected components of the local models. In the odd case, by subdividing each edge of $\mathcal{T}^{bas}$ that is contained in more than two projection paths as illustrated in Fig. 6b, the extended 2-level cactus $\mathcal{G}^+$ can be modified in such a way that the result is still a linear sized model for the minimum and minimum + 1 cuts of G, and has an embedding such that

(c) for each auxiliary edge e, the cycle on the edge set $e \cup p_e$ is a part of the boundary of some face f.

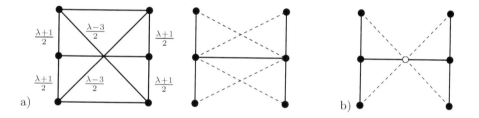

a)

b)

Fig. 6. a) A graph of odd connectivity $\lambda > 3$ and its 2-level cactus $\mathcal{G}$. Vertical edges refer to $\frac{\lambda+1}{2}$ edges, diagonal edges refer to $\frac{\lambda-3}{2}$ edges, and horizontal edges refer to one edge each. There is no embedding of $\mathcal{G}$ with the property that, for every simple cycle c that is generated by a dashed edge and its projection path, there is a face f such that c is part of the boundary of f. b) Splitting the edge that is contained in more than 2 generating paths achieves this property.

References

1. A. A. Benczúr. A representation of cuts within 6/5 times the edge connectivity with applications. In *Proceedings of the 36th Annual Symposium on Foundations of Computer Science (FOCS '95)*, pages 92–103. IEEE Computer Society Press, 1995.

2. U. Brandes, S. Cornelsen, and D. Wagner. How to draw the minimum cuts of a planar graph. In J. Marks, editor, *Proceedings of the 8th International Symposium on Graph Drawing (GD 2000)*, volume 1984 of *Lecture Notes in Computer Science*, pages 103–114. Springer, 2001.

3. Y. Dinitz. The 3-edge-components and a structural description of all 3-edge-cuts in a graph. In E. W. Mayr, editor, *Graph Theoretic Cocepts in Computer Science, 18th International Workshop, (WG '92)*, volume 657 of *Lecture Notes in Computer Science*, pages 145–157. Springer, 1993.

4. Y. Dinitz, A. V. Karzanov, and M. Lomonosov. On the structure of a family of minimal weighted cuts in a graph. In A. Fridman, editor, *Studies in Discrete Optimization*, pages 290–306. Nauka, 1976. (in Russian).

5. Y. Dinitz and Z. Nutov. A 2-level cactus model for the system of minimum and minimum+1 edge–cuts in a graph and its incremental maintenance. In *Proceedings of the 27th Annual ACM Symposium on the Theory of Computing (STOC '95)*, pages 509–518. ACM, The Association for Computing Machinery, 1995.

6. Y. Dinitz and Z. Nutov. A 2-level cactus tree model for the system of minimum and minimum+1 edge cuts of a graph and its incremental maintenance. Technical Report CS0915, Computer Science Department, Technion Haifa, 1997. 50 pages, available at `http://www.cs.technion.ac.il/users/wwwb/cgi-bin/tr-info.cgi?1997/CS/CS0%915`.

7. Y. Dinitz and J. Westbrook. Maintaining the classes of 4-edge-connectivity in a graph on-line. *Algorithmica*, 20:242–276, 1998.

8. Z. Galil and G. F. Italiano. Maintaining the 3-edge-connected components of a graph on-line. *SIAM Journal on Computing*, 22(1):11–28, 1993.

9. M. R. Henzinger and D. P. Williamson. On the number of small cuts in a graph. *Information Processing Letters*, 59(1):41–44, 1996.

10. C. Kuratowski. Sur le problème des courbes gauches en topologie. *Fundamenta Mathematicae*, 15:271–283, 1930.

11. S. MacLane. A combinatorial condition for planar graphs. *Fundamenta Mathematicae*, 28:22–32, 1937.

12. H. Nagamochi, K. Nishimura, and T. Ibaraki. Computing all small cuts in undirected networks. In D.-Z. Du, editor, *Proceedings of the 5th International Symposium on Algorithms and Computing (ISAAC '94)*, volume 834 of *Lecture Notes in Computer Science*, pages 190–198. Springer, 1994.

13. C. Thomassen. Planarity and duality of finite and infinite graphs. *Journal of Combinatorial Theory, Series B*, 29:244–271, 1980.

14. V. V. Vazirani and M. Yannakakis. Suboptimal cuts: Their enumeration, weight and number. In W. Kuich, editor, *Proceedings of the 19th International Colloquium on Automata, Languages and Programming*, volume 623 of *Lecture Notes in Computer Science*, pages 366–377. Springer, 1992.

Estimating All Pairs Shortest Paths in Restricted Graph Families: A Unified Approach

(Extended Abstract)

Feodor F. Dragan

Dept. of Computer Science, Kent State University, Kent, Ohio 44242, USA
dragan@cs.kent.edu

Abstract. In this paper we show that a very simple and efficient approach can be used to solve the *all pairs almost shortest path* problem on the class of weakly chordal graphs and its different subclasses. Moreover, this approach works well also on graphs with small size of largest induced cycle and gives a unified way to solve the *all pairs almost shortest path* and *all pairs shortest path* problems on different graph classes including chordal, strongly chordal, chordal bipartite, and distance-hereditary graphs.

1 Introduction

Let $G = (V, E)$ be a finite, unweighted, undirected, connected and simple (i.e., without loops and multiple edges) graph. Let also $|V| = n$ and $|E| = m$. The *distance* $d(v, u)$ between vertices v and u is the smallest number of edges in a path connecting v and u. The *all pairs shortest path problem* is to compute $d(v, u)$ for all pairs of vertices v and u in G.

The all pairs shortest path (APSP) problem is one of the most fundamental graph problems. There has been a renewed interest in it recently (see [1,2,4,6,12], [15,29] for general (unweighted, undirected) graphs, and [3,5,7,10,11,14,20,21,24], [28,32,33] for special graph classes). For general graphs, the best result known is by SEIDEL [29], who showed that the APSP problem can be solved in $O(M(n) \log n)$ time where $M(n)$ denotes the time complexity for matrix multiplication involving small integers only. The current best matrix multiplication algorithm is due to COPPERSMITH and WINOGRAD [13] and has $O(n^{2.376})$ time bound. In contrast, the naive algorithm for APSP performs breadth-first searches from each vertex, and requires time $\Theta(nm)$ which is $\Theta(n^3)$ for the dense graphs.

Given the fundamental nature of the APSP problem, it is important to consider the desirability of implementing the algorithms in practice. Unfortunately, fast matrix multiplication algorithms are far from being practical and suffer from large hidden constants in the running time bound. There is interest therefore in obtaining fast algorithms for the APSP problem that *do not use* fast matrix multiplication. The currently best *combinatorial* (i.e., in graph-theoretic terms) algorithm for the APSP problem is an $O(n^3 / \log n)$ time algorithm obtained

A. Brandstädt and V.B. Le (Eds.): WG 2001, LNCS 2204, pp. 103–116, 2001.
© Springer-Verlag Berlin Heidelberg 2001

by BASCH ET AL [6]. This offers only a marginal improvement over the naive algorithm.

Since an algorithm for the APSP problem would yield an algorithm with similar time bound for Boolean matrix multiplication, obtaining a combinatorial $O(n^{3-\epsilon})$ time algorithm for the APSP problem would be a major breakthrough. Nowadays, many researchers take approach to consider the all pairs *almost* shortest path (APASP) problem in general graphs or to design (optimal) $O(n^2)$ time algorithm for special well structured graph classes (which are interesting from practical point of view).

AWERBUCH ET AL [4] and COHEN [12] considered the problem of finding *stretch t all pairs paths*, where t is some fixed constant and a path is of stretch t if its length is at most t times the distance between the endvertices. They presented efficient algorithms for computing t-stretch paths for $t \geq 4$. A different approach was employed by AINGWORTH ET AL [1]. They described a simple and elegant $O(n^{5/2}\sqrt{\log n})$ time algorithm for finding all distances in unweighted, undirected graphs with an *additive one-sided error* of at most 2. They also make the very important observation that the small distances, and not the long distances, are the hardest to approximate. Recently, DOR ET AL [15] improved the result of AINGWORTH ET AL [1] and obtained an $\tilde{O}(\min\{n^{3/2}m^{1/2}, n^{7/3}\})$ time algorithm for finding all distances in unweighted, undirected graphs with an additive one-sided error of at most 2 (here, $\tilde{O}(f)$ means $O(f\,\text{polylog } n)$). A simple argument shows that computing all distances in G with an additive one-sided error of at most 1 is as hard as Boolean matrix multiplication (see [15]).

Besides, efficient algorithms have been developed for solving the APSP problem on some restricted classes of graphs. Optimal $O(n^2)$ time algorithms were proposed for interval graphs [3,24,28,33], circular arc graphs [3,33], permutation graphs [14], bipartite permutation graphs [10], strongly chordal graphs [5,14,20], chordal bipartite graphs [21], distance-hereditary graphs [14], and dually chordal graphs [7]. The optimal parallel algorithms for the APSP problem for some certain classes of graphs can be found in [10,14,20,32].

The approach to the shortest path and other optimization problems in metric spaces (e.g. in a simple rectilinear polygon endowed with some metric) taken by many researchers first finds a *visibility graph* of the metric space which contains a shortest path between any two given points. MOTWANI ET AL [25,26] studied two classes of rectilinear polygons and showed that the visibility graphs corresponding to them are *chordal* or *weakly chordal*, respectively. More recently, EVERETT and CORNEIL [18] have shown that the visibility graphs of spiral polygons are *interval graphs* (the definition of the various families of graphs will be presented in the next section).

This motivated researchers to look at distances in chordal graphs [8,11,20,32]. HAN ET AL showed in [20] that the APSP problem for a chordal graph G can be solved in $O(n^2)$ time if the *square G^2* of G is already given. They also note that computing G^2 for chordal graphs is as hard as for general graphs, and present few subclasses of chordal graphs for which G^2 can be computed more efficiently. From results in [8] it follows that within $O(n^2)$ time bound one can compute all distances in chordal graphs with an additive error of at most 2. Even more, given a pair of vertices $x, y \in V$, after a simple linear time preprocessing, in only $O(1)$ time one can compute $d(x, y)$ with an additive error of at most 2. SRIDHAR ET AL showed in [32] that using a more sophisticated linear time preprocessing of a chordal graph, the distance between any two vertices which is at most 1 greater than the shortest distance can be computed in $O(1)$ time.

In this paper we show that a very simple and efficient approach can be used to solve the APASP problem on the class of weakly chordal graphs and its different subclasses. Furthermore, we show that this approach works well also on graphs with small size of largest induced cycle and gives a unified way to solve the APASP and APSP problems on different graph classes including chordal, strongly chordal, chordal bipartite, and distance-hereditary graphs.

The rest of this paper is organized as follows. We begin in Section 2 by presenting some definitions and a simple $O(n^2)$ time algorithm for computing all pairs almost shortest path distances in graphs with special vertex orderings. In Section 3, we prove that this algorithm

– computes all distances
  • in House-Hole–free graphs with an additive one-sided error of at most 2,
  • in House-Hole-Domino–free graphs (and hence in chordal graphs) with an additive one-sided error of at most 1, and
– solves the APSP problem for distance-hereditary graphs, chordal bipartite graphs, strongly chordal (and hence interval) graphs.

In this section we also show that the APSP problem for a House-Hole–free graph G can be solved in $O(n^2)$ time, if the square G^2 of G is given. All these classes are subclasses of the weakly chordal graphs. Finally, in section 4 we consider graphs G with small size of largest induced cycle and show that our generic algorithm computes in $O(n^2)$ time all distances in G with an additive one-sided error of at most $k - 1$, where k is the size of largest induced cycle in G. Hence, for weakly chordal graphs (even for Hole-free graphs), all distances can be computed in $O(n^2)$ time with an additive one-sided error of at most 3.

2   Preliminaries and the Algorithm

The *(open) neighborhood* of a vertex v is the set $N(v) = \{u \in V : uv \in E\}$ and the *closed neighborhood* is $N[v] = N(v) \cup \{v\}$. A *path* is a sequence of vertices $(v_0, \ldots, v_l)$ such that $v_i v_{i+1} \in E$ for $i = 0, \ldots, l - 1$; its *length* is l. We say that this path connects vertices v_0 and v_l. An *induced path* is a path where $v_i v_j \in E$ if and only if $i = j - 1$ and $j = 1, \ldots, l$. A *cycle* is a sequence of vertices $(v_0, \ldots, v_k)$ such that $v_0 = v_k$ and $v_i v_{i+1} \in E$ for $i = 0, \ldots, k - 1$; its *length* is k. An *induced cycle* is a cycle where $v_i v_j \in E$ if and only if $|i - j| = 1$ (modulo k). A *hole* is an induced cycle of length at least 5.

We now define the various graph families that will be used in this paper. A graph G is *chordal* if there is no induced cycle of length greater than 3 in G. A *Hole-free* graph is a graph which contains no holes. A graph G is *weakly chordal* if both G and the complement of G are Hole–free graphs. A bipartite

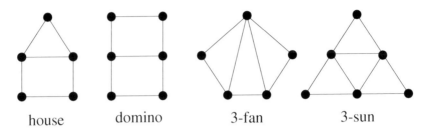

house   domino   3-fan   3-sun

Fig. 1.

weakly chordal graph is called *chordal bipartite*. An *interval* graph is the intersection graph of intervals of a line. A graph is called *AT–free* if it does not have an *asteroidal triple*, i.e. a set of three vertices such that there is a path between any pair of them avoiding the closed neighborhood of the third. A graph is *House-Hole–free (HH–free)* if it contains no induced houses or holes and is *House-Hole-Domino–free (HHD–free)* if it contains no induced houses, holes or dominos (see Figure 1). A *distance-hereditary graph* is a HHD–free graph which does not contain 3-fan as an induced subgraph. Finally, a graph $G = (V, E)$ is *strongly chordal* if it admits a *strong elimination ordering*, namely, an ordering $v_1, v_2, \ldots, v_n$ of V such that for each i, j, k and l, if $i < j, k < l, v_k, v_l \in N[v_i]$, and $v_k \in N[v_j]$, then $v_l \in N[v_j]$. For more information on these graph classes we refer the reader to [9,19]

For computing approximate distances in these graph classes we use the following simple $O(n^2)$ time algorithm. Let $G = (V, E)$ be a graph and $\sigma = [v_1, v_2, \ldots, v_n]$ be an ordering of the vertex set V of G. Let also $\sigma(v)$ be the number assigned to a vertex v in this ordering. Denote by $mn(v)$ the *maximum neighbor* of v, i.e., the vertex of $N[v]$ with maximum number in σ.

Algorithm APASP

    for $i = 1$ **to** n **do** $\hat{d}(v_i, v_i) := 0$;
    for $i = n - 1$ **downto** 1 **do**
        for $j = n$ **downto** $i + 1$ **do**
            if $v_i v_j \in E$ **then** $\hat{d}(v_i, v_j) := 1$ **else** $\hat{d}(v_i, v_j) := \hat{d}(mn(v_i), v_j) + 1$;
    return distances $\hat{d}$.

We assume that ordering σ used in the algorithm has the property that for all v_i with $i < n$, $mn(v_i) \neq v_i$ (otherwise algorithm will not work properly). This property holds for all orderings considered in this paper.

Note that this method is not new. It was already used to compute all pairs shortest path distances in strongly chordal graphs (see [14]) and in dually chordal graphs (see [7]). For both classes the characteristic orderings were employed. Here we show that this method gives the exact distances or a very good approximation of distances in many other graph classes.

Clearly, the algorithm can be modified to produce paths of length $\hat{d}$ rather than merely returning the approximate distances. Nevertheless, we present a separate procedure which for any two vertices u, v returns the path of length $\hat{d}(u, v)$ along which the APASP algorithm computes. Its complexity is $O(c \cdot \hat{d}(u, v))$, where c is the necessary time to verify the adjacency of two vertices).

Procedure (mn–path(u, v))

    if u and v are adjacent **then return** (u, v)
    else if $\sigma(u) < \sigma(v)$ **then return** $(u, \text{mn–path}(mn(u), v))$
        else return $(\text{mn–path}(u, mn(v)), v)$

The path between u and v returned by the procedure mn–path(u, v) will be called the *maximum-neighbor path*. Throughout this paper, by $d(u, v)$ we denote the (real) distance between u and v and by $\hat{d}(u, v)$ the approximate distance returned by the algorithm.

In this paper we will employ three types of vertex orderings: *breadth-first-search orderings, lexicographic-breadth-first-search orderings* of ROSE ET AL [31] and *lexical orderings* of Lubiw [23].

Let $\sigma = [v_1, v_2, \ldots, v_n]$ be any ordering of the vertex set of a graph G. We will write $a < b$ whenever in a given ordering σ vertex a has a smaller number than vertex b. Moreover, $\{a_1, \cdots, a_l\} < \{b_1, \cdots, b_k\}$ is an abbreviation for $a_i < b_j$ $(i = 1, \cdots, l; \ j = 1, \cdots, k)$. Let u be a vertex of G. We define the layers of G with respect to vertex u as follows: $L_i(u) = \{v : d(u, v) = i\}$ for $i = 0, 1, 2, \ldots$.

In a *breadth-first search (BFS)*, started at a vertex u, the vertices of a graph G with n vertices are numbered from n to 1 in decreasing order. The vertex u is numbered by n and put on an initially empty queue of vertices. Then a vertex v at the head of the queue is repeatedly removed, and neighbors of v that are still unnumbered are consequently numbered and placed onto the queue. Clearly, BFS operates by proceeding vertices in layers: the vertices closest to the start vertex are numbered first, and the most distant vertices are numbered last. BFS may be seen to generate a rooted tree T with vertex u as the root. A vertex v is the *father* in T of exactly those neighbors in G which are inserted into the queue when v is removed.

An ordering σ of the vertex set of a graph G generated by a BFS will be called a *BFS–ordering* of G. Denote by $f(v)$ the farther of a vertex v with respect to σ. The following properties of a BFS–ordering will be used in what follows. Since all layers of V considered here are with respect to u, we will frequently use notation L_i instead of $L_i(u)$.

(P1)   If $x \in L_i$, $y \in L_j$ and $i < j$, then $x > y$ in σ.

(P2)   If $v \in L_q$ $(q > 0)$ then $f(v) \in L_{q-1}$ and $f(v)$ is the vertex from $N(v) \bigcap L_{q-1}$ with the largest number in σ, i.e., $f(v) = mn(v)$.

(P3)   If $x > y$, then either $mn(x) > mn(y)$ or $mn(x) = mn(y)$.

(P4)   If $x, y, z \in L_j$, $x > y > z$ and $mn(x)z \in E$, then $mn(x) = mn(y) = mn(z)$ (in particular, $mn(x)y \in E$).

Lexicographic breadth-first search (LexBFS), started at a vertex u, orders the vertices of a graph by assigning numbers from n to 1 in the following way. The vertex u gets the number n. Then each next available number k is assigned to a vertex v (as yet unnumbered) which has lexically largest vector $(s_n, s_{n-1}, \ldots, s_{k+1})$, where $s_i = 1$ if v is adjacent to the vertex numbered i, and $s_i = 0$ otherwise. An ordering of the vertex set of a graph generated by LexBFS we will call a *LexBFS–ordering*. It is well–known that any LexBFS–ordering has property (P5) (cf.[22]). Moreover, any ordering fulfilling (P5) can be generated by LexBFS [17].

(P5)   If $a < b < c$ and $ac \in E$ and $bc \notin E$ then there exists a vertex d such that $c < d$, $db \in E$ and $da \notin E$.

LUBIW in [23] has shown that any graph admits a *lexical ordering*, namely an ordering of V such that

(P6)   If $a < b$ and $ac \in E$ and $bc \notin E$ then there exists a vertex d such that $c < d$, $d \in N[b]$ and $da \notin E$ ($d = b$ is allowed).

Clearly, any lexical ordering is a LexBFS–ordering, and any LexBFS–ordering is a BFS–ordering (but not conversely). Note also that for a given graph G, both a BFS–ordering and a LexBFS–ordering can be generated in linear time [19], while to date the fastest method – doubly lexical ordering of the (closed) neighborhood matrix of G [23] – producing a lexical ordering of G takes $O(m \log n)$ [27] or $O(n^2)$ [30] time.

3 Distances in Graphs from the Weakly Chordal Hierarchy

In this section, we consider different subclasses of weakly chordal graphs. The weakly chordal graphs itself will be considered in the next section. First we present the following simple but very useful lemma which holds for arbitrary graphs.

Lemma 1. *Let there exist integers $t \geq 2$ and $s \geq 0$ such that*

(1) for all $x, y \in V$ with $x < y$ and $d(x, y) \geq t$, $d(x, y) = d(mn(x), y) + 1$ holds,
(2) for all $x, y \in V$ with $d(x, y) < t$, $\hat{d}(x, y) \leq d(x, y) + s$ holds.

Then $\hat{d}(x, y) \leq d(x, y) + s$ holds for all $x, y \in V$.

Proof. The proof is by induction on $d(x, y)$. Assume that $\hat{d}(x, y) \leq d(x, y) + s$ holds for all $x, y \in V$ with $d(x, y) < k$ ($k \geq t$), and consider a pair x, y such that $d(x, y) = k$ and $x < y$. According to APASP algorithm we have $\hat{d}(x, y) = \hat{d}(mn(x), y) + 1$. Since $d(mn(x), y) = d(x, y) - 1$, by induction, $\hat{d}(mn(x), y) \leq d(mn(x), y) + s$ holds. Therefore, $\hat{d}(x, y) \leq d(mn(x), y) + s + 1 = d(x, y) + s$. □

Let $P = (x_0, x_1, \cdots, x_{k-1}, x_k)$ be an arbitrary path of G and σ be an ordering of the vertex set of this graph. The path P is *monotonic* (with respect to σ) if $x_0 < x_1 < \cdots < x_{k-1} < x_k$ holds whenever $x_0 < x_k$, and P is *convex* if there is an index i ($1 \leq i < k$) such that $x_0 < x_1 < \cdots < x_{i-1} < x_i > x_{i+1} > \cdots > x_{k-1} > x_k$. Vertex x_i is called the *switching point* of the convex path P. Let now

$P = (x_0, \cdots, x_k)$ be a shortest path of G connecting x_0 and x_k. We say that P is a *rightmost shortest path* if the sum $\sigma(x_0) + \sigma(x_1) + \cdots + \sigma(x_k)$ is largest among all shortest paths connecting x_0 and x_k.

3.1 Employing LexBFS–Orderings

Let G be a HH–free graph and σ be a LexBFS–ordering of G. By P_4 we denote a path on four vertices.

Lemma 2 ([16]). G has no induced $P_4 = (c, a, b, d)$ with $\{a, b\} < \{c, d\}$ in σ.

Corollary 1. *Let a, b, c be three distinct vertices of G such that $a < \{b, c\}$, $ab, ac \in E$ and $bc \notin E$. Then there is a vertex $d > \{b, c\}$ adjacent to b and c but not to a.*

Lemma 3 ([16]).
(1) Every rightmost shortest path of G is either monotonic or convex.
(2) Let $P = (x_0, \cdots, x_k)$ be a rightmost shortest path in G which is convex and x_i be the switching point of P. Then $d(x_0, x_i) \geq d(x_k, x_i)$ if $x_0 < x_k$.

Lemma 4. *Let x, y be a pair of vertices of G such that $x < y$ and $d(x, y) \geq 3$. Then $d(x, y) = d(mn(x), y) + 1$ holds.*

Proof. Let $P = (x, x_1, x_2, \cdots, y)$ be a rightmost shortest path in G connecting vertices x and y. Since $x < y$ and $d(x, y) \geq 3$, by Lemma 3, we have $x < x_1 < x_2$. Assume $x_1 \neq mn(x)$. Vertices $v = mn(x)$ and x_2 cannot be adjacent, since otherwise we can replace x_1 in P with v and get a shortest path between x and y with larger sum, contradicting the choice of P. For path (v, x, x_1, x_2) we have $\{x, x_1\} < \{v, x_2\}$. Hence, by Lemma 2, v and x_1 are adjacent. Now we apply Corollary 1 to $x_1 < \{v, x_2\}$, $vx_2 \notin E$ and get a vertex u such that $u > \{v, x_2\}$, $uv, ux_2 \in E$ and $x_1u \notin E$. Furthermore, x cannot be adjacent to u because $v = mn(x)$ and $u > v$. But then a house formed by x, v, x_1, u, x_2 is induced, that is impossible for a HH-free graph G. $\square$

Unfortunately, this result cannot be extended to weakly chordal graphs. For any even integer $k \geq 4$, there exist a weakly chordal graph $G = (V, E)$, a LexBFS–ordering σ of G, and a pair of vertices $x, y \in V$ such that $x < y$ in σ, $d(x, y) = k$ but $d(x, y) < d(mn(x), y) + 1$. Figure 2 gives such a weakly chordal graph for $k = 4$ (this example can easily be extended to a larger k).

Lemma 5. *Let x, y be a pair of vertices of G such that $x < y$ and $d(x, y) = 2$. Then*

(1) $d(x, y) = \hat{d}(x, y)$ if $mn(x)y \in E$ or there exists a vertex $z \in N(x) \bigcap N(y)$ such that $z < y$;
(2) $d(x, y) \geq \hat{d}(x, y) - 1$ if there exists a vertex $z \in N(x) \bigcap N(y)$ such that $mn(x)z \in E$.

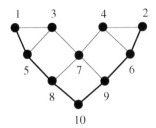

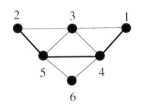

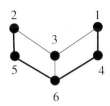

Fig. 2. A weakly chordal graph with a LexBFS–ordering; $d_{1,2} = d_{2,5} = 4$.

Fig. 3. A LexBFS–ordering of a 3-sun such that $\hat{d}_{1,2} - d_{1,2} = 1$.

Fig. 4. A LexBFS–ordering of a domino such that $\hat{d}_{1,2} - d_{1,2} = 2$.

Proof. (1) Clearly, if $mn(x)y \in E$, then $\hat{d}(mn(x), y) = 1$ and $\hat{d}(x, y) = \hat{d}(mn(x), y) + 1 = 2 = d(x, y)$. So, assume that $mn(x)y \notin E$ and there is a vertex $z \in N(x) \cap N(y)$ such that $z < y$. For path $(mn(x), x, z, y)$ we have $\{x, z\} < \{mn(x), y\}$. Hence, by Lemma 2, $mn(x)$ and z must be adjacent. Now we can proceed as in the proof of Lemma 4 and construct an induced house in HH-free graph G, which is impossible.

(2) We may assume that $mn(x)y \notin E$ and for the vertex $z \in N(x) \cap N(y)$ with $mn(x)z \in E$, $z > y$ holds. We have $d(mn(x), y) = 2$ and $y < z < mn(x)$. Hence, by (1), $\hat{d}(y, mn(x)) = d(y, mn(x))$, and therefore $\hat{d}(x, y) = \hat{d}(mn(x), y) + 1 = d(mn(x), y) + 1 = 3 = d(x, y) + 1$. □

Lemma 6. *Let x, y be a pair of vertices such that $x < y$, $d(x, y) = 2$, $mn(x)y \notin E$, and there exists a vertex $z \in N(x) \cap N(y)$ with $mn(x)z \notin E$ and $z > y$. Then G contains a domino as an induced subgraph and $d(x, y) = \hat{d}(x, y) - 2$ holds.*

Proof. Let $v = mn(x)$, $u = mn(z)$ and $w = mn(y)$ be the maximum neighbors in σ of x, z and y, respectively. We have $x < y < z < v$, $vz, vy \notin E$. By property (P3), we have also $u \geq w > v$ and hence $xu, xw \notin E$. Since $\{x, z\} < \{v, u\}$, by Lemma 2, vertices v and u must be adjacent. If $uy \in E$ then we get an induced house in G, which is impossible. Therefore, $d(y, u) = 2$ and we can apply Lemma 5 to $y < z < u$ and get $d(y, u) = \hat{d}(y, u) = 2$. That is, vertex $w = mn(y)$ is adjacent to u. If now $wz \in E$ or $wv \in E$ then vertices x, y, z, v, u, w induce either a house or a cycle of length 5. Since these induced subgraphs are forbidden, w is adjacent neither to v nor to z, i.e. vertices x, y, z, v, u, w form an induced domino.

Now consider maximum neighbors $h = mn(v)$, $s = mn(u)$ and $t = mn(w)$ of v, u and w, respectively. We had $d(v, w) = 2$, $v < w < u$ and $u \in N(v) \cap N(w)$. If also $hu, hw \notin E$ then, as we have shown above, vertices v, u, w, h, s, t will form an induced domino with edges $vu, vh, us, uw, wt, hs, st$. Since $v = mn(x)$, $u = mn(z)$, $w = mn(y)$ and $v < w < u < h < t < s$, there cannot be edges with one end in $\{x, z, y\}$ and the other end in $\{h, s, t\}$. That is, the cycle $(x, z, y, w, t, s, h, v, x)$ of G of length 8 is induced. Since such cycles are forbidden in G, we must have $hu \in E$ or $hw \in E$.

If $hu \in E$ then vertices x, z, v, u, h induce a house, which is impossible (note that h is adjacent neither to x nor to z because $h > u = mn(z) > v = mn(x)$).

So, it remains only to consider the case when $hw \in E$. In this case, according to the APASP algorithm, we have $\hat{d}(h, w) = 1$ and therefore $\hat{d}(x, y) = \hat{d}(v, y) + 1 = \hat{d}(v, w) + 1 + 1 = \hat{d}(h, w) + 1 + 1 + 1 = 4$. Finally, $d(x, y) = 2 = 4 - 2 = \hat{d}(x, y) - 2$.   □

Combining Lemmas 1, 4, 5, and 6 we obtain the following results for HH-free, HHD-free and chordal graphs.

Theorem 1. *Let G be a HH-free graph and σ be a LexBFS–ordering of G. Then, for all vertices $v, u \in V$, the distances returned in $\hat{d}$ by the algorithm APASP satisfy the inequality*

$$0 \leq \hat{d}(v, u) - d(v, u) \leq 2.$$

Corollary 2. *All distances in HH-free graphs with an additive one-sided error of at most 2 can be found in $O(n^2)$ time.*

Theorem 2. *Let G be a HHD-free graph and σ be a LexBFS–ordering of G. Then, for all vertices $v, u \in V$, the distances returned in $\hat{d}$ by the algorithm APASP satisfy the inequality*

$$0 \leq \hat{d}(v, u) - d(v, u) \leq 1.$$

Corollary 3. *All distances in HHD-free graphs with an additive one-sided error of at most 1 can be found in $O(n^2)$ time.*

Corollary 4 ([32]). *All distances in chordal graphs with an additive one-sided error of at most 1 can be found in $O(n^2)$ time.*

Note that the bounds given in Theorems 1 and 2 are tight. Figure 3 shows a chordal graph with a LexBFS–ordering where the difference between $\hat{d}$ and d can achieve 1, Figure 4 gives a similar example for HH-free graphs. In all figures dark edges indicate the maximum-neighbor path between vertices numbered 1 and 2.

For the distance-hereditary graphs this method gives the exact distances.

Theorem 3. *Let G be a distance-hereditary graph and σ be a LexBFS–ordering of G. Then the distances returned in $\hat{d}$ by the algorithm APASP are the real distances, i.e., $\hat{d}(v, u) = d(v, u)$ for any $v, u \in V$.*

Proof. According to Lemmas 1, 4, 5, and 6, we have to consider only the case $d(v, u) = 2$, $v < u$, $mn(v)u \notin E$ and there exists a vertex $z \in N(v) \bigcap N(u)$ such that $mn(v)z \in E$, $z > u$.

Consider the maximum neighbor $w = mn(u)$ of u. Since $v < u$ and $mn(v)u \notin E$, by property (P3), we have $mn(v) < w$, i.e., $v < u < z < mn(v) < w$, and $wv \notin E$. By Lemma 2, vertex w has to be adjacent to $mn(v)$ or z. If

$mn(v)w \in E$ then we get an induced house or an induced 3-fan depending on adjacency of w and z. Distance-hereditary graphs cannot contain such induced subgraphs. Hence, w is not adjacent to $mn(v)$ but is adjacent to z. Now we apply Corollary 1 to $z < \{mn(v), w\}$, $mn(v)w \notin E$ and get a vertex t such that $t > w$, $mn(v)t, tw \in E$ and $zt \notin E$. Since $t > w > mn(v)$, vertex v is adjacent neither to t nor to w. Thus, we have created an induced house in G formed by $v, z, mn(v), t, w$, contradicting the fact G is distance-hereditary. □

Corollary 5 ([14]). *The all pairs shortest path problem in distance-hereditary graphs can be solved in $O(n^2)$ time.*

Lemma 4 allows us also to state the following interesting result. All what one needs to compute the exact distances in a HH-free graph G in $O(n^2)$ time is to know in advance the square G^2 of G. Recall that the square G^2 of $G = (V, E)$ is the graph with the same vertex set V where two vertices u and v ($u \neq v$) are adjacent if their distance in G is at most 2. Unfortunately, computing G^2 even for split graphs (a subclass of chordal graphs) is as hard as for general graphs.

Theorem 4. *Let G be a HH-free graph and σ be a LexBFS–ordering of G. Let also the square G^2 of G is given. Then the all pairs shortest path problem on G can be solved in $O(n^2)$ time.*

Proof. Let $\sigma = [v_1, v_2, \ldots, v_n]$ be a LexBFS–ordering of G. Clearly (see Lemma 4), the following modification of APASP algorithm computes the exact distances in G.

> **for** $i = 1$ **to** n **do** $d(v_i, v_i) := 0$;
> **for** $i = n - 1$ **downto** 1 **do**
>   **for** $j = n$ **downto** $i + 1$ **do**
>   **if** $v_i v_j \in E(G)$ **then** $d(v_i, v_j) := 1$
>   **else if** $v_i v_j \in E(G^2) \setminus E(G)$ **then** $d(v_i, v_j) := 2$
>   **else** $d(v_i, v_j) := d(mn(v_i), v_j) + 1$;
> **return** distances d.

If the adjacency matrix of G^2 is given, this algorithm clearly runs in $O(n^2)$ time. □

Corollary 6. *The all pairs shortest path problem on a HHD-free graph G can be solved in $O(n^2)$ time if G^2 is known.*

Corollary 7 ([20]). *The all pairs shortest path problem on a chordal graph G can be solved in $O(n^2)$ time if G^2 is known.*

3.2   Employing Lexical Orderings

In this subsection we consider strongly chordal graphs and chordal bipartite graphs.

Theorem 5 (). *Let G be a strongly chordal graph and σ be a lexical ordering of G. Then the distances returned in $\hat{d}$ by the algorithm APASP are the real distances, i.e., $\hat{d}(v, u) = d(v, u)$ for any $v, u \in V$.*

Proof. Since any lexical ordering of a graph is a LexBFS–ordering and strongly chordal graphs are HH-free (even chordal) graphs, Lemmas 4 and 5 can be applied. According to those lemmas and Lemma 1, we have to consider only the case when $v < u$, $d(v, u) = 2$ and $mn(v)u \notin E$. We claim that this case is impossible.

It is well-known [23] that any lexical ordering of a strongly chordal graph is a strong elimination ordering. Consider a vertex $z \in N(v) \bigcap N(u)$. We have $zv, zu, mn(v)v \in E$, $z < mn(v)$, $v < u$. Since $mn(v)u \notin E$, a contradiction with σ is a strong elimination ordering arises.   □

Corollary 8 ([14,5]). *The all pairs shortest path problem in strongly chordal graphs can be solved in $O(n^2)$ time.*

Since interval graphs are strongly chordal, we immediately conclude.

Corollary 9 ([3,24,28,33]). *The all pairs shortest path problem in interval graphs can be solved in $O(n^2)$ time.*

In a similar way we can prove the following (proof is omitted).

Theorem 6. *Let G be a chordal bipartite graph and σ be a lexical ordering of G. Then the distances returned in $\hat{d}$ by the algorithm APASP are the real distances, i.e., $\hat{d}(v, u) = d(v, u)$ for any $v, u \in V$.*

Corollary 10 ([21]). *The all pairs shortest path problem in chordal bipartite graphs can be solved in $O(n^2)$ time.*

4   Distances in Graphs with Small Size of Largest Induced Cycle: Employing BFS–Orderings

In order to capture the notion of "small" induced cycles, we define a graph to be *k-chordal* if it has no induced cycles of size greater than k. Note that chordal graphs are precisely the 3-chordal graphs, weakly chordal graphs are 4-chordal graphs and AT-free graphs are 5-chordal.

For k-chordal graphs we obtained the following results, which we present here without proofs. It is interesting to note that even simple BFS–orderings succeed in getting a good approximation for distances in k-chordal graphs (for small $k \geq 4$).

Theorem 7. *Let G be a k-chordal graph and σ be a BFS–ordering of G. Then, for all vertices $v, u \in V$, the distances returned in $\hat{d}$ by the algorithm APASP satisfy the inequality*

$$0 \leq \hat{d}(v, u) - d(v, u) \leq k - 1.$$

Corollary 11. *All distances in k-chordal graphs with an additive one-sided error of at most $k - 1$ can be found in $O(n^2)$ time.*

Corollary 12. *All distances in hole-free graphs with an additive one-sided error of at most 3 can be found in $O(n^2)$ time.*

Corollary 13. *All distances in weakly chordal graphs with an additive one-sided error of at most 3 can be found in $O(n^2)$ time.*

This result can be strengthened further for AT-free graphs. We can prove the following.

Theorem 8. *Let G be an AT-free graph and σ be a BFS–ordering of G. Then, for all vertices $v, u \in V$, the distances returned in $\hat{d}$ by the algorithm APASP satisfy the inequality*

$$0 \leq \hat{d}(v, u) - d(v, u) \leq 2.$$

Corollary 14. *All distances in AT-free graphs with an additive one-sided error of at most 2 can be found in $O(n^2)$ time.*

In Figures 5, 6 and 7 we present three graphs with BFS–orderings which show that the bound stated in Theorem 7 is tight at least for 4-, 5-, and 6-chordal graphs. Since all orderings shown in figures are even LexBFS–orderings,

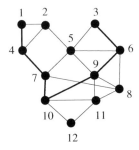

Fig. 5. A weakly chordal graph with a LexBFS–ordering; $\hat{d}_{1,3} - d_{1,3} = 3$.

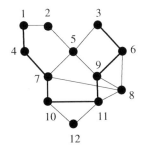

Fig. 6. A 5-chordal graph with a LexBFS–ordering; $\hat{d}_{1,3} - d_{1,3} = 4$.

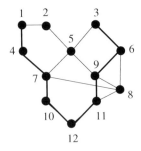

Fig. 7. A 6-chordal graph with a LexBFS–ordering; $\hat{d}_{1,3} - d_{1,3} = 5$.

considering LexBFS instead of BFS in Theorem 7 will not give any better bound (for $k \geq 4$). Furthermore, the domino with a LexBFS–ordering presented in Figure 4 shows that the bound in Theorem 8 for AT-free graphs is tight, too.

Note that distances in k-chordal graphs were already considered in [11]. It was shown that for any k-chordal graph $G = (V, E)$ there exists a tree $T = (V, F)$ such that $|d_G(u, v) - d_T(u, v)| \leq \lfloor k/2 \rfloor + \alpha$ for all $u, v \in V$, where $\alpha = 1$ if $k \neq 4, 5$, and $\alpha = 2$ otherwise. Such a tree T can be constructed in linear time $O(n + m)$.

Acknowledgements: This research was partially supported by the DFG.

References

1. D. AINGWORTH, C. CHEKURI, P. INDYK, and R. MOTWANI, Fast estimation of diameter and shortest paths (without matrix multiplications), *SIAM J. on Computing*, 28 (1999), 1167–1181.
2. N. ALON, Z. GALIL, and O. MARGALIT, On the exponent of the all pairs shortest path problem, Proceedings of the *32nd Annual IEEE Symposium on Foundations of Computer Science*, 1991, 569–575.
3. M. J. ATALLAH, D. Z. CHEN, and D. T. LEE, An optimal algorithm for shortest paths on weighted interval and circular-arc graphs, with applications, Proceedings of the *European Symposium on Algorithms, Lecture Notes in Computer Science*, 726 (1993), 13–24.
4. B. AWERBUCH, B. BERGER, L. COWEN, and D. PELEG, Near-linear cost sequential and distributed constructions of sparse neighborhood covers, Proceedings of the *34th Annual IEEE Symposium on Foundations of Computer Science*, 1993, 638–647.
5. V. BALACHANDHRAN, and C. PANDU RANGAN, All-pairs-shortest length on strongly chordal graphs, *Discrete Applied Math.*, 69 (1996), 169–182.
6. J. BASCH, S. KHANNA, and R. MOTWANI, On diameter verification and Boolean matrix multiplication, *Technical Report STAN-CS-95-1544*, Department of Computer Scince, Stanford University, 1995.
7. A. BRANDSTÄDT, V. D. CHEPOI, and F. F. DRAGAN, The algorithmic use of hypertree structure and maximum neighborhood orderings, *Discrete Applied Math.*, 82 (1998), 43–77.
8. A. BRANDSTÄDT, V. D. CHEPOI, and F. F. DRAGAN, Distance approximating trees for chordal and dually chordal graphs, *Journal on Algorithms*, 30 (1999), 166–184.
9. A. BRANDSTÄDT, V. B. LE, and J. P. SPINRAD, Graph classes: a survey, *SIAM monographs on discrete mathematics and applications*, 1999.
10. L. CHEN, Solving the shortest-paths problem on bipartite permutation graphs efficiently, *Information Processing Letters*, 55 (1995), 259–264.
11. V. D. CHEPOI, and F. F. DRAGAN, A note on distance approximating trees in graphs, *Europ. J. Combinatorics*, 21 (2000), 761–766.
12. E. COHEN, Fast algoruthms for constructing t-spanners and paths with stretch t (extended abstract), Proceedings of the *34th Annual IEEE Symposium on Foundations of Computer Science*, 1993, 648–658.
13. D. COPPERSMITH and S. WINOGRAD, Matrix multiplication via arithmetic progression, Proceedings of the *19th ACM Symposium on Theory of Computing*, 1987, 1–6.

14. E. DAHLHAUS, Optimal (parallel) algorithms for the all-to-all vertices distance problem for certain graph classes, Proceedings of the *International Workshop "Graph-Theoretic Concepts in Computer Science", Lecture Notes in Computer Science*, 657 (1992), 60–69.

15. D. DOR, S. HALPERIN, and U. ZWICK, All pairs almost shortest paths, Proceedings of the *37th Annual IEEE Symposium on Foundations of Computer Science*, 1996, 452–461.

16. F. F. DRAGAN, Almost diameter of a hose-hole free graph in linear time via LexBFS, *Discrete Applied Math.*, 95 (1999), 223-239.

17. F.F. DRAGAN, F. NICOLAI and A. BRANDSTÄDT, LexBFS–orderings and powers of graphs, Proceedings of the *International Workshop "Graph-Theoretic Concepts in Computer Science", Lecture Notes in Computer Science*, 1197 (1997), 166–180.

18. H. EVERETT AND D.G. CORNEIL, Recognizing visibility graphs of spiral polygons, *Journal on Algorithms*, 11 (1990), 1–26.

19. M. C. GOLUMBIC, Algorithmic Graph Theory and Perfect Graphs, *Academic Press*, New York, 1980.

20. K. HAN, CHANDRA N. SEKHARAN and R. SRIDHAR, Unified all-pairs shortest path algorithms in the chordal hierarchy, *Discrete Applied Math.*, 77 (1997), 59–71.

21. CHIN-WEN HO and JOU-MING CHANG, Solving the all-pairs-shortest-length problem on chordal bipartite graphs, *Information Processing Letters*, 69 (1999), 87–93.

22. B. JAMISON and S. OLARIU, On the semi–perfect elimination, *Advances in Applied Math.*, 9 (1988), 364–376.

23. A. LUBIW, Doubly lexical orderings of matrices, *SIAM J. on Computing*, 16 (1987), 854–879.

24. P. MIRCHANDANI, A simple $O(n^2)$ algorithm for the all-pairs shortest path problem on an interval graph, *Networks*, 27 (1996), 215–217.

25. R. MOTWANI, A. RAGUNATHAN AND H. SARAN, Perfect graphs and ortogonally convex covers, *SIAM J. Discrete Math.*, 2 (1989), 371–392.

26. R. MOTWANI, A. RAGUNATHAN AND H. SARAN, Covering orthogonal polygons with star polygons: the perfect graphs approach, *J. of Computer and System Sciences*, 40 (1990), 19–48.

27. R. PAIGE and R.E. TARJAN, Three partition refinement algorithms, *SIAM J. on Computing*, 16 (1987), 973–989.

28. R. RAVI, M. V. MARATHE, and C. PANDU RANGAN, An optimal algorithm to solve the all-pair shortest path problem on interval graphs, *Networks*, 22 (1992), 21–35.

29. R. SEIDEL, On the all-pair-shortest-path problem, Proceedings of the *24th ACM Symposium on Theory of Computing*, 1992, 745–749.

30. J.P. SPINRAD, Doubly lexical ordering of dense 0–1– matrices, *Information Processing Letters*, 45 (1993), 229–235.

31. D. ROSE, R.E. TARJAN and G. LUEKER, Algorithmic aspects on vertex elimination on graphs, *SIAM J. on Computing*, 5 (1976), 266–283.

32. R. SRIDHAR, K. HAN, and N. CHANDRASEKHARAN, Efficient algorithms for shortest distance queries on special classes of polygons, *Theoretical Computer Science*, 140 (1995), 291–300.

33. R. SRIDHAR, D. JOSHI, and N. CHANDRASEKHARAN, Efficient algorithms for shortest distance queries on interval, directed path and circular arc graphs, Proceedings of the *5th International Conference on Computing and Information*, 1993, 31–35.

How to Solve NP-hard Graph Problems
on Clique-Width Bounded Graphs
in Polynomial Time

Wolfgang Espelage, Frank Gurski[*], and Egon Wanke

Department of Computer Science, D-40225 Düsseldorf, Germany
{espelage,gurski,wanke}@cs.uni-dueseldorf.de

Abstract. We show that many non-MSO$_1$ NP-hard graph problems can be solved in polynomial time on clique-width and NLC-width bounded graphs using a very general and simple scheme. Our examples are partition into cliques, partition into triangles, partition into complete bipartite subgraphs, partition into perfect matchings, partition into forests, cubic subgraph, Hamiltonian path, minimum maximal matching, and vertex/edge separation problems.

1 Introduction

Graphs of bounded clique-width are defined by three operations for vertex-labeled graphs, the vertex disjoint union $G \oplus H$ of two graphs, the addition of edges $\eta_{i,j}(G)$ between vertices labeled by i and vertices labeled by j, and the relabeling $\rho_{i \to j}(G)$ of vertices labeled by i into vertices labeled by j, see [3]. Graphs of bounded NLC-width are defined by two similar operations, the union $G \times_S H$ of two graphs that additionally creates all edges between vertices from G labeled by i and vertices from H labeled by j for every pair $(i,j) \in S$, and the relabeling $\circ_R(G)$ of vertices labeled by i into vertices labeled by $R(i)$, see [11]. The clique-width or NLC-width of a graph is the minimum number of labels needed to define it. Every graph of clique-width at most k has NLC-width at most k and every graph of NLC-width at most k has clique-width at most $2k$, see [8].

Clique-width and NLC-width bounded graphs are especially interesting from an algorithmic point of view. A lot of NP-hard graph problems can be solved in polynomial time for graphs of bounded clique-width or bounded NLC-width if an expression for the graph is explicitly given. For example, all graph properties which are expressible in monadic second order logic with quantifications over vertices and vertex sets (MSO$_1$-logic) are decidable in linear time on clique-width and NLC-width bounded graphs, see [2]. The MSO$_1$-logic has been extended by counting mechanisms which allow the expressibility of optimization problems concerning maximal or minimal vertex sets, see [2]. All these graph problems

---

[*] The work of the second author was supported by the German Research Association (DFG) grant WA 674/9-1.

A. Brandstädt and V.B. Le (Eds.): WG 2001, LNCS 2204, pp. 117–128, 2001.
© Springer-Verlag Berlin Heidelberg 2001

expressible in extended MSO_1-logic can also be solved in linear time on clique-width and NLC-width bounded graphs.

If a graph G has clique-width or NLC-width at most k then the edge complement $\overline{G}$ has clique-width at most $2k$, see [3], and NLC-width at most k, see [11]. The set of all graphs of clique-width at most 2 or NLC-width 1 is the set of all labeled cographs. The clique-width and NLC-width of permutation graphs, interval graphs, grids and planar graphs are not bounded by some fixed integer k, see [6]. An arbitrary graph with n vertices has clique-width at most $n - r$, if $2^r < n - r$, and NLC-width at most $\lceil \frac{n}{2} \rceil$, see [8]. Every graph of tree-width at most k has clique-width at most $3 \cdot 2^{k-1}$, see [5], and NLC-width at most $2^{k+1} - 1$, see [11]. The graphs of clique-width at most 2 or NLC-width 1 do not have bounded tree-width. In [7], it is shown that every graph of clique-width or NLC-width at most k which does not contain the complete bipartite graph $K_{n,n}$ for some $n > 1$ as a subgraph has tree-width at most $3k(n - 1) - 1$. The recognition problem for graphs of clique-width or NLC-width at most k is still open for $k \geq 4$ and $k \geq 3$, respectively. Clique-width of at most 3 is decidable in time $O(n^2 m)$, where n is the number of vertices and m is the number of edges of the input graph, see [1]. NLC-width of at most 2 is decidable in time $O(n^4 \log(n))$, see [9].

There are many NP-hard graph problems which are not expressible in extended MSO_1-logic, but which are, nevertheless, solvable in polynomial time on clique-width and NLC-width bounded graphs, see also [11]. In this paper, we extend this problem list by partition into independent sets/cliques (chromatic number), partition into independent sets/cliques of bounded size (partition into triangles), partition into complete bipartite subgraphs, partition into perfect matchings, partition into forests, degree bounded subgraph problems (the cubic subgraph problem), the Hamiltonian path problem, the minimum maximal matching problem, and various vertex/edge separation problems. All these problems are not expressible in extended MSO_1-logic. The input of our algorithms is always a clique-width or NLC-width expression. Note that clique-width and NLC-width expressions can simply be transformed into each other.

The aim of this paper is to illustrate the power of our simple scheme to design such polynomial time algorithms. We also demonstrate that it is sometimes much more convenient to use an NLC-width expression instead of a clique-width expression. This is especially the case for degree bounded subgraph problems and vertex/edge separation problems. The proofs of the correctness are more or less straight forward and are omitted in this version due to space limitations. We are not interested in bounding the constants in the running times of the algorithms which are sometimes more than double exponential in k (the clique-width of the graphs). Such considerations seem to be only interesting if the time complexity can be reduced to $O(n^c)$, where c is independent of k. Up to now, no such algorithms are known for the examples of this paper.

2   Preliminaries

Let $[k] := \{1, \ldots, k\}$ be the set of all integers between 1 and k. We work with finite undirected labeled *graphs* $G = (V, E, \text{lab})$, where V is a finite set of *vertices* labeled by some mapping $\text{lab} : V \to [k]$ and $E \subseteq \{\{u, v\} \mid u, v \in V, \ u \neq v\}$ is a finite set of *edges*. For a vertex set $U \subseteq V$ let $\text{lab}(U) = \{\text{lab}(u) \mid u \in U\}$ and $\text{lab}^2(U)$ be the set of all labels i such that U has at least two vertices labeled by i. A labeled graph $J = (V', E', \text{lab}')$ is a subgraph of G if $V' \subseteq V$, $E' \subseteq E$ and $\text{lab}'(u) = \text{lab}(u)$ for all $u \in V'$. J is an *induced subgraph* of G if additionally $E' = \{\{u, v\} \in E \mid u, v \in V'\}$. The labeled graph consisting of a single vertex labeled by $i \in [k]$ is denoted by $\bullet_i$.

The notion of clique-width for labeled graphs is defined by Courcelle and Olariu in [3].

Definition 1 (Clique-width, [3]). *Let k be some positive integer. The class CW_k of labeled graphs is recursively defined as follows.*

1. *The single vertex graph $\bullet_i$ for some $i \in [k]$ is in CW_k.*
2. *Let $G = (V_G, E_G, \text{lab}_G) \in CW_k$ and $J = (V_J, E_J, \text{lab}_J) \in CW_k$ be two vertex disjoint labeled graphs. Then $G \oplus J := (V', E', \text{lab}')$ defined by $V' := V_G \cup V_J$, $E' := E_G \cup E_J$, and*

$$\text{lab}'(u) := \begin{cases} \text{lab}_G(u) \ \text{if } u \in V_G \\ \text{lab}_J(u) \ \text{if } u \in V_J \end{cases}, \ \forall u \in V'$$

   is in CW_k.
3. *Let $i, j \in [k]$ be two distinct integers and $G = (V, E, \text{lab}) \in CW_k$ be a labeled graph then*
   (a) $\eta_{i,j}(G) := (V, E', \text{lab})$ defined by
       $E' := E \cup \{\{u, v\} \mid u, v \in V, \ \text{lab}(u) = i, \ \text{lab}(v) = j\}$ is in CW_k and
   (b) $\rho_{i \to j}(G) := (V, E, \text{lab}')$ defined by

$$\text{lab}'(u) := \begin{cases} \text{lab}(u) \ \text{if } \text{lab}(u) \neq i \\ j \qquad \ \text{if } \text{lab}(u) = i \end{cases}, \ \forall u \in V$$

   is in CW_k.

The notion of NLC-width[1] of labeled graphs is defined in [11].

Definition 2 (NLC-width, [11]). *Let k be some positive integer. The class NLC_k of labeled graphs is recursively defined as follows.*

1. *The single vertex graph $\bullet_i$ for some $i \in [k]$ is in NLC_k.*
2. *Let $G = (V_G, E_G, \text{lab}_G) \in NLC_k$ and $J = (V_J, E_J, \text{lab}_J) \in NLC_k$ be two vertex disjoint labeled graphs and $S \subseteq [k]^2$ be a relation, then $G \times_S J :=$*

---

[1] The abbreviation NLC results from the *node label controlled* embedding mechanism originally defined for graph grammars.

(V', E', lab') defined by $V' := V_G \cup V_J$, $E' := E_G \cup E_J \cup \{\{u, v\} \mid u \in V_G,\ v \in V_J,\ (lab_G(u), lab_J(v)) \in S\}$, and

$$lab'(u) := \begin{cases} lab_G(u) \text{ if } u \in V_G \\ lab_J(u) \text{ if } u \in V_J \end{cases},\ \forall u \in V'$$

is in NLC_k.

3. Let $G = (V, E, lab) \in NLC_k$ and $R : [k] \to [k]$ be a function, then $\circ_R(G) := (V, E, lab')$ defined by $lab'(u) := R(lab(u))$, $\forall u \in V$ is in NLC_k.

The *clique-width* (*NLC-width*) of a labeled graph G is the least integer k such that $G \in CW_k$ ($G \in NLC_k$, respectively). An expression X built with the operations $\bullet_i, \oplus, \rho_{i \to j}, \eta_{i,j}$ for integers $i, j \in [k]$ is called a *clique-width k-expression*. An expression X built with the operations $\bullet_i, \times_S, \circ_R$ for $i \in [k]$, $S \subseteq [k]^2$, and $R : [k] \to [k]$ is called an *NLC-width k-expression*. The graph defined by expression X is denoted by $val(X)$.

The polynomial time algorithms we want to introduce for NP-hard graph problems on clique-width and NLC-width bounded graphs are based on the following scheme, see also [11]. Let Π be a graph problem for labeled graphs. If there is a mapping F that maps every clique-width k-expression X onto some abstract structure $F(X)$ such that for all k-expressions X, Y (1.) the size of $F(X)$ is polynomially bounded in the size of X, (2.) the answer to Π for $val(X)$ is computable in polynomial time from $F(X)$, (3.) $F(\bullet_i)$ is computable in time $O(1)$, (4.) $F(X \oplus Y)$ is computable in polynomial time from $F(X)$ and $F(Y)$, and (5.) $F(\eta_{i,j}(X))$ and $F(\rho_{i \to j}(X))$ are computable in polynomial time from $F(X)$ then for every clique-width k-expression X, the answer to Π for $val(X)$ is computable in polynomial time from X. This works also for NLC-width k-expressions built with the operations $\bullet_i, \times_S$, and $\circ_R$ instead of $\bullet_i, \oplus, \eta_{i,j}$, and $\rho_{i \to j}$. Then $F(X \times_S Y)$ has to be computable in polynomial time from $F(X)$ and $F(Y)$, and $F(\circ_R(X))$ has to be computable in polynomial time from $F(X)$.

In the next section, we frequently use the notion of a *multi set*, i.e., a set that may have several equal elements. For a multi set $\mathcal{M}$ with elements $x_1, \ldots, x_n$ we write $\mathcal{M} = \langle x_1, \ldots, x_n \rangle$. There is no order on the elements of $\mathcal{M}$. The number how often an element x occurs in $\mathcal{M}$ is denoted by $\psi(\mathcal{M}, x)$. Two multi sets $\mathcal{M}_1$ and $\mathcal{M}_2$ are *equal* if for each element $x \in \mathcal{M}_1 \cup \mathcal{M}_2$, $\psi(\mathcal{M}_1, x) = \psi(\mathcal{M}_2, x)$, otherwise they are called *different*. The empty multi set is denoted by $\langle \rangle$.

For two distinct integers i, j let $R_{i \to j} : [k] \to [k]$ be defined by $R_{i \to j}(t) = t$ if $t \neq i$, and $R_{i \to j}(t) = j$ if $t = i$. For $R : [k] \to [k]$ and $L \subseteq [k]$ let $R(L) = \{R(i) \mid i \in L\}$ and $\rho_{i \to j}(L) = R_{i \to j}(L)$. For $L \subseteq [k]^r$, $r > 1$, let $\rho_{i \to j}(L) = \{(R(i_1), \ldots, R(i_r)) \mid (i_1, \ldots, i_r) \in L\}$ and $\rho_{i \to j}(L) = R_{i \to j}(L)$.

3   Examples

Problem 1 (partition into independent sets (into cliques)).
INSTANCE: Graph $G = (V, E)$, positive integer $r \leq |V|$.
QUESTION: Is there a partition of V into r disjoint sets $V_1, \ldots, V_r$ such that

$V_1 \cup \cdots \cup V_r = V$ and no set V_t, $1 \leq t \leq r$, has two vertices joined by an edge (every set V_t, $1 \leq t \leq r$, induces a complete subgraph, respectively)?

Let $G = (V, E, \text{lab}) = \text{val}(X)$ for some clique-width k-expression X. For a disjoint partition of V into independent sets $V_1, \ldots, V_r$ let $\mathcal{M}$ be the multi set $\langle \text{lab}(V_1), \ldots, \text{lab}(V_r) \rangle$. Let $F(X)$ be the set of all multi sets $\mathcal{M}$ for all disjoint partitions of vertex set V into independent sets. Then $F(X)$ is polynomially bounded in the size of X, because $F(X)$ has at most $(|V| + 1)^{2^k - 1}$ multi sets each with at most $|V|$ nonempty subsets of $[k]$. The following observations show that $F(\bullet_i)$ is computable in time $O(1)$, $F(X \oplus Y)$ is computable in polynomial time from $F(X)$ and $F(Y)$, and $F(\eta_{i,j}(X))$, and $F(\rho_{i \to j}(X))$ are computable in polynomial time from $F(X)$.

1. $F(\bullet_i) = \{\langle \{i\} \rangle\}$
2. Starting with set $D = \{\langle \rangle\} \times F(X) \times F(Y)$, extend D by all triples that can be obtained from some triple $(\mathcal{M}, \mathcal{M}', \mathcal{M}'') \in D$ by removing a set L' from $\mathcal{M}'$ or a set L'' from $\mathcal{M}''$ and inserting it into $\mathcal{M}$, or by removing both sets and inserting $L' \cup L''$ into $\mathcal{M}$.

   D gets at most $(|V| + 1)^{3(2^k - 1)}$ triples and thus is computable in polynomial time.

   $F(X \oplus Y) = \{\mathcal{M} \mid (\mathcal{M}, \langle \rangle, \langle \rangle) \in D\}$
3. $F(\eta_{i,j}(X)) = \{\langle L_1, \ldots, L_r \rangle \in F(X) \mid \{i, j\} \nsubseteq L_t \text{ for } t = 1, \ldots, r\}$
4. $F(\rho_{i \to j}(X)) = \{\langle \rho_{i \to j}(L_1), \ldots, \rho_{i \to j}(L_r) \rangle \mid \langle L_1, \ldots, L_r \rangle \in F(X)\}$

There is a partition of the vertex set of $\text{val}(X)$ into r independent sets if and only if there is some $\mathcal{M} \in F(X)$ consisting of r label sets. A partition of $G = (V, E, \text{lab})$ into cliques corresponds to a partition of its edge complement $\overline{G} = (V, \overline{E}, \text{lab})$, $\overline{E} = \{\{u, v\} \mid u, v \in V, u \neq v, \{u, v\} \notin E\}$, into independent sets. A $2k$-expression $\overline{X}$ for $\text{val}(X)$ can easily be obtained from X in polynomial time, see [3]. This proves the following result.

Theorem 1. *Given a clique-width k-expression X for some fixed integer k, then the problems partition into independent sets and partition into cliques for graph $\text{val}(X)$ are solvable in polynomial time in the size of X.*

The problem partition into independent sets (chromatic number) is also shown in [10] to be solvable in polynomial time on clique-width bounded graphs. Due to its simplicity, we have used the chromatic number problem as our introducing example. Note that our solution differs from that given in [10].

Problem 2 (partition into independent sets (into cliques) of bounded size).
INSTANCE: Graph $G = (V, E)$, positive integer $s \leq |V|$.
QUESTION: Is there a partition of V into disjoint sets $V_1, \ldots, V_r$ of size s such that $V_1 \cup \cdots \cup V_r = V$ and no set V_t, $1 \leq t \leq r$, has two vertices joined by an edge (every set V_t, $1 \leq t \leq r$, induces a complete graph of G, respectively)?

Let s be some fixed integer. For a disjoint partition of V into independent sets $V_1, \ldots, V_r$ of size $\leq s$ let $\mathcal{M}$ be the multi set $\langle (\text{lab}(V_1), |V_1|), \ldots, (\text{lab}(V_r), |V_r|) \rangle$. Let $F(X)$ be the set of all multi sets $\mathcal{M}$ for all disjoint partitions of vertex set

V into independent sets of size $\leq s$. Then $F(X)$ has at most $(|V| + 1)^{(2^k - 1) \cdot s}$ multi sets each with at most $|V|$ tuples (L, x), where L is a nonempty subset of $[k]$ and $1 \leq x \leq s$.

$F(\bullet_i)$, $F(X \oplus Y)$, $F(\eta_{i,j}(X))$, and $F(\rho_{i \rightarrow j}(X))$ are computable in the same way as for the problem partition into independent sets. The only difference worth to emphasize is the computation of $F(X \oplus Y)$ in the case where we remove both tuples, the tuple (L', x') from $\mathcal{M}'$ and the tuple (L'', x'') from $\mathcal{M}''$. In this case, we insert $(L' \cup L'', x' + x'')$ into $\mathcal{M}$ only if $x' + x'' \leq s$.

There is a partition of $\mathrm{val}(X)$ into independent sets of size s if and only if there is some $\mathcal{M} \in F(X)$ such that for all $(L, x) \in \mathcal{M}$, $x = s$. This and the edge complement property of clique-width bounded graphs proves the following theorem.

Theorem 2. *Given a clique-width k-expression X for some fixed integer k, then the problems partition into independent sets of bounded size and partition into cliques of bounded size for graph $\mathrm{val}(X)$ are solvable in polynomial time.*

The problem partition into cliques of size three is also called partition into triangles.

Problem 3 (partition into complete bipartite graphs).
INSTANCE: Graph $G = (V, E)$, positive integer $r \leq |V|$.
QUESTION: Is there a partition of V into r disjoint sets $V_1, \ldots, V_r$ such that $V_1 \cup \cdots \cup V_r = V$ and each set V_t, $1 \leq t \leq r$, induces a complete bipartite graph?

Let $V_{1,1}, V_{1,2}, \ldots, V_{r,1}, V_{r,2}$ be a disjoint partition of vertex set V into independent sets such that the subgraphs G_t induced by the vertices from $V_{t,1} \cup V_{t,2}$ are bipartite (not necessarily complete bipartite) for $t = 1, \ldots, r$. Let $\mathcal{M}$ be the multi set that has a triple $(L_1, L_2, L_{\overline{E}})$ for every pair $(V_{t,1}, V_{t,2})$ defined as follows. $L_1 = \mathrm{lab}(V_{t,1})$, $L_2 = \mathrm{lab}(V_{t,2})$, and $L_{\overline{E}}$ is the set of all label pairs (i, j) such that there are two nonadjacent vertices $u \in V_{t,1}$, $v \in V_{t,2}$ labeled by $\mathrm{lab}(u) = i$, $\mathrm{lab}(v) = j$. Let $F(X)$ be the set of all multi sets $\mathcal{M}$ defined as above for all disjoint partitions of vertex set V into bipartite graphs. Then $F(X)$ has at most $(|V| + 1)^{2^k \cdot 2^k \cdot 2^{k^2}}$ multi sets each with at most $|V|$ triples $(L_1, L_2, L_{\overline{E}})$, where L_1 and L_2 are subsets of $[k]$ and $L_{\overline{E}}$ is a subset of $[k] \times [k]$.

$F(\bullet_i)$, $F(X \oplus Y)$, $F(\eta_{i,j}(X))$, and $F(\rho_{i \rightarrow j}(X))$ are computable as follows.
1. $F(\bullet_i) = \{\langle(\{i\}, \emptyset, \emptyset), (\emptyset, \{i\}, \emptyset)\rangle\}$
2. Starting with set $D = \{\langle\rangle\} \times F(X) \times F(Y)$ extend D by all triples that can be obtained from some triple $(\mathcal{M}, \mathcal{M}', \mathcal{M}'') \in D$ by removing a triple $(L_1', L_2', L_{\overline{E}}')$ from $\mathcal{M}'$ or a triple $(L_1'', L_2'', L_{\overline{E}}'')$ from $\mathcal{M}''$ and inserting it into $\mathcal{M}$, or by removing both triples and inserting $(L_1' \cup L_1'', \; L_2' \cup L_2'', \; L_{\overline{E}}' \cup L_{\overline{E}}'' \cup L_1' \times L_2'' \cup L_1'' \times L_2')$ into $\mathcal{M}$.
   $F(X \oplus Y) = \{\mathcal{M} \mid (\mathcal{M}, \langle\rangle, \langle\rangle) \in D\}$
3. We say $\mathcal{M} \in F(X)$ is feasible for operation $\eta_{i,j}$ if for every $(L_1, L_2, L_{\overline{E}}) \in \mathcal{M}$, $\{i, j\} \not\subseteq L_1$ and $\{i, j\} \not\subseteq L_2$.
   If $\mathcal{M}$ is feasible for operation $\eta_{i,j}$ then let $\eta_{i,j}(\mathcal{M})$ be the multi set obtained from $\mathcal{M}$ by removing (i, j) and (j, i) from all sets $L_{\overline{E}}$ for all $(L_1, L_2, L_{\overline{E}}) \in \mathcal{M}$.

$$F(\eta_{i,j}(X)) = \{\eta_{i,j}(\mathcal{M}) \mid \mathcal{M} \in F(X), \mathcal{M} \text{ feasible for } \eta_{i,j}\}$$

4. For operation $\rho_{i\to j}$ and $\mathcal{M} \in F(X)$, let $\rho_{i\to j}(\mathcal{M})$ be the multi set obtained from $\mathcal{M}$ by substituting every triple $(L_1, L_2, L_{\overline{E}}) \in \mathcal{M}$ by $(\rho_{i\to j}(L_1), \rho_{i\to j}(L_2), \rho_{i\to j}(L_{\overline{E}}))$.
$$F(\rho_{i\to j}(X)) = \{\rho_{i\to j}(\mathcal{M}) \mid \mathcal{M} \in F(X)\}$$

There is a partition of $\mathrm{val}(X)$ into r complete bipartite graphs if and only if there is some $\mathcal{M} \in F(X)$ consisting of r triples $(L_1, L_2, \emptyset)$ such that L_1 and L_2 are nonempty.

Theorem 3. *Given a clique-width k-expression X for some fixed integer k, then the problem partition into complete bipartite graphs for $\mathrm{val}(X)$ is solvable in polynomial time.*

Problem 4 (partition into perfect matchings).
INSTANCE: Graph $G = (V, E)$, positive integer $r \leq |V|$.
QUESTION: Is there a partition of V into r disjoint sets $V_1, \ldots, V_r$ such that $V_1 \cup \cdots \cup V_r = V$ and each set V_t, $1 \leq t \leq r$, induces a perfect matching?

A vertex set $V' \subseteq V$ is called *edge independent* if the subgraph G' of G induced by V' does not contain two incident edges. Let $V_1, \ldots, V_r$ be a disjoint partition of vertex set V into edge independent sets. For such a partition of V let $\mathcal{M}$ be the multi set that has a 4-tuple $(L_{\geq 1}, L_{\geq 2}, L_E, L_U)$ for every vertex set V_t, $1 \leq t \leq r$, defined as follows. $L_{\geq 1} = \mathrm{lab}(V_t)$, $L_{\geq 2} = \mathrm{lab}^2(V_t)$, L_E is the set of all label sets $\{i, j\}$ such that the subgraph G_t of G induced by the vertices of V_t has two adjacent vertices labels by i and j, respectively, and L_U is the set of all labels i such that G_t has a vertex labeled by i that is not incident to any edge of G_t. Let $F(X)$ be the set of all multi sets $\mathcal{M}$ defined as above for all disjoint partitions of vertex set V into edge independent sets. Then $F(X)$ has at most $(|V| + 1)^{(2^k-1)\cdot 2^k \cdot 2^{k^2} \cdot 2^k}$ multi sets each with at most $|V|$ 4-tuples $(L_{\geq 1}, L_{\geq 2}, L_E, L_U)$, where $L_{\geq 1}$ is a nonempty subset of $[k]$, $L_{\geq 2}$ and L_U are subsets of $[k]$, and L_E is a subset of $\{\{i, j\} \mid i, j \in [k]\}$.
$F(\bullet_i)$, $F(X \oplus Y)$, $F(\eta_{i,j}(X))$, and $F(\rho_{i\to j}(X))$ are computable as follows.

1. $F(\bullet_i) = \{\langle(\{i\}, \emptyset, \emptyset, \{i\})\rangle\}$
2. Starting with set $D = \{\langle\rangle\} \times F(X) \times F(Y)$ extend D by all triples that can be obtained from some triple $(\mathcal{M}, \mathcal{M}', \mathcal{M}'') \in D$ by removing a 4-tuple $(L'_{\geq 1}, L'_{\geq 2}, L'_E, L'_U)$ from $\mathcal{M}'$ or a 4-tuple $(L''_{\geq 1}, L''_{\geq 2}, L''_E, L''_U)$ from $\mathcal{M}''$ and inserting it into $\mathcal{M}$, or by removing both 4-tuples and inserting

$$(L'_{\geq 1} \cup L''_{\geq 1}, \ L'_{\geq 2} \cup L''_{\geq 2} \cup (L'_{\geq 1} \cap L''_{\geq 1}), \ L'_E \cup L''_E, \ L'_U \cup L''_U)$$

into $\mathcal{M}$.
$F(X \oplus Y) = \{\mathcal{M} \mid (\mathcal{M}, \langle\rangle, \langle\rangle) \in D\}$
3. We say $\mathcal{M} \in F(X)$ is feasible for operation $\eta_{i,j}$ if for every 4-tuple $(L_{\geq 1}, L_{\geq 2}, L_E, L_U) \in \mathcal{M}$ the following holds. If $\{i, j\} \subseteq L_{\geq 1}$ then $\{i, j\} \subseteq L_{\geq 1} - L_{\geq 2}$ and either $\{i, j\} \in L_E$ or $i, j \in L_U$. If $\mathcal{M}$ is feasible for operation

$\eta_{i,j}$ then let $\eta_{i,j}(\mathcal{M})$ be the multi set obtained from $\mathcal{M}$ by the following modification of the 4-tuples $(L_{\geq 1}, L_{\geq 2}, L_E, L_U) \in \mathcal{M}$. If $i, j \in L_{\geq 1} - L_{\geq 2}$ and $i, j \in L_U$ then remove i, j from L_U and insert $\{i, j\}$ into L_E.
$$F(\eta_{i,j}(X)) = \{\eta_{i,j}(\mathcal{M}) \mid \mathcal{M} \in F(X), \mathcal{M} \text{ feasible for } \eta_{i,j}\}$$

4. For operation $\rho_{i \rightarrow j}$ let $\rho_{i \rightarrow j}(\mathcal{M})$ be the multi set obtained from $\mathcal{M}$ by the following modification of the 4-tuples $(L_{\geq 1}, L_{\geq 2}, L_E, L_U) \in \mathcal{M}$. If $i, j \in L_{\geq 1}$ then insert j into $L_{\geq 2}$. Then relabel every label i into label j in every set $L_{\geq 1}, L_{\geq 2}, L_U$, and L_E.
$$F(\rho_{i \rightarrow j}(X)) = \{\rho_{i \rightarrow j}(\mathcal{M}) \mid \mathcal{M} \in F(X)\}$$

There is a partition of $\mathrm{val}(X)$ into r perfect matchings if and only if there is some $\mathcal{M} \in F(X)$ consisting of r 4-tuples $(L_{\geq 1}, L_{\geq 2}, L_E, L_U)$ such that L_U is empty.

Theorem 4. *Given a clique-width k-expression X for some fixed integer k, then the problem partition into perfect matchings for $\mathrm{val}(X)$ is solvable in polynomial time.*

Problem 5 (partition into forests).
INSTANCE: Graph $G = (V, E)$, positive integer $r \leq |V|$.
QUESTION: Is there a partition of V into r disjoint sets $V_1, \ldots, V_r$ such that $V_1 \cup \cdots \cup V_r = V$ and each set V_t, $1 \leq t \leq r$, induces a forest?

Let $V_1, \ldots, V_r$ be a disjoint partition of the vertex set V such that the subgraphs G_t of G induced by the vertices from V_t are forests for $t = 1, \ldots, r$. For such a partition of V let $\mathcal{M}$ be the multi set that has a triple $(L_{\geq 1}, L_{\geq 2}, L_T)$ for every vertex set V_t, $1 \leq t \leq r$, defined as follows. $L_{\geq 1} = \mathrm{lab}(V_t)$, $L_{\geq 2} = \mathrm{lab}^2(V_t)$, and L_T is the set of all triples $(K_{\geq 1}, K_{\geq 2}, K_{\overline{E}})$ such that G_t has a maximal connected component (a tree) $T = (V_T, E_T, \mathrm{lab}_T)$ and $K_{\geq 1} = \mathrm{lab}(V_T)$, $K_{\geq 2} = \mathrm{lab}^2(V_T)$, and $K_{\overline{E}}$ is the set of all label sets $\{i, j\}$ such that V_T has two nonadjacent vertices labeled by i and j, respectively. Let $F(X)$ be the set of all multi sets $\mathcal{M}$ defined as above for all disjoint partitions of vertex set V into forests.

The size of $F(X)$ is polynomially bounded in the size of X, because $F(X)$ has at most $(|V| + 1)^{(2^k - 1) \cdot 2^k \cdot 2^{2^k \cdot 2^k \cdot 2^{k^2}}}$ multi sets each with at most $|V|$ triples $(L_{\geq 1}, L_{\geq 2}, L_T)$, where $L_{\geq 1}$ is a nonempty subset of $[k]$, $L_{\geq 2}$ is a subset of $[k]$, and L_T is a set of triples $(K_{\geq 1}, K_{\geq 2}, K_{\overline{E}})$ where $K_{\geq 1}$ is a nonempty subset of $[k]$, $K_{\geq 2}$ is a subset of $[k]$, and $K_{\overline{E}}$ is a subset of $\{\{i, j\} \mid i, j \in [k]\}$.

$F(\bullet_i)$, $F(X \oplus Y)$, $F(\eta_{i,j}(X))$, and $F(\rho_{i \rightarrow j}(X))$ are computable as follows.

1. $F(\bullet_i) = \{\langle (\{i\}, \emptyset, \{(\{i\}, \emptyset, \emptyset)\}) \rangle\}$
2. Starting with set $D = \{\langle \rangle\} \times F(X) \times F(Y)$ extend D by all triples that can be obtained from some triple $(\mathcal{M}, \mathcal{M}', \mathcal{M}'') \in D$ by removing a triple $(L'_{\geq 1}, L'_{\geq 2}, L'_T)$ from $\mathcal{M}'$ or a triple $(L''_{\geq 1}, L''_{\geq 2}, L''_T)$ from $\mathcal{M}''$ and inserting it into $\mathcal{M}$, or by removing both triples and inserting $(L'_{\geq 1} \cup L''_{\geq 1}, L'_{\geq 2} \cup L''_{\geq 2} \cup (L'_{\geq 1} \cap L''_{\geq 1}), L'_T \cup L''_T)$ into $\mathcal{M}$.
$$F(X \oplus Y) = \{\mathcal{M} \mid (\mathcal{M}, \langle \rangle, \langle \rangle) \in D\}$$

3. We say $\mathcal{M} \in F(X)$ is feasible for operation $\eta_{i,j}$ if for every triple $(L_{\geq 1}, L_{\geq 2}, L_T) \in \mathcal{M}$, $\{i,j\} \not\subseteq L_{\geq 2}$ and for every triple $(K_{\geq 1}, K_{\geq 2}, K_{\overline{E}}) \in L_T$, $\{i,j\} \not\subseteq K_{\overline{E}}$, and if $i \in L_{\geq 1} - K_{\geq 1}$ then $j \notin K_{\geq 2}$, and if $j \in L_{\geq 1} - K_{\geq 1}$ then $i \notin K_{\geq 2}$. If $\mathcal{M}$ is feasible for operation $\eta_{i,j}$ then let $\eta_{i,j}(\mathcal{M})$ be the multi set obtained from $\mathcal{M}$ by the following modification of the triples $(L_{\geq 1}, L_{\geq 2}, L_T) \in \mathcal{M}$. Assume $i \in L_{\geq 1} - L_{\geq 2}$, for $j \in L_{\geq 1} - L_{\geq 2}$ change i and j. Then let

$$(K_{1,\geq 1}, K_{1,\geq 2}, K_{1,\overline{E}}), \ldots, (K_{s,\geq 1}, K_{s,\geq 2}, K_{s,\overline{E}}) \in L_T$$

be all triples such that $i \in K_{1,\geq 1}$ and $j \in K_{t,\geq 1}$ for $t = 2, \ldots, s$. Remove these s triples from L_T and insert a new triple $(K'_{\geq 1}, K'_{\geq 2}, K'_{\overline{E}})$ into L_T defined as follows.

$$K'_{\geq 1} = \bigcup_{1 \leq t \leq s} K_{t,\geq 1},$$

$$K'_{\geq 2} = \bigcup_{1 \leq t \leq s} K_{t,\geq 2} \cup \bigcup_{1 \leq t_1 < t_2 \leq s} K_{t_1,\geq 1} \cap K_{t_2,\geq 1}, \text{ and}$$

$$K'_{\overline{E}} = \bigcup_{1 \leq t \leq s} K_{t,\overline{E}} \cup \bigcup_{1 \leq t_1 < t_2 \leq s} \{\{i',j'\} \mid i' \in K_{t_1,\geq 1}, \; j' \in K_{t_2,\geq 1}\} - \{i,j\}.$$

$$F(\eta_{i,j}(X)) = \{\eta_{i,j}(\mathcal{M}) \mid \mathcal{M} \in F(X), \; \mathcal{M} \text{ feasible for } \eta_{i,j}\}$$

4. For operation $\rho_{i \to j}$ let $\rho_{i \to j}(\mathcal{M})$ be the multi set obtained from $\mathcal{M}$ by the following modification of the triples $(L_{\geq 1}, L_{\geq 2}, L_T) \in \mathcal{M}$. If both labels i, j are in $L_{\geq 1}$ or in $K_{\geq 1}$ of some triple $(K_{\geq 1}, K_{\geq 2}, K_{\overline{E}}) \in L_T$ then insert j into $L_{\geq 2}$ or $K_{\geq 2}$, respectively.
Then relabel every i into j in every set of $(L_{\geq 1}, L_{\geq 2}, L_T)$.
$F(\rho_{i \to j}(X)) = \{\rho_{i \to j}(\mathcal{M}) \mid \mathcal{M} \in F(X)\}$

There is a partition of $\mathrm{val}(X)$ into r forests if and only if there is some $\mathcal{M} \in F(X)$ consisting of r triples.

Theorem 5. *Given a clique-width k-expression X for some fixed integer k, then the problem partition into forests for $\mathrm{val}(X)$ is solvable in polynomial time.*

Problem 6 (cubic subgraph).
INSTANCE: Graph $G = (V, E)$.
QUESTION: Is there a nonempty subset $E' \subseteq E$ such that in graph (V, E') every vertex has either degree 3 or degree 0?

We illustrate this example with an NLC-width k-expression instead of a clique-width k-expression. For some fixed integer s and a subset of edges $E' \subseteq E$ such that in $G' = (V, E', \mathrm{lab})$ all vertices have degree $\leq s$, let $\mathcal{M}$ be the following multi set. $\mathcal{M}$ has for every vertex $u \in V$ a tuple $(\mathrm{lab}(u), \deg(u))$, where $\deg(u)$ is the vertex degree of u in G'. Let $F(X)$ be the set of all multi sets $\mathcal{M}$ for integer s and all such subsets E'. Then $F(X)$ has at most $(|V| + 1)^{k \cdot (s+1)}$ multi sets each with $|V|$ tuples (l, d), where l is a label from $[k]$ and d is an integer between 0 and s.
$F(\bullet_i)$, $F(X \times_S Y)$, and $F(\circ_R(X))$ are computable as follows.

1. $F(\bullet_i) = \{\langle\langle i, 0\rangle\rangle\}$
2. Starting with set $D = \{\langle\rangle\} \times F(X) \times F(Y)$ extend D by all triples that can be obtained from some triple $(\mathcal{M}, \mathcal{M}', \mathcal{M}'') \in D$ by
   (a) removing a tuple (l', d') from $\mathcal{M}'$ or (l'', d'') from $\mathcal{M}''$ and inserting it into $\mathcal{M}$ or by
   (b) removing a tuple (l', d') from $\mathcal{M}'$ and choosing at most r, $1 \le r \le s - d'$, tuples $(l''_1, d''_1), \ldots, (l''_r, d''_r)$ from $\mathcal{M}''$ such that $(l', l''_t) \in S$ and $d''_t < s$ for $t = 1, \ldots, r$, and inserting $(l', d' + r)$ into $\mathcal{M}$ and changing every (l''_t, d''_t) into $(l''_t, d''_t + 1)$ without removing them from $\mathcal{M}''$.
   $$F(X \times_S Y) = \{\mathcal{M} \mid (\mathcal{M}, \langle\rangle, \langle\rangle) \in D\}$$
3. $F(\circ_R(X)) = \{\langle\langle(R(l_1), d_1), \ldots, (R(l_t), d_t)\rangle\rangle \mid \langle\langle(l_1, d_1), \ldots, (l_t, d_t)\rangle\rangle \in \mathcal{M}\}$

There is a subgraph of $\mathrm{val}(X)$ with at least one edge such that all vertices have degree 3 or 0 if and only if there is some $\mathcal{M} \in F(X)$ such that for all $(l, d) \in \mathcal{M}$, $d = 3$ or $d = 0$ and there is at least one $(l, d) \in \mathcal{M}$ with $d = 3$.

Theorem 6. *Given an NLC k-expression X for some fixed integer k, then the cubic subgraph problem for $\mathrm{val}(X)$ is solvable in polynomial time.*

Problem 7 (Hamiltonian path).
INSTANCE: Graph $G = (V, E)$.
QUESTION: Is there a simple path of length $|V|$ in G?

For a subset of edges $E' \subseteq E$ such that subgraph $G' = (V, E', \mathrm{lab})$ contains no cycles and every vertex in G' has degree at most 2, let $\mathcal{M}$ be the following multi set. $\mathcal{M}$ has a multi set $\langle \mathrm{lab}(v_1), \mathrm{lab}(v_r)\rangle$ for every path $p = v_1, \ldots, v_r$, $r \ge 1$, in G', where v_1, v_r have degree ≤ 1 in G'. Let $F(X)$ be the set of all multi sets $\mathcal{M}$ for all such subsets $E' \subseteq E$. Then $F(X)$ has at most $(|V| + 1)^{k^2}$ multi sets each with at most $|V|$ multi sets of size two.
$F(\bullet_i)$, $F(X \oplus Y)$, $F(\eta_{i,j}(X))$, and $F(\rho_{i \to j}(X))$ are computable as follows.

1. $F(\bullet_i) = \{\langle\langle i, i\rangle\rangle\}$
2. $F(X \oplus Y) = \{\mathcal{M}' \cup \mathcal{M}'' \mid \mathcal{M}' \in F(X), \mathcal{M}'' \in F(Y)\}$
3. $F(\eta_{i,j}(X))$ can be computed from $F(X)$ as follows. Starting with set $D = F(X)$, extend D by all multi sets that can be obtained from a multi set $\mathcal{M}$ of D by removing two multi sets $\langle i_1, j_1\rangle, \langle i_2, j_2\rangle$, where $i = i_1$ and $j = j_2$, and inserting the new multi set $\langle i_2, j_1\rangle$. The resulting set D is $F(\eta_{i,j}(X))$.
4. $F(\rho_{i \to j}(X)) = \{\langle\langle\rho_{i \to j}(M)\rangle\rangle \mid M \in \mathcal{M}\rangle \mid \mathcal{M} \in F(X)\}$

Graph $\mathrm{val}(X)$ has a Hamiltonian path if and only if $F(X)$ has a multi set $\mathcal{M}$ consisting of exactly one multi set.

Theorem 7. *Given a clique-width k-expression X for some fixed integer k, then the Hamiltonian path problem for $\mathrm{val}(X)$ is computable in polynomial time.*

A simple modification leads to a polynomial time solution for the Hamiltonian circuit problem as in [11].

Problem 8 (minimum maximal matching).
INSTANCE: Graph $G = (V, E)$, positive integer $r \leq |E|$.
QUESTION: Is there a subset $E' \subseteq E$ with $|E'| \leq r$ such that $|E'|$ is a maximal matching, i.e. no two edges in E' share a common vertex and every edge in $E - E'$ shares a common vertex with some edge in E'?

For a subset of edges $E' \subseteq E$ and a disjoint partition of V into three sets V_1, V_2, V_3 such that V_1 is an independent set and E' is a perfect matching for the vertices of V_2, we define a triple $(\text{lab}(V_1), |E'|, \langle \text{lab}(u_1), \ldots, \text{lab}(u_s) \rangle)$, where $V_3 = \{u_1, \ldots, u_s\}$. Let $F(X)$ be the set of all triples defined as above for all such edge sets and partitions.

Then $F(X)$ has at most $2^k \cdot \frac{|V|}{2} \cdot (|V| + 1)^k$ triples of size at most $k + 1 + |V|$. $F(\bullet_i)$, $F(X \oplus Y)$, $F(\eta_{i,j}(X))$, and $F(\rho_{i \to j}(X))$ are computable as follows.

1. $F(\bullet_i) = \{(\emptyset, 0, \langle i \rangle), (\{i\}, 0, \langle \rangle)\}$
2. $F(X \oplus Y) = \{(L \cup L', t + t', \mathcal{M} \cup \mathcal{M}') \mid (L, t, \mathcal{M}) \in F(X), (L', t', \mathcal{M}') \in F(Y)\}$
3. $F(\eta_{i,j}(X))$ can be computed from $F(X)$ as follows. Starting with set $D = F(X)$, remove all triples $(L, t, \mathcal{M})$ from D for which $\{i, j\} \subseteq L$, and then extend D by all triples $(L, t + 1, \mathcal{M} - \langle \{i\}, \{j\} \rangle)$ for all triples $(L, t, \mathcal{M})$ in D, where $\langle \{i\}, \{j\} \rangle \subseteq \mathcal{M}$. The resulting set D is $F(\eta_{i,j}(X))$.
4. $F(\rho_{i \to j}(X)) = \{(\rho_{i \to j}(L), t, \rho_{i \to j}(\mathcal{M})) \mid (L, t, \mathcal{M}) \in F(X)\}$

There is a subset $E' \subseteq E$ with $|E'| = r$ such that $|E'|$ is a maximal matching in G if and only if there is some triple $(L, r, \langle \rangle) \in F(X)$.

Theorem 8. *Given a clique-width k-expression X for some fixed integer k, then the minimum maximal matching problem for* $\text{val}(X)$ *is computable in polynomial time.*

Let us finally consider some vertex and edge separation problems. Let $F(X)$ be the set of all 5-tuples $(\text{lab}(V_1), |V_1|, |V_2|, \text{lab}(V_3), |V_3|)$ for all disjoint partitions of vertex set V into three sets V_1, V_2, V_3 such that G has no edge between a vertex from V_1 and a vertex from V_3. Then $F(X)$ has at most $2^k \cdot (|V| + 1) \cdot (|V| + 1) \cdot 2^k \cdot (|V| + 1)$ 5-tuples. $F(\bullet_i) = \{(\{i\}, 1, 0, \emptyset, 0), (\emptyset, 0, 1, \emptyset, 0), (\emptyset, 0, 0, \{i\}, 1)\}$. $F(X \times_S Y)$ is the set of all 5-tuples $(L_1, x_1, x_2, L_3, x_3)$ obtained from two 5-tuples $(L_1', x_1', x_2', L_3', x_3') \in F(X)$ and $(L_1'', x_1'', x_2'', L_3'', x_3'') \in F(Y)$ such that $L_1 = L_1' \cup L_1''$, $x_1 = x_1' + x_1''$, $x_2 = x_2' + x_2''$, $L_3 = L_3' \cup L_3''$, $x_3 = x_3' + x_3''$, if there is no pair $(i, j) \in S$ such that $i \in L_1'$ and $j \in L_3''$ or $i \in L_3'$ and $j \in L_1''$. $F(\circ_R(X)) = \{(R(L_1), x_1, x_2, R(L_3), x_3) \mid (L_1, x_1, x_2, L_3, x_3) \in F(X)\}$.

The vertex separation problem whether $G = \text{val}(X)$ has a vertex set V_2 of size $\leq r$ such that G without the vertices of V_2 can be partitioned into two disjoint sets V_1, V_3 of size $\leq \alpha|V|$ such that there is no edge between vertices from V_1 and vertices from V_3, for some given r and α, can now be solved in polynomial time.

Let $F(X)$ be the set of all triples

$$(\langle \text{lab}(u_1), \ldots, \text{lab}(u_r) \rangle, x, \langle \text{lab}(u_{r+1}), \ldots, \text{lab}(u_n) \rangle)$$

for all disjoint partitions of $V = \{u_1, \ldots, u_n\}$ into $V_1 = \{u_1, \ldots, u_r\}$ and $V_2 = \{u_{r+1}, \ldots, u_n\}$ such that x is the number of edges between the vertices from V_1 and V_2. Then $F(X)$ has at most $(|V| + 1)^k \cdot (|E| + 1) \cdot (|V| + 1)^k$ triples.

$F(\bullet_i) = \{(\langle i \rangle, 0, \langle \rangle), (\langle \rangle, 0, \langle i \rangle)\}$. $F(X \times_S Y)$ is the set of all triples $(\mathcal{M}_1, x, \mathcal{M}_2)$ obtained from two triples $(\mathcal{M}_1', x', \mathcal{M}_2') \in F(X)$ and $(\mathcal{M}_1'', x'', \mathcal{M}_2'') \in F(Y)$ such that $\mathcal{M}_1 = \mathcal{M}_1' \cup \mathcal{M}_1''$, $\mathcal{M}_2 = \mathcal{M}_2' \cup \mathcal{M}_2''$, and

$$x = x' + x'' + \sum_{(i,j) \in S} \psi(\mathcal{M}_1', i) \cdot \psi(\mathcal{M}_2'', j) + \sum_{(i,j) \in S} \psi(\mathcal{M}_2', i) \cdot \psi(\mathcal{M}_1'', j).$$

$F(\circ_R(X)) = \{(R(\mathcal{M}_1), x, R(\mathcal{M}_2)) \mid (\mathcal{M}_1, x, \mathcal{M}_2) \in F(X)\}$.

The edge separation problem whether there is a partition of the vertex set of $G = \text{val}(X)$ into two disjoint sets V_1, V_2 of size $\leq \alpha|V|$ such that there are at most r edges between vertices from V_1 and vertices from V_2, for some given r and α, can now be solved in polynomial time.

References

1. D.G. Corneil, M. Habib, J.M. Lanlignel, B. Reed, and U. Rotics. Polynomial time recognition of clique-width at most three graphs. In *Proceedings of Latin American Symposium on Theoretical Informatics (LATIN '2000)*, volume 1776 of *LNCS*. Springer-Verlag, 2000.
2. B. Courcelle, J.A. Makowsky, and U. Rotics. Linear time solvable optimization problems on graphs of bounded clique width. *Theory of Computing Systems*, 33(2):125–150, 2000.
3. B. Courcelle and S. Olariu. Upper bounds to the clique width of graphs. *Discrete Applied Mathematics*, 101:77–114, 2000.
4. D.G. Corneil, Y. Perl, and L.K. Stewart. A linear recognition algorithm for cographs. *SIAM Journal on Computing*, 14(4):926–934, 1985.
5. D.G. Corneil and U. Rotics. On the relationship between clique-width and tree-width. In *Proceedings of Graph-Theoretical Concepts in Computer Science, LNCS*, Springer-Verlag, 2001. to appear
6. M.C. Golumbic and U. Rotics. On the clique-width of perfect graph classes. *IJFCS: International Journal of Foundations of Computer Science*, 11(3):423–443, 2000.
7. F. Gurski and E. Wanke. The tree-width of clique-width bounded graphs without $K_{n,n}$. In *Proceedings of Graph-Theoretical Concepts in Computer Science*, volume 1938 of *LNCS*, pages 196–205. Springer-Verlag, 2000.
8. Ö. Johansson. Clique-decomposition, NLC-decomposition, and modular decomposition - relationships and results for random graphs. *Congressus Numerantium*, 132:39–60, 1998.
9. Ö. Johansson. NLC2 decomposition in polynomial time. In *Proceedings of Graph-Theoretical Concepts in Computer Science*, volume 1665 of *LNCS*, pages 110–121. Springer-Verlag, 1999.
10. D. Kobler and U. Rotics. Polynomial algorithms for partitioning problems on graphs with fixed clique-width. In *Proceedings of the ACM-SIAM Symposium on Discrete Algorithms*, pages 468–476 ACM-SIAM, 2001.
11. E. Wanke. k-NLC graphs and polynomial algorithms. *Discrete Applied Mathematics*, 54:251–266, 1994. Revised version, "http://www.cs.uni-duesseldorf.de/~wanke".

$(g, f)-$Factorizations Orthogonal to k Subgraphs*

Haodi Feng

Department of Computer Science
City University of HongKong
Hong Kong
fenghd@cs.cityu.edu.hk

Abstract. We establish two results for $(g, f)-$factorizations:
1. every $(mg + m - 1, mf - m + 1)-$graph with $9k/4 \leq g \leq f$ has a $(g, f)-$factorization orthogonal to any set of k given edge disjoint $m-$subgraphs of G;
2. every $(0, mf - m + 1)-$graph G with $f \geq k + 2$ has a $(0, f)-$factorization orthogonal to any set of k given vertex disjoint $m-$subgraphs of G. Polynomial-time algorithms to find the claimed orthogonal factorizations under the above conditions are presented.

1 Introduction

Consider a simple graph $G = (V, E)$. Let $d_G(x)$ be the degree of $x \in V$. For any two integer-valued functions g and f defined on V such that $\forall x \in V : g(x) \leq f(x)$, a $(g, f)-$factor of G is a spanning subgraph F of G satisfying $g(x) \leq d_F(x) \leq f(x)$ for all $x \in V(F)$. In particular, if G is a $(g, f)-$factor itself, then G is called a $(g, f)-$graph. A $(g, f)-$factorization $\mathcal{F} = \{F_1, F_2, \ldots, F_m\}$ of G is a partition of E into edge disjoint $(g, f)-$factors $F_1, F_2, ..., F_m$. Let H be a subgraph of G with m edges (called an $m-$subgraph of G). A factorization $\mathcal{F} = \{F_1, F_2, ..., F_m\}$ of G is orthogonal to H if $|E(H) \cap E(F_i)| = 1, 1 \leq i \leq m$.

The factorization problem can be viewed as a generalization of the edge coloring problem. Imposing the condition orthogonal to a given subgraph (or k subgraphs) introduces to the study an appealing feature for application areas. The study of orthogonal factorization problems was first proposed by Alspach et al.[1], to formulate practical problems that found applications in combinatorial design, network design, circuit layout, etc.. The decision problem of the orthogonal factorization problem is, in general, NP-complete[1].

Early studies in this area have been focused on the factorization orthogonal to one subgraph. Recently, Lam, et al.[6], studied the more general problem of

---

* Research is partially supported by a grant from the Research Grants Council of Hong Kong SAR (CityU 1074/00E) and a grant from CityU of Hong Kong (Project No.7001215).

[1] Reduction from the edge coloring problem to it can be done by setting H as the set of edges incident to a maximum degree node, and setting g to be all zeros and f to be all ones.

A. Brandstädt and V.B. Le (Eds.): WG 2001, LNCS 2204, pp. 129–139, 2001.
© Springer-Verlag Berlin Heidelberg 2001

finding factorizations orthogonal to each of k subgraphs. More precisely, they consider the following problem: given subgraphs $H_1, H_2, ..., H_k$ of G, does there exist a factorization orthogonal to each $H_i, 1 \le i \le k$? They proved that, for any vertex-disjoint m−subgraphs $H_1, H_2, ..., H_k$ of an $(mg+m-1, mf-m+1)$−graph G, there exists a (g, f)−factorization of G orthogonal to every $H_i, 1 \le i \le k$, where $k \le g(x) \le f(x)$ for every $x \in V$. The requirement of vertex-disjoint subgraphs is rather stringent in practical applications. As an open problem, Lam, et al., asked that, if $H_1, H_2, ..., H_k$ are mutually edge-disjoint m−subgraphs of G, will the above conclusion still hold? In this paper, we answer the question in the affirmative under the condition $9k/4 \le g \le f$.

Another question that left open in their work is when g is allowed to be all zeros. We also study this for $(0, mf - m + 1)$−graphs and prove that: for any vertex-disjoint m−subgraphs $H_1, H_2, ..., H_k$ of a $(0, mf - m + 1)$−graph G, there exists a $(0, f)$−factorization of G orthogonal to each $H_i, 1 \le i \le k$, where $k + 2 \le f(x)$ for all $x \in V$.

In Section 2, we introduce some preliminary results necessary for our study. In Section 3, we present our work on factorizations orthogonal to k edge disjoint graphs. In Section 4, we prove our extension for (g, f)-factorization with $g = 0$ (for k vertex disjoint graphs). In Section 5, we conclude our paper with discussion on open problems and on polynomial-time algorithms for finding the above (g, f)−factorizations.

2   Preliminaries

For $S \subset V$, let $E(S) = \{xy \in E(G) : x, y \in S\}$. Suppose $S, T \subset V$ with $S \cap T = \emptyset$, and $E_1, E_2 \subset E$ with $E_1 \cap E_2 = \emptyset$. Let $E_G(S, T) = \{(s, t) \in E : s \in S, t \in T\}$ and $e_G(S, T) = |E_G(S, T)|$. Let $D = V \setminus (S \cup T)$.

Let

$$
\begin{aligned}
E_1' &= E_1 \cap E(S), & E_1'' &= E_1 \cap E(S, D); \\
E_2' &= E_2 \cap E(T), & E_2'' &= E_2 \cap E(T, D); \\
H_1 &= G[E_1' \cup E_1''], & H_2 &= G[E_2' \cup E_2''].
\end{aligned}
\tag{1}
$$

And let

$$
\begin{aligned}
\alpha_G(S, T; E_1, E_2) &= 2|E_1'| + |E_1''| = \sum_{x \in S} d_{H_1}(x), \\
\beta_G(S, T; E_1, E_2) &= 2|E_2'| + |E_2''| = \sum_{x \in T} d_{H_2}(x), \\
\Delta_G(S, T; E_1, E_2) &= \alpha_G(S, T; E_1, E_2) + \beta_G(S, T; E_1, E_2).
\end{aligned}
\tag{2}
$$

For simplicity, we use α, β, Δ for $\alpha_G(S, T; E_1, E_2)$, $\beta_G(S, T; E_1, E_2)$ and $\Delta_G(S, T; E_1, E_2)$.

Moreover, for any function u and any set S, we denote $\sum_{x \in S} u(x)$ by $u(S)$.

We define auxiliary functions u, w and p, q on V. For each $x \in V$, let:

$$
\begin{aligned}
u(x) &= \max\{g(x), d_G(x) - (m - 1)f(x) + m - 2\}, \\
w(x) &= \min\{f(x), d_G(x) - (m - 1)g(x) - m + 2\}. \\
p(x) &= \max\{0, d_G(x) - (m - 1)f(x) + m - 2\}, \\
q(x) &= \min\{f(x), d_G(x)\}.
\end{aligned}
\tag{3}
$$

From the definitions of u, w and p, q it is easy to see:

Lemma 1. *For any integer $m \geq 2$, and for each $x \in V$:*

$$
g(x) \leq u(x) < w(x) \leq f(x).
\tag{4}
$$

Lemma 2. *For any integer $m \geq 2$, and for each $x \in V$:*

$$
0 \leq p(x) < q(x) \leq f(x).
\tag{5}
$$

For all $x \in V$, define:

$$
\Omega_1(x) = d_G(x)/m - u(x), \quad \Omega_2(x) = w(x) - d_G(x)/m;
\tag{6}
$$

$$
\Delta_1(x) = d_G(x)/m - p(x), \quad \Delta_2(x) = q(x) - d_G(x)/m.
\tag{7}
$$

Lemma 3. *For any disjoint subsets S, T of V, we have([6]):*

$$
\delta_G(S, T; u, w) = \Omega_1(T) + \Omega_2(S) + d_{G-S}(T)(m - 1)/m + d_{G-T}(S)/m.
\tag{8}
$$

$$
\delta_G(S, T; p, q) = \Delta_1(T) + \Delta_2(S) + d_{G-S}(T)(m - 1)/m + d_{G-T}(S)/m.
\tag{9}
$$

We should also make use of the following lemma which gives a sufficient and necessary condition for a graph to have a (g, f)−factor that includes some given edges and excludes some other given edges [6].

Lemma 4. *Let $G = (V, E)$ be a graph, g and f be integer-valued functions defined on V such that for each $x \in V$, $0 \leq g(x) < f(x) \leq d_G(x)$. Let E_1 and E_2 be two disjoint subsets of E. Then G has a (g, f)−factor F such that $E_1 \subseteq E(F)$ and $E_2 \cap E(F) = \emptyset$ if and only if $\delta_G(S, T; g, f) = d_G(T) - e_G(S, T) - g(T) + f(S) \geq \Delta$ for all disjoint subsets S and T of V.*

3   Factorization Orthogonal to k Edge Disjoint Sets

Theorem 1. *Let G be an $(mg + m - 1, mf - m + 1)$−graph, where g and f be integer-valued functions defined on V satisfying $9k/4 \leq g \leq f$. Let $H_1, H_2, ..., H_k$ be edge disjoint m−subgraphs of G. Then there exists a (g, f)−factorization of G orthogonal to each $H_i, 1 \leq i \leq k$.*

Proof (of Theorem 1). Choose $u_i v_i \in E(H_i), 1 \leq i \leq k$. Let $E_1 = \{u_i v_i : 1 \leq i \leq k\}$, $E_2 = \bigcup_{i=1}^{k} E(H_i) \setminus E_1$. Then $|E_1| = k$, $|E_2| = (m-1)k$. For any two disjoint subsets S and T of V, E_1', E_1'', E_2', E_2'', α, β and Δ are defined as before. According to the definitions of α and β, we have:

$$\alpha \leq \min\{2k, k|S|\}, \quad \beta \leq \min\{2(m-1)k, (m-1)k|T|\}. \tag{10}$$

Define functions u, w and Ω_1, Ω_2 on V as before. We have the following claim.

Claim 1. *For every $x \in V$ and $m \geq 2$, we have*

$$\Omega_1(x) \geq \begin{cases} 1/m & : \quad \text{if } u(x) > g(x) \text{ and } d_G(x) = mf(x) - m + 1, \\ 1 - 1/m, & : \quad \text{else;} \end{cases} \tag{11}$$

$$\Omega_2(x) \geq \begin{cases} 1/m, & : \quad \text{if } w(x) < f(x) \text{ and } d_G(x) = mg(x) + m - 1, \\ 1 - 1/m, & : \quad \text{else.} \end{cases} \tag{12}$$

Proof (of Claim 1). If $u(x) > g(x)$ and $d_G(x) = mf(x) - m + 1$, then

$$\begin{aligned} \Omega_1(x) &= d_G(x)/m - u(x) = d_G(x)/m - d_G(x) + (m-1)f(x) - (m-2) \\ &= (mf(x) - m + 1)(1-m)/m + (m-1)f(x) - m + 2 = 1/m. \end{aligned} \tag{13}$$

If $u(x) > g(x)$ and $d_G(x) \leq mf(x) - m$, then

$$\begin{aligned} \Omega_1(x) &= d_G(x)/m - u(x) = d_G(x)/m - d_G(x) + (m-1)f(x) - (m-2) \\ &\geq (mf(x) - m)(1-m)/m + (m-1)f(x) - m + 2 = 1. \end{aligned} \tag{14}$$

If $u(x) = g(x)$, then

$$\Omega_1(x) = d_G(x)/m - g(x) \geq (mg(x) + m - 1)/m - g(x) = 1 - 1/m. \tag{15}$$

The second inequality can be proven similarly. □

Lemma 5. *Let G be an $(mg + m - 1, mf - m + 1)$-graph, where $m \geq 2$. Then G has a (u, w)-factor F such that $E_1 \subseteq E(F)$ and $E_2 \cap E(F) = \emptyset$.*

Proof (of Lemma 5). According to Lemma 4, it is enough to show for all disjoint subsets S, T of V, $\delta_G(S, T; u, w) \geq \alpha + \beta$.

If $S = \emptyset$, according to Lemma 3 and Claim 1, we have

$$\begin{aligned} \delta_G(S, T; u, w) &= d_G(T)(m-1)/m + \Omega_1(T) \\ &\geq (mg(T) + (m-1)|T|)(m-1)/m + |T|/m \\ &\geq (m-1)k|T| \geq \beta = \alpha + \beta. \end{aligned} \tag{16}$$

Similarly when $T = \emptyset$, we have

$$\begin{aligned} \delta_G(S, T; u, w) &= d_G(S)/m + \Omega_2(S) \geq (mg(S) + (m-1)|S|)/m + |S|/m \\ &\geq k|S| \geq \alpha = \alpha + \beta. \end{aligned} \tag{17}$$

Therefore, we should assume $S \neq \emptyset$ and $T \neq \emptyset$.
Let

$$
\begin{aligned}
S_0 &= \{x \in S : w(x) < f(x) \text{ and } d_G(x) = mg(x) + m - 1\}, \\
T_0 &= \{x \in T : u(x) > g(x) \text{ and } d_G(x) = mf(x) - m + 1\}.
\end{aligned}
\tag{18}
$$

Let $S_1 = S \setminus S_0$ and $T_1 = T \setminus T_0$.

If $|S| \geq 2$ and $|T| \geq 2$, then choose $x \in T$ such that $|E_2 \cap E(x, D)|$ is as small as possible and choose $y \in S$ such that $|E_1 \cap E(y, D)|$ is as small as possible. By Lemma 3 and Claim 1, we have

$$
\begin{aligned}
&\delta_G(S, T; u, w) \\
&= \Omega_1(T) + \Omega_2(S) + d_{G-S}(T)(m-1)/m + d_{G-T}(S)/m \\
&\geq (|T_0| + (m-1)|T_1|)/m + (|S_0| + (m-1)|S_1|)/m \\
&\quad + d_{G-S}(T)(m-1)/m + d_{G-T}(S)/m \\
&= ((m-2)(|S_1| + |T_1|))/m + (|S| + d_{G-S}(T))/m \\
&\quad + (|T| + d_{G-T}(S))/m + d_{G-S}(T)(m-2)/m \\
&\geq ((m-2)(|S_1| + |T_1|))/m + (mg(x) + m - 1 + \beta - (m-1)k/2)/m \\
&\quad + (mg(y) + m - 1 + \alpha - k/2)/m + \beta(m-2)/m \\
&\geq ((m-2)(|S_1| + |T_1|))/m + (4mk + 2m - 2)/m + (\alpha + (m-1)\beta)/m \\
&\geq ((m-2)(|S_1| + |T_1|))/m + (2(m-1)k + (m-1)2k + 4k + 2m - 2)/m \\
&\quad + (\alpha + (m-1)\beta)/m \\
&\geq (m\alpha + m\beta)/m = \alpha + \beta.
\end{aligned}
\tag{19}
$$

If $|S| = 1$ or $|T| = 1$, we choose $x \in T$, $y \in S$. According to Lemma 3 and Claim 1, we have

$$
\begin{aligned}
&\delta_G(S, T; u, w) \\
&= \Omega_1(T) + \Omega_2(S) + d_{G-S}(T)(m-1)/m + d_{G-T}(S)/m \\
&\geq ((m-2)(|S_1| + |T_1|))/m + (mg(x) + m - 1 + \beta - (m-1)k)/m \\
&\quad + (mg(y) + m - 1 + \alpha - k)/m + \beta(m-2)/m \\
&\geq (7mk/2 + 2m - 2)/m + (\alpha + (m-1)\beta)/m \\
&= (2(m-1)k + 3mk/2 + 2m - 2)/m + (\alpha + (m-1)\beta)/m.
\end{aligned}
\tag{20}
$$

If $|S| = 1$, then $\alpha \leq k|S| = k$, and subsequently we have

$$
\begin{aligned}
\delta_G(S, T; u, w) &\geq (m\beta + m\alpha)/m + (mk/2 + k + 2m - 2)/m \\
&> \alpha + \beta.
\end{aligned}
\tag{21}
$$

If $|T| = 1$, then $\beta \leq (m-1)k|T| = (m-1)k$, and subsequently we have

$$
\begin{aligned}
\delta_G(S, T; u, w) &\geq (m\alpha + m\beta)/m + (mk/m + k + 2m - 2)/m \\
&> \alpha + \beta.
\end{aligned}
\tag{22}
$$

Therefore, Lemma 5 is proven.   □

Now we prove Theorem 1.

We apply induction on m. The claim is trivial when $m = 1$. Suppose the theorem is true for $m - 1$, where $m \geq 2$. We will prove the theorem is also true for m. According to Lemma 5, G has a (u, w)−factor F such that $E_1 \subseteq E(F)$ and $E_2 \cap E(F) = \emptyset$. From Lemma 1, F is also a (g, f)−factor. In addition, for all $x \in V$, we have:

$$d_{G-F}(x)$$
$$= d_G(x) - d_F(x) \geq d_G(x) - w(x)$$
$$\geq d_G(x) - d_G(x) + (m-1)g(x) + (m-2) = (m-1)g(x) + (m-1) - 1,$$
$$d_{G-F}(x)$$
$$= d_G(x) - d_F(x) \leq d_G(x) - u(x)$$
$$\leq d_G(x) - d_G(x) + (m-1)f(x) - (m-2) = (m-1)f(x) - (m-1) + 1.$$
$$(23)$$

Hence, $G - F$ is an $((m-1)g + (m-1) - 1, (m-1)f - (m-1) + 1)$−graph. Let $H_i' = H_i \setminus \{u_i v_i\}$, then $H_1', H_2', ..., H_k'$ are $(m-1)$−subgraphs of $G-F$. According to induction hypothesis, $G-F$ has a (g, f)−factorization $\mathcal{F}' = \{F_1, F_2, ..., F_{m-1}\}$ orthogonal to $H_1', H_2', ..., H_k'$. Then G has a (g, f)−factorization $\mathcal{F} = \mathcal{F}' \cup \{F\}$ orthogonal to $H_1, H_2, ..., H_k$. Therefore, the theorem is proven.   □

Substitute $g - 1$ and $f + 1$ for g and f in Theorem 1, we derive the following corollary.

Corollary 1. *Let G be an $(mg - 1, mf + 1)$−graph, where $9k/4 + 1 \leq g \leq f$. Let $H_1, H_2, ..., H_k$ be edge disjoint m−subgraphs of G. Then G has a $(g - 1, f + 1)$−factorization orthogonal to each $H_i, 1 \leq i \leq k$.*

4   Some Extension to Factorization Orthogonal to k Vertex Disjoint Sets

Theorem 2. *Let k be a positive integer, G be a $(0, mf - m + 1)$−graph, where f is an integer-valued function defined on V satisfying for each $x \in V$, $f(x) \geq k + 2$. Let $H_1, H_2, ..., H_k$ be vertex disjoint m−subgraphs of G. Then there exists a $(0, f)$−factorization of G orthogonal to each $H_i, 1 \leq i \leq k$.*

Proof (of Theorem 2). We apply induction on m. When $m = 1$, the theorem is trivial. Assume that the theorem is true for $m - 1$, where $m \geq 2$. We will prove it is also true for m.

Choose $u_i v_i \in H_i, 1 \leq i \leq k$. Let

$$E_1 = \{u_i v_i, 1 \leq i \leq k\}, E_2 = (\bigcup_{i=1}^{k} E(H_i)) \setminus E_1. \tag{24}$$

We have $|E_1| = k$ and $|E_2| = (m-1)k$. For two disjoint subsets S, T of V, E_1', E_1'', E_2', E_2'', α, β and Δ are defined as before. It is easy to see from the definitions of α and β that:

$$\alpha \leq \min\{2k, |s|\}, \beta \leq \min\{2(m-1)k, (m-1)|T|\}. \tag{25}$$

Next, we choose S, T satisfying the following conditions 1, 2 and 3:

1. $\delta_G(S, T; p, q) - \Delta_G(S, T; E_1, E_2)$ is as small as possible, where p, q are defined as before;
2. $|T|$ is as small as possible subject to 1;
3. $|S|$ is as small as possible subject to 1 and 2.

Claim 2. *If $T \neq \emptyset$, then for every $x \in T$, $p(x) \geq 1$ (and $p(x) = d_G(x) - ((m - 1)f(x) - (m - 1) + 1)$ and $d_G(x) \geq (m - 1)f(x) - m + 3$).*

Proof (of Claim 2). Otherwise, $T^* = \{x \in T; p(x) = 0\} \neq \emptyset$. Let $T_0 = T \setminus T^*$. Then

$$
\begin{aligned}
\delta_G(S, T; p, q) &= d_G(T) - e_G(S, T) - p(T) + q(S) \\
&= \delta_G(S, T_0; p, q) + d_{G-S}(T^*) - p(T^*) \\
&= \delta_G(S, T_0; p, q) + d_{G-S}(T^*).
\end{aligned} \tag{26}
$$

On the other hand,

$$
\Delta_G(S, T; E_1, E_2) \leq \Delta_G(S, T_0; E_1, E_2) + \beta_G(S, T^*; E_1, E_2). \tag{27}
$$

Therefore,

$$
\begin{aligned}
&\delta_G(S, T; p, q) - \Delta_G(S, T; E_1, E_2) \\
&\geq \delta_G(S, T_0; p, q) - \Delta_G(S, T_0; E_1, E_2) + d_{G-S}(T^*) - \beta_G(S, T^*; E_1, E_2).
\end{aligned} \tag{28}
$$

Notice that $d_{G-S}(T^*) \geq \beta_G(S, T^*; E_1, E_2)$, so we have

$$
\delta_G(S, T; p, q) - \Delta_G(S, T; E_1, E_2) \geq \delta_G(S, T_0; p, q) - \Delta_G(S, T_0; E_1, E_2), \tag{29}
$$

a contradiction to the choice of T. Therefore, $T^* = \emptyset$ □

Claim 3. *If $S \neq \emptyset$, then for every $x \in S$, $q(x) \leq d_G(x) - 1$ (and $q(x) = f(x)$ and $d_G(x) \geq f(x) + 1$).*

The proof of Claim 3 is similar to that of Claim 2 and hence omitted.
    For all $x \in V$, Let $\Delta_1(x)$ and $\Delta_2(x)$ be defined as in the section 2.

Claim 4. *If $S \neq \emptyset$ and $T \neq \emptyset$, then for all $x \in T$,*

$$
\Delta_1(x) \geq \begin{cases} 1/m, & : \quad \text{if } d_G(x) = mf(x) - m + 1, \\ 1, & : \quad else; \end{cases} \tag{30}
$$

for all $x \in S$,

$$
\Delta_2(x) \geq \begin{cases} 1 - 1/m, & : \quad \text{if } d_G(x) = mf(x) - m + 1, \\ 1, & : \quad else. \end{cases} \tag{31}
$$

Proof (of Claim 4). According to Claim 2, for every $x \in T$, $p(x) = d_G(x) - ((m - 1)f(x) - (m - 1) + 1)$. Therefore,

$$
\begin{aligned}
\Delta_1(x) &= d_G(x)/m - p(x) \\
&= d_G(x)/m - d_G(x) + (m - 1)f(x) - (m - 2) \\
&= d_G(x)(1 - m)/m + (m - 1)f(x) - (m - 2).
\end{aligned} \tag{32}
$$

If $d_G(x) = mf(x) - m + 1$, then

$$\Delta_1(x) = (mf(x) - m + 1)(1 - m)/m + (m - 1)f(x) - (m - 2) = 1/m. \quad (33)$$

If $d_G(x) \leq mf(x) - m$, then

$$\Delta_1(x) \geq (mf(x) - m)(1 - m)/m + (m - 1)f(x) - (m - 2) = 1. \quad (34)$$

According to Claim 3, for all $x \in S$, $q(x) = f(x)$. Therefore,

$$\Delta_2(x) = q(x) - d_G(x)/m = f(x) - d_G(x)/m. \quad (35)$$

If $d_G(x) = mf(x) - m + 1$, then

$$\Delta_2(x) = f(x) - (mf(x) - m + 1)/m = 1 - 1/m, \quad (36)$$

else,

$$\Delta_2(x) \geq f(x) - (mf(x) - m)/m = 1. \quad (37)$$

$\square$

In order to prove Theorem 2, we will first prove that G has a (p, q)–factor F such that $E_1 \subseteq E(F)$ and $E_2 \cap E(F) = \emptyset$. According to Lemma 4 and the choice of S, T, it suffices to prove that $\delta_G(S, T; p, q) \geq \alpha + \beta$.

If $S = \emptyset$ and $T \neq \emptyset$, then $\alpha = 0$. According to Claim 2,

$$
\begin{aligned}
\delta_G(S, T; p, q) &= d_G(T) - p(T) \\
&= d_G(T) - (d_G(T) - ((m - 1)f(T) - (m - 2)|T|)) \\
&= (m - 1)f(T) - (m - 2)|T| \\
&\geq (m - 1)(k + 2)|T| - (m - 2)|T| \\
&> (m - 1)|T| \geq \beta = \beta + \alpha.
\end{aligned}
\quad (38)
$$

If $T = \emptyset$ and $S \neq \emptyset$, then $\beta = 0$. According to Claim 3,

$$\delta_G(S, T; p, q) = q(S) = f(S) \geq (k + 2)|S| > |S| \geq \alpha = \alpha + \beta. \quad (39)$$

If $S = \emptyset$ and $T = \emptyset$, then $\alpha = 0$, $\beta = 0$. And subsequently, $\delta_G(S, T; p, q) = 0 \geq \alpha + \beta$. We should therefore assume that $S \neq \emptyset$ and $T \neq \emptyset$ in the rest of proof of Theorem 2.

According to Lemma 3,

$$\delta_G(S, T; p, q) = \Delta_1(T) + \Delta_2(S) + d_{G-S}(T)(m - 1)/m + d_{G-T}(S)/m. \quad (40)$$

Let

$$
\begin{aligned}
T_0 &= \{x \in T | d_G(x) = mf(x) - m + 1\}, \ T_1 = T \setminus T_0; \\
S_0 &= \{x \in S | d_G(x) = mf(x) - m + 1\}, \ S_1 = S \setminus S_0.
\end{aligned}
\quad (41)
$$

We have several cases to discuss:

Case 1. $T_0 = \emptyset$. Now $T_1 = T$. Notice that $d_{G-S}(T) \geq \beta$ and that $d_{G-T}(S) \geq \alpha$. According to Claim 4,

$$
\begin{aligned}
\delta_G(S,T;p,q) &= \Delta_1(T) + \Delta_2(S) + d_{G-S}(T)(m-1)/m + d_{G-T}(S)/m \\
&\geq |T|m/m + |S|(m-1)/m + \beta(m-1)/m + \alpha/m \\
&\geq (\beta + |T|)/m + \alpha(m-1)/m + \beta(m-1)/m + \alpha/m \\
&> \alpha + \beta.
\end{aligned}
\tag{42}
$$

Case 2. $T_0 \neq \emptyset$ and $|T_1| \geq k$. According to Claim 2, Claim 3 and Claim 4,

$$
\begin{aligned}
&\delta_G(S,T;p,q) \\
&= \Delta_1(T) + \Delta_2(S) + d_{G-S}(T)(m-1)/m + d_{G-T}(S)/m \\
&\geq (|T_0| + m|T_1|)/m + |S|(m-1)/m + d_{G-S}(T)(m-1)/m + d_{G-T}(S)/m \\
&= |T_1|(m-1)/m + (|T| + d_{G-T}(S))/m + (|S| + d_{G-S}(T))/m \\
&\quad + |S|(m-2)/m + d_{G-S}(T)(m-2)/m \\
&\geq |T_1|(m-1)/m + (d_G(x) + \alpha - 1)/m + (d_G(y) + \beta - (m-1))/m \\
&\quad + \alpha(m-2)/m + \beta(m-2)/m \\
&\geq k(m-1)/m + (f(x) + 1 + \alpha - 1)/m + (mf(y) - m + 1 + \beta - (m-1))/m \\
&\quad + \alpha(m-2)/m + \beta(m-2)/m \\
&\geq k(m-1)/m + (k+2)/m + (m(k+2) - 2m + 2)/m \\
&\quad + \alpha(m-1)/m + \beta(m-1)/m \\
&= 2mk/m + 4/m + \alpha(m-1)/m + \beta(m-1)/m \\
&> (m\alpha + m\beta)/m = \alpha + \beta,
\end{aligned}
\tag{43}
$$

where $x \in S$, $y \in T_0$.

Case 3. $T_0 \neq \emptyset$, $|T_1| \leq k-1$ and $|T| \leq k$. In this case,

$$
\begin{aligned}
&\delta_G(S,T;p,q) \\
&= d_G(T) - e_G(S,T) - p(T) + q(S) \\
&= d_G(T) - (d_G(T) - ((m-1)f(T) - (m-2)|T|)) - e_G(S,T) + f(S) \\
&\geq (m-1)(k+2)|T| - (m-2)|T| - |S||T| + (k+2)|S| \\
&\geq (m-1)|T| + |S| + (k - |T|)|S| \geq \alpha + \beta,
\end{aligned}
\tag{44}
$$

where the second equality comes from Claim 2.

Case 4. $T_0 \neq \emptyset$, $|T_1| \leq k-1$ and $|T| \geq k+1$. We have $|T_0| \geq 2$ in this case.

If $S_0 = \emptyset$, then $S_1 = S$. Choose $x, y \in T_0$, $z \in S$. According to Claim 2, Claim 3 and Claim 4,

$$
\begin{aligned}
&\delta_G(S, T; p, q) \\
&= \Delta_1(T) + \Delta_2(S) + d_{G-S}(T)(m-1)/m + d_{G-T}(S)/m \\
&\geq (|T| + (m-1)|T_1|)/m + |S|m/m + d_{G-S}(T)(m-1)/m + d_{G-T}(S)/m \\
&= |T_1|(m-1)/m + (|T| + d_{G-T}(S))/m + (2|S| + d_{G-S}(T))/m \\
&\quad + |S|(m-2)/m + d_{G-S}(T)(m-2)/m \\
&\geq |T_1|(m-1)/m + (d_G(z) + \alpha - 1)/m + (d_G(x) + d_G(y) + \beta - 2(m-1))/m \\
&\quad + (\alpha + \beta)(m-2)/m \\
&\geq |T_1|(m-1)/m + (f(z) + \alpha)/m \\
&\quad + (m(f(x) + f(y)) - 2(m-1) + \beta - 2m + 2)/m + (\alpha + \beta)(m-2)/m \\
&\geq (k+2)/m + (2m(k+2) - 2m + 2 - 2m + 2)/m + (\alpha + \beta)(m-1)/m \\
&\geq (2(m-1)k + 2k + k + 6)/m + (\alpha + \beta)(m-1)/m \\
&\geq (m\alpha + m\beta)/m = \alpha + \beta.
\end{aligned}
\tag{45}
$$

If $S_0 \neq \emptyset$, then there exists a $z \in S_0$ such that $d_G(z) = mf(z) - m + 1$. Choose an $x \in T_0$. According to Claim 2, Claim 3 and Claim 4, and similar as the discussion above, we have

$$
\begin{aligned}
\delta_G(S, T; p, q) &= \Delta_1(T) + \Delta_2(S) + d_{G-S}(T)(m-1)/m + d_{G-T}(S)/m \\
&\geq (mf(z) - m + 1 + \alpha - 1)/m + (mf(x) - m + 1 + \beta - m + 1)/m \\
&\quad + \alpha(m-2)/m + \beta(m-2)/m \\
&\geq (m(k+2) - m)/m + (m(k+2) - 2m + 2)/m + (\alpha + \beta)(m-1)/m \\
&> \alpha + \beta.
\end{aligned}
\tag{46}
$$

From the above discussion, for the chosen S, T, we always have $\delta_G(S, T; p, q) \geq \alpha + \beta$. According to Lemma 4, G has a (p, q)-factor F such that $E_1 \subseteq E(F)$ and $E_2 \cap E(F) = \emptyset$. From Lemma 2, F is also a $(0, f)$-factor. According to the definitions of p, q, we have

$$
\begin{aligned}
d_{G-F}(x) &= d_G(x) - d_F(x) \geq d_G(x) - q(x) \geq d_G(x) - d_G(x) = 0, \\
d_{G-F}(x) &= d_G(x) - d_F(x) \leq d_G(x) - p(x) \\
&\leq d_G(x) - (d_G(x) - ((m-1)f(x) - (m-1) + 1)) \\
&= (m-1)f(x) - (m-1) + 1.
\end{aligned}
\tag{47}
$$

Therefore, $G - F$ is a $(0, (m-1)f(x) - (m-1) + 1)$-graph. Let $H_i' = H_i \setminus \{u_i v_i\}$. Then $H_i', 1 \leq i \leq k$ are vertex disjoint $(m-1)$-subgraphs of $G - F$. According to induction hypothesis, $G - F$ has a $(0, f)$-factorization $\mathcal{F}' = \{F_1, F_2, ..., F_{m-1}\}$ orthogonal to $H_1', H_2', ..., H_k'$. It is obvious that $\mathcal{F} = \mathcal{F}' \cup \{F\}$ is a $(0, f)$-factorization of G that is orthogonal to $H_1, H_2, ..., H_k$.

Thus completes the proof of Theorem 2.   □

Corollary 2. *Let G be a $(0, mf - m + 1)$-graph, where $f \geq k+2$ is an integer-valued function defined on V. Then for any km-matching M of G, G has a $(0, f)$-factorization $\mathcal{F} = \{F_1, F_2, ..., F_m\}$ satisfying that each $F_i, 1 \leq i \leq m$ shares exactly k edges with M.*

Remark 1. Let m, k and l be integers such that $k \leq m - 1, l \leq m$. Let G be a $(0, mf - k)$−graph, where f is an integer-valued function defined on V satisfying $f \geq l$. Let $H_1, H_2, ..., H_l$ be m−subgraphs of G. Then G has a subgraph R such that R has a $(0, f)$−factorization orthogonal to $H_1, H_2, ..., H_l$.

Proof. Without lose of generality, we suppose $E(H_i) = \{e_{i_1}, e_{i_2}, ..., e_{i_m}\}, 1 \leq i \leq l$. Let $F_j = \{e_{i_j}, 1 \leq i \leq l\}, 1 \leq j \leq m$. Let $R = \bigcup_{j=1}^{m} F_j$. Then $\mathcal{F} = \{F_1, F_2, ..., F_m\}$ is a $(0, f)$−factorization of R that is orthogonal to $H_1, H_2, ..., H_l$.
□

Remarks and Discussion Anstee [2] gave a polynomial-time algorithm which either finds a (g, f)−factor or shows that one does not exist in $O(|V|^3)$ operations. Hell and Kirkpatrick [5] presented an $O(\sqrt{(g(V))}|E|)$ algorithm for the general (g, f)−factor problem. If $g(x) \neq f(x)$ for every $x \in V$, it was shown that there is a (g, f)−factor algorithm of time complexity $O(g(V)|E|)$ by finding alternating paths in [4]. The value $p(x), q(x)$ and the set E_1 can easily be found by $O(|V|)$ operations. Let $G_1 = G - E_2$. Set $p'(x) = p(x) - 1$ and $q'(x) = q(x) - 1$ when $x = u_i, v_i, 1 \leq i \leq k$, otherwise, set $p'(x) = p(x), q'(x) = q(x)$. Then we can find a (p', q')−factor F in G_1 by the algorithm in [5] or [2]. It is easy to see that $F \cup E_1$ is a $(0, f)$−factor of G containing E_1 and excluding E_2. It follows from the proof of Theorem 2 that $G - F$ is a $(0, (m - 1)f - (m - 1) + 1)$−graph. Repeating the above procedure, we can find after at most $m - 1$ operations a $(0, f)$−factorization $\mathcal{F} = \{F_1, F_2, ..., F_m\}$ orthogonal to mutually vertex-disjoint subgraphs $H_1, H_2, ..., H_k$ in G.

Our work on the factorization orthogonal to edge disjoint subgraphs is the first result in this direction. It would be interesting to know the tight conditions for this problem. For vertex disjoint subgraphs, we completely relax the restriction of the previous work on the lower bound g (to the all zero function) and conclude the study in this direction.

Acknowledgments. The author is grateful to Dr. Xiaotie Deng and Professor Guizhen Liu for their helpful comments and advice on this paper.

References

1. B.Alspach, K.Heinrich, G.Z.Liu: Contemporary Design Theory-A collection of surveys. John Wiley and Sons, New York (1992) 13–37
2. R.P.Anstee: An algorithm proof of Tutte's f−factor theorem. J ALg **6**(1985) 112–131
3. J.A.Bondy, U.S.R.Murty: Graph Theory with Applications. Elsevier-North Holland, New York (1976)
4. K.Heinrich, P.Hell, D.G.Kirkpatrick, G.Z.Liu: A simple existence criterion for $(g < f)$-factors. Discrete Math. **85** (3)(1990) 313–317
5. P.Hell, D.G.Kirkpatrick: Algorithms for degree constrained graph factors of minimum deficiency. J. Algorithms **14**(1)(1993)115–138
6. P.C.B.Lam, G.Z.Liu, G.J.Li, W.C.Shiu: Orthogonal (g, f)−Factorizations in NetWorks. NetWorks **35** (4) (2000) 274–278

On Star Coloring of Graphs

Guillaume Fertin[1], André Raspaud[2], and Bruce Reed[3]

[1] IRIN UPRES-EA 2157, Université de Nantes
2 rue de la Houssinière - BP 92208 - F44322 Nantes Cedex 3, France
fertin@irin.univ-nantes.fr
[2] LaBRI U.M.R. 5800, Université Bordeaux 1
351 Cours de la Libération - F33405 Talence Cedex, France
raspaud@labri.u-bordeaux.fr
[3] Univ. Paris 6 - Equipe Combinatoire - Case 189 -
4 place Jussieu - F75005 Paris, France
reed@ecp6.jussieu.fr

Abstract. In this paper, we deal with the notion of star coloring of graphs. A *star coloring* of an undirected graph G is a proper vertex coloring of G (i.e., no two neighbors are assigned the same color) such that any path of length 3 in G is not bicolored.

We give the exact value of the star chromatic number of different families of graphs such as trees, cycles, complete bipartite graphs, outerplanar graphs and 2-dimensional grids. We also study and give bounds for the star chromatic number of other families of graphs, such as hypercubes, tori, d-dimensional grids, graphs with bounded treewidth and planar graphs.

Keywords : graphs, vertex coloring, proper coloring, star coloring, acyclic coloring, treewidth.

1 Introduction

All graphs considered are undirected. In the following definitions (and in the whole paper), the term *coloring* will be used to define *vertex coloring* of graphs. A **proper** coloring of a graph G is a labeling of the vertices of G such that no two neighbors in G are assigned the same label. Usually, the labeling (or coloring) of vertex x is denoted by $c(x)$. In the following, all the colorings that we will define and use are proper colorings.

Definition 1 (Star coloring). *A* **star coloring** *of a graph G is a proper coloring of G such that no path of length 3 in G is bicolored.*

We also introduce here the notion of *acyclic coloring*, that will be useful for our purpose.

Definition 2 (Acyclic coloring). *An* **acyclic coloring** *of a graph G is a proper coloring of G such that no cycle in G is bicolored.*

A. Brandstädt and V.B. Le (Eds.): WG 2001, LNCS 2204, pp. 140–153, 2001.
© Springer-Verlag Berlin Heidelberg 2001

For any of the above colorings, we define by **chromatic number** of a graph G the minimum number of colors which are necessary to color G according to the definition of the given coloring. Depending on the coloring, the notation for the chromatic number of G differs ; it is denoted $\chi(G)$ for the *proper* coloring, $\chi_s(G)$ for the *star* coloring and $a(G)$ for the *acyclic* coloring.

By extension, the chromatic number of a family $\mathcal{F}$ of graphs is the minimum number of colors that are necessary to color any member of $\mathcal{F}$ according to the definition of the given coloring. Depending on the coloring, this will be denoted either by $\chi(\mathcal{F})$, $\chi_s(\mathcal{F})$ or $a(\mathcal{F})$.

The purpose of this paper is to determine and give properties on $\chi_s(\mathcal{F})$ for a large number of families of graphs. In Section 2, we motivate the problem and present general properties for the star chromatic number of graphs. In the following sections (Section 3 to 6), we determine precisely $\chi_s(\mathcal{F})$ for trees, cycles, complete bipartite graphs and 2-dimensional grids and we give bounds on $\chi_s(\mathcal{F})$ for other families of graphs, such as hypercubes, graphs with bounded treewidth, outerplanar graphs, d-dimensional grids, 2-dimensional tori.

2   Generalities

We note that for any graph G, a star coloring of G is also an acyclic coloring of G : indeed, a cycle in G can be bicolored if and only if it is of even length, that is of length greater than or equal to 4. However, by definition of a star coloring, no path of length 3 in G can be bicolored. Thus the following observation.

Observation 1 *For any graph G, $a(G) \le \chi_s(G)$.*

Actually we can remark that the star coloring is an acyclic coloring such that if we take two classes of colors then the induced subgraph is a bipartite graph composed only of stars. Star coloring was introduced in 1973 by Grünbaum [8]. He linked star coloring to *acyclic coloring* by showing that any planar graph has an acyclic chromatic number less than or equal to 9, and by suggesting that this implies that any planar graph has a star chromatic number less than or equal to $9 \cdot 2^8 = 2304$.

However, this property can be generalized for any given graph G, as mentioned in [3,2]. In [3,2], the result was just stated, but no proof was given. We detail the proof here for completeness.

Theorem 1 (Relation acyclic/star coloring). *For any graph G, if the acyclic chromatic number of G satisfies $a(G) \le k$, then the star chromatic number of G satisfies $\chi_s(G) \le k \cdot 2^{k-1}$.*

In order to prove the theorem, we shall use the following proposition.

Proposition 1. *Let T be a tree and V_1 and V_2 be the bipartition of its set of vertices, then there exists a star coloring of T $c : V(T) \to \{0, 1, 2, 3\}$ such that if $v \in V_1$ then $c(v) \in \{0, 2\}$ and if $v \in V_2$ then $c(v) \in \{1, 3\}$.*

Proof. Let v be any vertex belonging to V_1, we give the following coloring c of the vertices of T : $c(v) = 0$ and for $x \in V(T) \setminus \{v\}$, $c(x) = d(v,x) \bmod 4$. It is easy to see that this is a star coloring and that it has the required properties of the proposition.

Now we prove the theorem.

Proof. Let $G = (V, E)$ be a graph with $a(G) \leq k$, and let $V_1, V_2, \dots, V_k$ be the color classes of an acyclic coloring of the vertices of G with k colors. The color classes form a partition of the set V and for any $i \neq j$ belonging to $\{1, \dots, k\}$, the subgraph induced by $V_i \cup V_j$, denoted by $G[V_i \cup V_j]$, is a forest. For any $G[V_i \cup V_j]$ $(i < j)$, we denote by $c_{i,j}$ a star coloring of the vertices with 4 colors such that if $v \in V_i$ then $c_{i,j}(v) \in \{0,2\}$, and if $v \in V_j$ then $c(v) \in \{1,3\}$. We define the following coloring c of the vertices of G : let $u \in V$, then there is a unique $i \in \{1, \dots, k\}$ such that $v \in V_i$, let $c(u) = (c_{1,i}(u), \dots, c_{i-1,i}(u), *, c_{i,i+1}(u), \dots, c_{i,k}(u))$. We have to notice that the terms before the star belong to $\{0,2\}$ and that the terms after the star belong to $\{1,3\}$. By this definition we use at most $k2^{k-1}$ colors. It is clear that this is a proper coloring, indeed if $xy \in E$ then x and y belong to different sets V_i, and by definition $c(x) \neq c(y)$. Now we have to prove that there is no bicolored path of length 3. By contradiction let us assume that there is such a path P with $V(P) = \{u, v, w, t\}$ and $E(P) = \{uv, vw, wt\}$. By definition of the coloring the bicolored path is between V_i and V_j for some $i < j$. W.l.o.g., assume that $u, w \in V_i$ and $v, t \in V_j$; if $c(u) = c(w)$ and $c(v) = c(t)$, then we have $c_{i,j}(u) = c_{i,j}(w)$ and $c_{i,j}(v) = c_{i,j}(t)$, which is impossible because there is no bicolored path in $G[V_i \cup V_j]$ with the coloring $c_{i,j}$. This completes the proof.

In the following, let $\mathcal{P}$ denote the family of planar graphs. We deduce the following easy result.

Theorem 2. $7 \leq \chi_s(\mathcal{P}) \leq 80$.

Proof. Grünbaum showed that any planar graph has an acyclic coloring using at most 9 colors, but conjectured that the exact answer was 5. Moreover, he gave an example of a planar graph G for which $a(G) = 5$. This conjecture was solved in 1979 by Borodin [6]. This result, combined with Theorem 1 gives the upper bound for the theorem.

Moreover, there exists a planar graph G_2 for which any star coloring needs 7 colors. This graph is shown in Figure 1 (right). Thanks to the computer, we know that $\chi_s(G_2) = 7$.

The girth g of a graph G is the length of its shortest cycle. In [3], it is proved that if G is planar with girth $g \geq 5$ (resp. $g \geq 7$), then $a(G) \leq 4$ (resp. $a(G) \leq 3$). Together with Theorem 1, we deduce :

Corollary 1. *If G is a planar graph with girth $g \geq 5$, then $\chi_s(G) \leq 32$. If G is a planar graph with girth $g \geq 7$, then $\chi_s(G) \leq 12$.*

Several graphs are cartesian product of graphs (Grid, Hypercube, Tori), so it is interesting to have an upper bound for the star chromatic number of cartesian

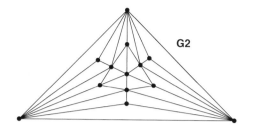

Fig. 1. Planar graph : $\chi_s(G_2) = 7$

product of graphs. We recall that the cartesian product of two graphs $G = (V, E)$ and $G' = (V', E')$, denoted by $G\Box G'$, is the graph such that the set of vertices is $V \times V'$ and two vertices (x, x') and (y, y') are linked if and only if $x = y$ and $x'y'$ is an edge of G' or $x' = y'$ and xy is an edge of G.

Theorem 3. *For any two graphs G and H, $\chi_s(G\Box H) \leq \chi_s(G) \cdot \chi_s(H)$.*

Proof. Suppose that $\chi_s(G) = g$ and $\chi_s(H) = h$, and let c_G (resp. c_H) be a star coloring of G (resp. H) using g (resp. h) colors. In that case, we assign to any vertex (u, v) of $G\Box H$ color $[c_g(u), c_h(v)]$. This coloring uses $g \cdot h$ colors, and this defines a star coloring. Indeed, suppose that there exists a path P of length 3 that is bicolored in $G\Box H$, with $V(P) = \{x, y, z, t\}$ and $E(P) = \{xy, yz, zt\}$. Depending on the composition of the ordered pairs corresponding to the vertices of the path, we have 8 possible paths. We will only consider 4 of them, because by permuting the first and second component of each ordered pairs, we obtain the others. The 4 possible paths are :

(1) $x = (u, v), y = (u, v_1), z = (u, v_2), t = (u, v_3)$
(2) $x = (u, v), y = (u, v_1), z = (u, v_2), t = (u_1, v_2)$
(3) $x = (u, v), y = (u, v_1), z = (u_1, v_1), t = (u_1, v_2)$
(4) $x = (u, v), y = (u, v_1), z = (u_2, v_1), t = (u_3, v_1)$

Clearly, in the first case P cannot be bicolored, since the path v, v_1, v_2, v_3 is not bicolored in H. For the second case : y and t have different colors ($v_1 v_2$ is an edge of H). For the third case : x and z have different colors (vv_1 is an edge of H). The same argument works for the last case.

Observation 2 *For any graph G and for any $1 \leq \alpha \leq |V(G)|$, let $G_1, \ldots, G_p$ be the p connected components obtained by removing α vertices from G. In that case, $\chi_s(G) \leq \max_i\{\chi_s(G_i)\} + \alpha$.*

Proof. Star color each G_i, and reconnect them by adding the α vertices previously deleted, using a new color for each of the α vertices. Any path of length 3 within a G_i will be star colored by construction, and if this path begins in G_i and ends in G_j with $i \neq j$, then it contains at least one of the α vertices, which has a unique color. Thus the path of length 3 cannot be bicolored, and we get a star coloring of G.

Remark 1. For any $\alpha \geq 1$, the above result is optimal for complete bipartite graphs $K_{n,m}$. W.l.o.g., suppose $n \leq m$ and let $\alpha = n$. Remove the $\alpha = n$ vertices of partition V_n. We then get m isolated vertices, which can be independently colored with a single color. Then, give a unique color to the $\alpha = n$ vertices. We then get a star coloring with $n + 1$ colors ; this coloring can be shown to be optimal by Theorem 6.

Observation 3 *For any graph G that can be partitioned into p stables $S_1, \ldots, S_p$, $\chi_s(G) \leq 1 + |V(G)| - \max_i\{|S_i|\}$.*

Proof. Let $1 \leq j \leq p$ such that S_j is of maximum cardinality. Color each vertex of S_j with a single color c, and give new pairwise distinct different colors to all the others vertices. This coloring holds the desired number of colors. It is clearly a proper coloring, and it is also a star coloring, because there is only one color which is used at least twice.

Remark 2. The above result is optimal for complete p-partite graphs $K_{s_1, s_2, \ldots, s_p}$.

3   Trees, Cycles, Complete Bipartite Graphs, Hypercubes

Theorem 4. *Let $\mathcal{F}_d$ be the family of forests such that d is the maximum depth over all the trees contained in $\mathcal{F}_d$. In that case, $\chi_s(\mathcal{F}_d) = \min\{3, d+1\}$.*

Proof. Let F be a forest contained in $\mathcal{F}_d$. When $d = 0$, the result is trivial (F holds no edge). When $d = 1$, we color each root of each tree contained in F with color 1, and all the remaining vertices by color 2. This is obviously a proper coloring, and since in that case there is no path of length 3, it is consequently a star coloring as well. Now we assume $d \geq 2$. We then color each vertex v, of depth d_v in F, as follows : $c(v) = d_v \bmod 3$. Clearly, this is a proper coloring of F and it is easy to see that it is a star coloring.

Theorem 5. *Let C_n be the cycle with $n \geq 3$ vertices.*

$$\chi_s(C_n) = \begin{cases} 4 \text{ when } n = 5 \\ 3 \text{ otherwise} \end{cases}$$

Proof. It can be easily checked that $\chi_s(C_5) = 4$. Now let us assume $n \neq 5$. Clearly, 3 colors at least are needed to star color C_n. We now distinguish 3 cases : first, if $n = 3k$, we color alternatively the vertices around the cycle by colors 1,2 and 3. Thus, for any vertex u, its two neighbors are assigned distinct colors, and consequently this is a valid star coloring. Hence $\chi_s(C_{3k}) \leq 3$. Suppose now $n = 3k+1$. In that case, let us color $3k$ vertices of C_n consecutively, by repeating the sequence of colors 1,2 and 3. There remains 1 uncolored vertex, to which we assign color 2. One can check easily that this is also a valid star coloring, and thus $\chi_s(C_{3k+1}) \leq 3$. Finally, let $n = 3k + 2$. Since the case $n = 5$ is excluded here, we can assume $k \geq 2$. Thus $n = 3(k-1) + 5$, with $k - 1 \geq 1$. In that case, let us color $3(k-1)$ consecutive vertices along the cycle, alternating colors 1,2 and 3. For the 5 remaining vertices, we give the following coloring : $2, 1, 2, 3, 2$. It can be checked that this is a valid star coloring, and thus $\chi_s(C_{3k+2}) \leq 3$ for any $k \geq 2$. Globally, we have $\chi_s(C_n) = 3$ for any $n \neq 5$, and the result is proved.

Theorem 6. *Let $K_{n,m}$ be the complete bipartite graph with $n+m$ vertices. Then $\chi_s(K_{n,m}) = \min\{m, n\} + 1$.*

Proof. W.l.o.g., let $n \leq m$. The upper bound of $n+1$ immediately derives from Observation 2 (cf. Remark 1 for a detailed proof).

Now let us prove that $\chi_s(K_{n,m}) \geq n+1$. For this, let us show that any coloring with n colors will give us at least one bicolored cycle of length 4. Let S_n (resp. S_m) be the set of colors used to color the vertices of V_n (resp. V_m). Clearly, since all possible edges exist between vertices of V_n and vertices of V_m, we must have $S_n \cap S_m = \emptyset$ in order to achieve a proper coloring. Suppose then that we use $n - x$ colors for the vertices of V_n, and x colors for the vertices of V_m, with $1 \leq x \leq n - 1$. In that case, there exists at least 2 vertices a_n and b_n in V_n (resp. a_m and b_m in V_m) that are given the same color c_1 (resp. c_2). By definition of $K_{n,m}$, there exists a cycle of length 4 going through the vertices a_n, b_m, b_n, a_m, and this cycle is bicolored with colors c_1 and c_2. Thus, no coloring that uses n colors can be a star coloring, and $\chi_s(K_{n,m}) \geq n+1$.

For our purpose we introduce a new definition here : the k *distance coloring* of graphs.

Definition 3 (k distance coloring). *A k distance coloring of a graph G is a coloring of G such that any two vertices at distance at most k are assigned different colors.*

Remark 3. For any $k \geq 1$, k distance coloring is a generalization of proper coloring, since the latter can be seen as a k distance coloring where $k = 1$.

We define, as in Section 1, the k distance chromatic number of a graph G as being the minimum number of colors necessary to get a k distance coloring of G. It is noted $\chi_{\bar{k}}(G)$.

In this section, we will only focus on the case $k = 2$. Indeed, it is easy to see that for any graph G, a 2 distance coloring of G is also a star coloring of G. Hence the following observation.

Observation 4 *For any graph G and for any k ($k \geq 2$), $\chi_s(G) \leq \chi_{\bar{k}}(G)$.*

2 distance coloring (and, more generally, k distance coloring) of hypercubes has already been studied [12,9]. The main result concerning 2 distance coloring is the following.

Theorem 7. *[12] Let H_n denote the hypercube of dimension n. For any $n \geq 1$, $n + 1 \leq \chi_{\bar{2}}(H_n) \leq 2^{\lceil \log_2(n+1) \rceil}$.*

Wan conjectured that the above upper bound is tight. Note that Theorem 7 above implies that $\chi_{\bar{2}}(H_n)$ lies between $n + 1$ and $2n$, depending on the value of n ; this also shows that there is an infinite number of cases for which the result is optimal : indeed, when $n = 2^k - 1$ the lower and upper bounds coincide, and in that case $\chi_{\bar{2}}(H_{2^k-1}) = 2^k = n + 1$.

Thanks to this theorem, and thanks to Observation 4, we get the following property.

Property 1. For any n, $\chi_s(H_n) \leq 2^{\lceil \log_2(n+1) \rceil}$.

However, as can be seen in Table 1, this upper bound is not tight. All results given in bold characters in Table 1 are optimal ; the ones concerning star coloring are discussed below.

Table 1. Star and 2 distance coloring of H_n

n	$\chi_s(H_n)$	$\chi_{\bar{2}}(H_n)$	n	$\chi_s(H_n)$	$\chi_{\bar{2}}(H_n)$
2	**3**	**4**	5	**6**	**8**
3	**4**	**4**	6	**6**	**8**
4	**5**	**8**	7	$6 \leq \ \leq 8$	**8**

Property 2. $\chi_s(H_2) = 3$, $\chi_s(H_3) = 4$ and $\chi_s(H_4) = 5$.

Proof. The first result is trivial. It can also be easily seen that $\chi_s(H_3) \geq 4$. Since H_3 is a subgraph of H_4, we deduce that $\chi_s(H_4) \geq 4$.

The upper bounds are given, in each case, by a star coloring with the appropriate number of colors. These are given in Figure 2.

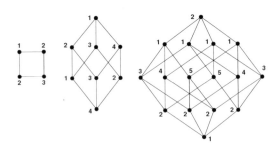

Fig. 2. Star coloring of H_2, H_3 and H_4

Property 3. $\chi_s(H_5) = 6$, $\chi_s(H_6) = 6$ and $\chi_s(H_7) \leq 8$.

Proof. The lower bound of 6 for $\chi_s(H_5)$ and $\chi_s(H_6)$ is given by computer : there is no possible star coloring of those two graphs with 5 colors.

The upper bounds are given in each of the 3 cases by an appropriate star coloring using the required number of colors. Those colorings are detailed in the appendix of [7], where the result is given as a list of correspondences between each vertex and its color.

4   *d*-Dimensional Grids

In this section, we study the star chromatic number of grids. More precisely, we give the star chromatic number of 2 dimensional grids, and we extend this result in order to get bounds on the star chromatic number of grids of dimension d.

We recall that the grid $G(n, m)$ is the cartesian product of two paths of length $n - 1$ and $m - 1$. Due to symmetries, we will always consider in the following that $m \geq n$. A summary of the results is given in Table 2 ; those results are detailed below.

Table 2. Star coloring of 2-dimensional grids $G(n, m)$ $(n \leq m)$

	$m = 2$	$m = 3$	$m \geq 4$
$n = 2$	3	4	4
$n = 3$	XXX	4	4
$n \geq 4$	XXX	XXX	5

Proposition 2. $\chi_s(G(2, 2)) = 3$, and for any $m \geq 4$, $\chi_s(G(2, m)) = \chi_s(G(3, m)) = 4$.

Proof. The first result is trivial.

It is easily checked that $\chi_s(G(2, m)) \geq 4$ for any $m \geq 3$, since star coloring of $G(2, 3)$ requires at least 4 colors, and since $G(2, 3)$ is subgraph of $G(2, m)$ for any $m \geq 3$. Moreover, as can be seen in Figure 3 (left), it is possible to find a 4 star coloring of $G(2, m)$. Thus $\chi_s(G(2, m)) \leq 4$, and altogether we have $\chi_s(G(2, m)) = 4$.

The fact that $\chi_s(G(3, m)) \geq 4$ for any $m \geq 3$ is trivial, since $G(2, 3)$ is a subgraph of $G(3, m)$. Moreover, Figure 3 (right) shows a 4 star coloring of $G(3, m))$ for any $m \geq 3$.

The main theorem of this section is the following.

Theorem 8. *For any n and m such that $\min\{n, m\} \geq 4$, $\chi_s(G(n, m)) = 5$.*

Proof. By a rather tedious case by case analysis (confirmed by the computer), it is possible to show that 5 colors at least are needed to color $G(4, 4)$. Hence, for any n and m such that $\min\{n, m\} \geq 4$, $\chi_s(G(n, m)) \geq 5$.

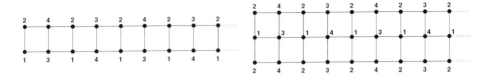

Fig. 3. 4 star colorings of $G(2, m)$ (left) and $G(3, m)$ (right)

Now let us show that 5 colors are sufficient to get a star coloring of $G(n, m)$. Our aim is to have colors such that no two vertices at distance 2 are assigned the same color (except for the color 1): for this, we will identify any vertex v_i of $G(n, m)$ by its coordinates (x_i, y_i), $0 \le x_i \le n - 1$ and $0 \le y_i \le m - 1$. We now describe the coloring scheme :

- Any vertex $v = (x, y)$ with $x + y \equiv 0 \bmod 2$ is assigned color 1 ;
- Any vertex $v = (4p, 4q + 1)$ and $v = (4p + 2, 4q + 3)$ is assigned color 2 ;
- Any vertex $v = (4p, 4q + 3)$ and $v = (4p + 2, 4q + 1)$ is assigned color 3 ;
- Any vertex $v = (4p + 1, 4q)$ and $v = (4p + 3, 4q + 2)$ is assigned color 4 ;
- Any vertex $v = (4p + 1, 4q + 2)$ and $v = (4p + 3, 4q)$ is assigned color 5.

An example of this coloring is shown in Figure 4, with $n = 5$ and $m = 9$.

It can be easily checked that the above coloring assigns a color to each vertex, and that this coloring is proper ; moreover, this is a valid 5 star coloring of $G(n, m)$, since two vertices u and v at distance 2 are assigned the same color iff $c(u) = c(v) = 1$. In that case, there cannot exist a path of length 3 that is bicolored. Thus $\chi_s(G(n, m)) \le 5$, and the result is proved.

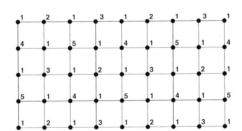

Fig. 4. A 5 star coloring of $G(5, 9)$

We know that d-dimensional grids G_d are isomorphic to the cartesian product of d paths. Hence, by Theorems 3 and 4, we get an upper bound for $\chi_s(G_d)$: $\chi_s(G_d) \le 3^d$ for any $d \ge 1$. Due to the result of Theorem 8 above, we can slightly improve this bound to $5 \cdot 3^{d-2}$, for any $d \ge 2$. However, it is even possible to do better : for this, we generalize the star coloring of paths and 2-dimensional grids. This is the purpose of the following theorem.

Theorem 9. Let G_d be any d-dimensional grid, $d \ge 1$. Then $\chi_s(G_d) \le 2^d + 1$.

Proof. The idea here is to design a star coloring for G_d that generalizes the one given for $d = 2$ in Proof of Theorem 8. More precisely, we want to define a coloring of G_d such that, for any fixed $d-2$ dimensions, the induced 2-dimensional grid has a coloring similar to the one of Figure 4. Let us now detail the coloring in itself : first, G_d being bipartite, we color every vertex of one of the two partitions of G_d by color 1. More formally, if every vertex $v \in G_d$ is defined by its d coordinates, that is $v = (x_1, x_2 \ldots x_d)$, then $c(v) = 1 \ \forall \ x_1, x_2 \ldots x_d$ s.t. $x_1 + x_2 + \ldots + x_d \equiv 0 \bmod 2$. Now, we need to assign the 2^d remaining colors to the remaining vertices. Again, we will do it depending on the coordinates $x_1, x_2 \ldots x_d$ of the considered vertex v. For this, we will use a set S_d of 2^d vertices, chosen thanks to their coordinates, and for which every vertex of S_d will be assigned a unique color $2 \leq c \leq 2^d + 1$. Starting from the vertices of S_d we will then apply a rule to color the still uncolored vertices. First, let us detail the construction of S_d. Let V_0 be the set of vertices of G_d holding an even number (resp. odd number) of coordinates equal to 0 if d is odd (resp. if d is even). Let now $V_{0,1}$ be the subset of V_0 such that all the non-zero coordinates are all equal to 1. It is not difficult to see that $|V_{0,1}| = 2^{d-1}$.

Now let $V_{0,1,3}$ be the subset of V_0 such that all but one of the non-zero coordinates are equal to 1, the last one being equal to 3. Finally, let $V'_{0,1,3}$ be the maximum subset of $V_{0,1,3}$ such that any two vertices u and v in $V'_{0,1,3}$ have at least one zero-coordinate location which differs. In that case, it is clear that $|V'_{0,1,3}| = 2^{d-1}$. We now define S_d as follows : $S_d = V_{0,1} \cup V'_{0,1,3}$.

Example : in the case $d = 4$, $V_{0,1} = \{(1,0,0,0), (0,1,0,0), (0,0,1,0), (0,0,0,1), (1,1,1,0), (1,1,0,1), (1,0,1,1), (0,1,1,1)\}$, while $V'_{0,1,3} = \{(3,0,0,0), (0,3,0,0), (0,0,3,0), (0,0,0,3), (3,1,1,0), (3,1,0,1), (3,0,1,1), (0,3,1,1)\}$.

We assign to each vertex of S_d a unique color $2 \leq c \leq 2^d + 1$. Now, let us take a vertex $u \in S_d$ with color $c(u)$; we then assign $c(u)$ to any vertex u' obtained from u by any combination of the two following rules :

- **(Rule 1)** : add ± 4 to exactly one of the coordinates ;
- **(Rule 2)** : add ± 2 to exactly 2 of the coordinates.

In that case, it can be seen that all the vertices of the d-dimensional grid have been assigned a color : indeed, every vertex v such that the sum of its coordinates is even is assigned color 1 ; moreover, for every other vertex, S_d constitues a basis of colored vertices, from which, thanks to Rules 1 and 2, every vertex in G_d is assigned a color.

This coloring is clearly a proper coloring of G_d (no two neighbors have been assigned the same color). Moreover, this is also a valid star coloring, since every vertex u s.t. $c(u) \neq 1$ is such that any vertex v with color $c(v) = c(u)$ is at least at distance 4 from u (by Rules 1 and 2). The above-mentioned coloring using $2^d + 1$ colors, we conclude that $\chi_s(G_d) \leq 2^d + 1$.

Remark 4. We note that for dimensions 1 and 2, the upper bound given by Theorem 9 for d-dimensional grids is tight (cf. Theorem 4 when $d = 1$ and Theorem 8 when $d = 2$).

5   2-Dimensional Tori

In the following, for any $n, m \geq 3$, we denote the toroidal 2-dimensional grid by $TG(n, m)$. We recall that $TG(n, m)$ is the cartesian product of two cycles of length n and m. Our main result here is the following.

Theorem 10. *For any* $3 \leq n \leq m$, $5 \leq \chi_s(TG(n, m)) \leq 6$.

Proof. The proof is given in [7]. More precisely, we show that there is an infinite number of cases of $TG(n, m)$, for which $\chi_s(TG(n, m)) = 5$. The fact that 5 colors are always needed, and that 6 colors always suffice for the star coloring of $TG(n, m)$, $n, m \geq 3$ derives from several propositions (see [7]).

In some cases, we have been unable to determine precisely the number of colors necessary to star color $TG(n, m)$, $n, m \geq 3$ (though we know it up to an additive constant of 1). However, we pose the following conjecture.

Conjecture 1. For any $n \geq m \geq 3$

$$\chi_s(TG(n, m)) = \begin{cases} 6 \text{ when } n = m = 3 \text{ or when } n = 3 \text{ and } m = 5 \\ 5 \text{ otherwise} \end{cases}$$

6   Graphs with Bounded Treewidth

The notion of treewidth was introduced by Robertson and Seymour [11]. A *tree decomposition* of a graph $G = (V, E)$ is a pair $(\{X_i | i \in I\}, T = (I, F))$ where $\{X_i | i \in I\}$ is a family of subsets of V, one for each node of T, and T a tree such that :

(1) $\bigcup_{i \in I} X_i = V$
(2) For all edges $vw \in E$, there exists an $i \in I$ with $v \in X_i$ and $w \in X_i$
(3) For all $i, j, k \in I$: if j is on the path from i to k in T, then $X_i \cap X_k \subseteq X_j$

The *width* of a tree decomposition $(\{X_i | i \in I\}, T = (I, F))$ is $\max_{i \in I} |X_i| - 1$. The *treewidth* of a graph G is the minimum width over all possible tree decomposition of G.

We will prove the following theorem.

Theorem 11. *If a graph G has a treewidth at most k, then $\chi_s(G) \leq k(k + 3)/2 + 1$*

Actually we will prove Theorem 11 for k-trees, because it is well known that the treewidth of a graph G is at most k $(k > 0)$ if and only if G is a partial k-tree [5].

We recall the definition of a k-tree [4] :
(1) a clique with k-vertices is a k-tree
(2) If $T = (V, E)$ is a k-tree and C is a clique of T with k vertices and $x \notin V$, then $T' = (V \cup \{x\}, E \cup \{cx : c \in C\})$ is a k-tree.

If a k-tree holds exactly k vertices, then it is a clique by definition. If not, it contains at least a $k + 1$ clique ; moreover, it is easy to see that the greedy coloring with $k + 1$ colors of a k-tree gives an acyclic coloring. Hence we can deduce the following observation.

Observation 5 *For any $k \geq 1$ and any k-tree T_k :*

- $\chi_s(T_k) = k$ *if* $|V(T_k)| = k$;
- $k + 1 \leq \chi_s(T_k) \leq (k+1) \cdot 2^k$ *otherwise.*

Theorem 11 is in fact a corollary of the following result, which gives a much tighter bound than the one of Observation 5.

Theorem 12. $k + 1 \leq \chi_s(T_k) \leq k(k+3)/2 + 1$

Proof. Consider a k-tree G. We recall that a k-tree G is an intersection graph [10] and can be represented by a tree T and a subtree S_v for each v in G s.t. :

(1) $uv \in E(G) \Longleftrightarrow S_u \cap S_v \neq \emptyset$
(2) for any $t \in T$, $|\{v : t \in S_v\}| = k + 1$

We can see that by considering the tree decomposition of the k-tree. The tree T is the one of the tree decomposition and the subtree S_v for $v \in V(G)$ is exactly the subtree of T containing the nodes of T corresponding to the subsets of the tree decomposition containing v.

We root T at some node r and for each vertex v of G let $t(v)$ bet the first node of S_v obtained when traversing T in pre-order (i.e. $t(v)$ is the "highest" node of S_v). We choose some fixed preorder and order the nodes as $v_1, v_2, \ldots, v_n$ so that for $i < j$, $t(v_i)$ is considered in the preorder before $t(v_j)$. We $k(k+3)/2+1$ color the nodes in this order.

For each i we let :

$$X_{v_i} = \bigcup_{S_{v_j} \ni t(v_i)} \{v_l \neq v_i : S_{v_l} \ni t(v_j)\}$$

We show now that $|X_{v_i}| \leq k(k+3)/2$ and for any $v_j \in X_{v_i}$, $j < i$. Indeed let $A = \{a_1, a_2, \cdots, a_k, v_i\}$ be the subset of vertices corresponding to $t(v_i)$. We assume that $t(a_i)$ is before $t(a_{i+1})$ ($i \in \{1, \ldots, k-1\}$) in the preorder. We first give an upper bound for the number of subtrees S_{v_l} which contain $t(a_i)$ ($v_l \notin A$). The number of S_{v_l} ($v_l \notin A$) for $i \in \{1, \cdots, k\}$ which contain $t(a_1)$ is at most k, because the corresponding subset can intersect A only in a_1. The number of S_{v_l} ($v_l \notin A$) which contain $t(a_2)$ is at most $k-1$, a_1 is in the subset corresponding to $t(a_2)$ and we do not count S_{a_1}. It is easy to see that the number of S_{v_l} ($v_l \notin A$) which contain $t(a_i)$ is at most $k+1-i$, because the subset corresponding to $t(a_i)$ contains $\{a_1, a_2, \cdots, a_i\}$. In total we have at most $\sum_{i=1}^{i=k} i = k(k+1)/2$ subtrees S_{v_l} with $v_l \notin A$ containing $t(a_i)$ $i \in \{1, \cdots, k\}$. Now we have to add the number of S_{a_i}, this gives $|X_{v_i}| \leq k(k+3)/2$, without counting v_i. We color v_i with any color not yet used on X_{v_i}. This clearly yields a proper coloring, indeed if xy is an edge of G then $S_x \cap S_y \neq \emptyset$ and either $t(x) \in S_y$ or $t(y) \in S_x$, hence by construction x and y have different colors. We claim it also yields a coloring with no bichromatic P_4 (a path of length 3) : assume the contrary, and let $\{x, y, z, w\}$ be this P_4 labelled so that

(1) $t(x)$ is the first of $t(x), t(y), t(z), t(w)$ considered in the preorder,
(2) x and y have the same color,
(3) xz and yz are in $E(G)$.

We have to notice that x, z, y are not in the same clique K_{k+1} of the graph G corresponding to a node of T, because by construction this would imply that the colors are different. Now $S_x \cap S_y = \emptyset$, $S_z \cap S_x \neq \emptyset$, $S_z \cap S_y \neq \emptyset$, so by (1) $t(y)$ is in S_z. Further since $zx \in E(G)$, by (1) we have $t(z)$ is in S_x. So $x \in X_y$, contradicting the fact that x and y get the same color.

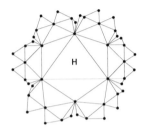

Fig. 5. A 2-tree H that satisfies $\chi_s(H) = 6$

We can notice that for 1-trees (i.e. the usual trees), the upper bound we obtain matches the one given by Theorem 4. For 2-trees, the upper bound is optimal because of the graph in Figure 5, which has been shown by computer to have a star chromatic number equal to 6.

In the following, we will denote by $\mathcal{O}$ the family of outerplanar graphs.

Corollary 2. $\chi_s(\mathcal{O}) = 6$.

Proof. It is well-known that any outerplanar graph is a partial 2-tree, then 6 colors suffice ; moreover the graph in Figure 5 which is also outerplanar, needs 6 colors.

7   Conclusion

In this paper, we have provided many new results concerning the star chromatic number of different families of graphs. In particular, we have provided exact results for paths and trees, cycles, complete bipartite graphs, 2-dimensional grids and outerplanar graphs. We have also determined bounds for the chromatic number in several other families of graphs, such as tori, d-dimensional grids, graphs with bounded treewidth, planar graphs and hypercubes. We have also determined several more general properties concerning the star chromatic number.

Using the techniques of [1], we can show that the star chromatic number of a graph of maximum degree Δ is $O(\Delta^{\frac{3}{2}})$ and that for every Δ, there exists a graph of maximum degree Δ whose star chromatic number exceeds $\epsilon.\frac{\Delta^{3/2}}{\log\Delta}$ for some positive absolute constant ϵ. Details will appear in the journal version of the paper.

A large number of problems remain open here, such as getting optimal results for other families of graphs, or refining our non optimal bounds ; getting one or several methods to provide good lower bounds for the star chromatic number is also another challenging problem.

References

1. N. Alon, C. McDiarmid, and B. Reed. Acyclic colourings of graphs. *Random Structures and Algorithms*, 2:277–288, 1990.
2. O.V. Borodin, A.V. Kostochka, A. Raspaud, and E. Sopena. Acyclic k-strong coloring of maps on surfaces. *Mathematical Notes*, 67 (1):29–35, 2000.
3. O.V. Borodin, A.V. Kostochka, and D.R. Woodall. Acyclic colourings of planar graphs with large girth. *J. London Math. Soc.*, 60 (2):344–352, 1999.
4. A. Brandstädt, V.B. Le, and J.P. Spinrad. *Graph Classes A survey*. SIAM Monographs on D.M. and Applications, 1999.
5. H.L. Bodlander. A partial k-arboretum of graphs with bounded treewidth. *Theoretical Computer Science*, 209:1–45, 1998.
6. O.V. Borodin. On acyclic colorings of planar graphs. *Discrete Mathematics*, 25:211–236, 1979.
7. G. Fertin, A. Raspaud, and B. Reed. On star coloring of graphs. *Technical report*, 2001.
8. B. Grünbaum. Acyclic colorings of planar graphs. *Israel J. Math.*, 14(3):390–408, 1973.
9. D.S. Kim, D.-Z. Du, and P.M. Pardalos. A coloring problem on the n-cube. *Discrete Applied Mathematics*, 103:307–311, 2000.
10. T.A. McKee and F.R. McMorris. *Topics in intersection graph theory*. SIAM Monographs on D.M. and Applications, 1999.
11. N. Robertson and P.D. Seymour. Graph minors. 1. excluding a forest. *J. Combin. Theory Ser. B*, 35:39–61, 1983.
12. P.-J. Wan. Near-optimal conflict-free channel set assignments for an optical cluster-based hypercube network. *J. Combin. Optim.*, 1:179–186, 1997.

Graph Subcolorings: Complexity and Algorithms
(Extended Abstract)[*]

Jiří Fiala[1,2], Klaus Jansen[1], Van Bang Le[3,**], and Eike Seidel[1]

[1] Institut für Informatik und Praktische Mathematik,
Christian-Albrechts-Universität zu Kiel, 24098 Kiel, Germany
{jfi,kj,ese}@informatik.uni-kiel.de
[2] Institute of Theoretical Computer Science,
Charles University, Prague, Czech Republic [†]
[3] Institut für Theoretische Informatik, Fachbereich Informatik,
Universität Rostock, 18051 Rostock, Germany
le@informatik.uni-rostock.de

Abstract. In a graph coloring, each color class induces a disjoint union of isolated vertices. A graph subcoloring generalizes this concept, since here each color class induces a disjoint union of complete graphs. Erdős and independently Albertson et al. proved that every graph of maximum degree at most 3 has a 2-subcoloring. We point out in this paper that this fact is best possible with respect to degree-constraints by showing that the problem of recognizing 2-subcolorable graphs with maximum degree 4 is NP-complete, even when restricted to triangle-free planar graphs. Moreover, in general, for fixed k, recognizing k-subcolorable graphs is NP-complete on graphs with maximum degree at most k^2. In contrast, we show that, for arbitrary k, k-SUBCOLORABILITY can be computed efficiently on graphs of bounded treewidth and on cographs.

1 Introduction and Results

A k-*coloring* of a graph G is a partition of the vertices into k pairwise disjoint sets $V_1, \ldots, V_k$ such that for every $i = 1, 2, \ldots, k$, each color class V_i consists of isolated vertices, i.e. it forms a stable set. The smallest k for which the graph G has a k-coloring is called the *chromatic number* of G, denoted by $\chi(G)$. Graph coloring is a well studied topic, both for its theoretic and algorithmic aspects. It is well-known that testing 3-COLORABILITY is NP-complete for triangle-free graphs with maximum degree 4 ([14]) and for planar graphs with maximum degree 4 ([12]). Testing 3-COLORABILITY is easy for graphs with maximum degree 3 (Brooks theorem) and for triangle-free planar graphs (Grötzsch theorem).

Graph coloring has been generalized in several ways and by a number of authors; see [2] for a comprehensive survey. In this paper we address one of

[*] Research supported in part by EU ARACNE project HPRN-CT-1999-00112 and EU APPOL project IST-1999-14084
[**] Financial support by Deutsche Forschungsgemeinschaft is gratefully acknowledged.
[†] Supported by the Ministry of Education of the Czech Republic as project LN00A056

A. Brandstädt and V.B. Le (Eds.): WG 2001, LNCS 2204, pp. 154–165, 2001.
© Springer-Verlag Berlin Heidelberg 2001

these generalized colorings. A partition $V_1, \ldots, V_k$ of vertex set of a graph G is called a k-*subcoloring* of G if each color class V_i induces in G a disjoint union of complete subgraphs (of various sizes). The *subchromatic number* $\chi_s(G)$ of G is the smallest integer k for which G has a k-subcoloring. Subcolorings have been discussed in [2,4,5,15]. It turns out that subcolorings have many interesting properties similar to colorings, and every k-coloring is also a k-subcoloring, hence $\chi(G) \geq \chi_s(G)$. Among other results we would like to mention the following properties of graph subcolorings:

(1) For every $k \geq 1$, there is a triangle-free graph G_k with $\chi_s(G_k) = k$ [2,15].
(2) For every planar graph G, $\chi_s(G) \leq 4$. In addition, if G is outerplanar, $\chi_s(G) \leq 3$. These bounds are tight [4].
(3) For every graph G with maximum degree Δ, $\chi_s(G) \leq \lfloor \frac{\Delta}{2} \rfloor + 1$ [2].

By (3), every graph with maximum degree at most 3 is 2-subcolorable. Actually, this fact follows also from a theorem due to Erdős [10] saying that every graph G has a bipartite spanning subgraph H such that the degree in H of every vertex is at least one half of its degree in G. Thus, if the maximum degree in G is at most 3, every bipartition of H defines a sub-bipartition of G. Moreover, such a bipartite spanning subgraph H of G can be found in polynomial time, easily by a "local improvement" technique.

Albertson et al. ([2]) then pointed out the difficulties involved in characterizing 2-subcolorable graphs by giving a number of examples. In this paper we prove the following theorems (Sect. 2.1 and 2.2):

Theorem 1. *Recognizing 2-subcolorable graphs of maximum degree 4 is NP-complete, even on triangle-free planar graphs.*

Albertson informed us that, independently, Gimbel also proved Theorem 1 with a completely different reduction

Notice that the NP-completeness of k-SUBCOLORABILITY for the class of *all* graphs follows from a theorem by Achlioptas [1]. We prove in Sect. 2.3 that:

Theorem 2. *For every fixed $k \geq 2$, k-SUBCOLORABILITY is NP-complete for graphs of maximum degree at most k^2.*

For fixed k, k-SUBCOLORABILITY can be expressed as a monadic second order logic formula, and hence can be tested in linear time for graphs with bounded treewidth and cographs. Due to the fact that for these graphs $\chi_s(G) \leq c$ for a constant c, we get that there exists an algorithm that in linear time determine $\chi_s(G)$ for graphs with bounded treewidth and cographs. On the other side, the general algorithm is unnecessarily complicated for our purpose and we present two simple algorithms solving this problem. (see Sect. 3 and 4)

Theorem 3. *For every k, k-SUBCOLORABILITY can be efficiently decided on graphs of bounded treewidth, and on cographs.*

Various proofs are omitted in this paper due to space restrictions. In the full version of this paper we present a general algorithm for graphs with bounded cliquewidth and also NP-completeness of 3-SUBCOLORABILITY of planar graphs.

2   The NP-Completeness of the Subcoloring Problem

We prove Theorem 1 by reducing the Not-all-equal 3-satisfiability problem, which was proven NP-complete by Schaefer [17] (see also [11, Problem LO3]).
Problem: *Let C be a Boolean formula consisting of of m clauses such that every clause has exactly three distinct literals. The the decision problem whether there exists a satisfying assignment for C such that each clause in C has at least one false (and at least one true) literal is NP-complete.*

We denote the class of all formulas that allow such assignment by NAE3SAT.

2.1   Triangle-Free Graphs of Maximum Degree 4

We first prove in this section the NP-completeness of 2-SUBCOLORABILITY for non-planar triangle-free graphs of maximum degree 4 (the problem is clearly in NP). Let $C = \{C_1, C_2, \ldots, C_m\}$ be a Boolean formula in normal form of m clauses with variable set $V = \{v_1, v_2, \ldots, v_n\}$ such that every clause C_j of C contains exactly three literals, $C_j = \{c_1^{(j)}, c_2^{(j)}, c_3^{(j)}\}$. We will construct a triangle-free graph $G = G(C)$ of maximum degree 4 such that G has a 2-subcoloring if and only if $C \in \text{NAE3SAT}$.

Before describing the construction, we consider the graph F shown in Figure 1 (left); its symbolic representation will be used when larger portions of the graph are drawn.

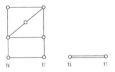

Fig. 1. The graph F (left) and its symbolic representation (right)

Lemma 1. *The graph F is 2-subcolorable and in every 2-subcoloring of F, the labeled vertices u and v belong to different color classes.*

Now, for each clause $C_j = \{c_1^{(j)}, c_2^{(j)}, c_3^{(j)}\}$ we consider the graph $G(C_j)$ shown in Figure 2.

Fig. 2. The graph $G(C_j)$

The following Lemma indicates the meaning of $G(C_j)$.

Lemma 2. *In every 2-subcoloring of $G(C_j)$, not all of c_{j1}, c_{j2}, c_{j3} belong to the same color class and each c_{jk} has a neighbor in $G(C_j)$ that has the same color as c_{jk}. Moreover, every bipartition of $\{c_{j1}, c_{j2}, c_{j3}\}$ into two nonempty sets can be extended to a 2-subcoloring of $G(C_j)$.*

For each variable $v_i \in V$ let $G(v_i)$ be the graph shown in Figure 3.

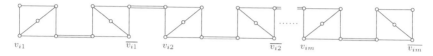

Fig. 3. The graph $G(v_i)$

Lemma 3. *The graph $G(v_i)$ is 2-subcolorable. Any its 2-subcoloring contains vertices $v_{i1}, v_{i2}, \ldots, v_{im}$ in the same color class while vertices $\overline{v_{i1}}, \overline{v_{i2}}, \ldots, \overline{v_{im}}$ are colored by the other color.*

We complete the construction of the graph G by taking m disjoint copies of $G(C_j)$ (one copy for each clause $C_j : 1 \leq j \leq n$) by n disjoint copies of $G(v_i)$ (each copy correspond to one variable v_i) and we connect these graphs by adding additional edges connecting variables to the corresponding clauses $\{(c_{jk}, v_{ij}) : c_k^{(j)} = v_i\}$ and also the negated variables to clauses $\{(c_{ik}, \overline{v_{ij}}) : c_k^{(j)} = \overline{v_i}\}$.

Observe that graphs $G(v_i)$ and $G(C_j)$ are triangle free and the additional edges form a matching, hence G is triangle-free and moreover every its vertex is of degree at most 4.

For each $1 \leq j \leq m$ and each $1 \leq k \leq 3$, the vertex c_{jk} in $G(C_j)$ has exactly one neighbor v outside $G(C_j)$, more formally $v \in \{v_{ij}, \overline{v_{ij}}\}$ if v_i or $\overline{v_i}$ is the literal $c_k^{(j)}$ in C_j. We call v the *personal neighbor* of c_{jk}.

Lemma 4. *In every 2-subcoloring of G, c_{jk} and its personal neighbor belong to different color classes.*

Proof. By Lemma 2, c_{jk} has a neighbor in $G(C_j)$ that has the same color as c_{jk}. Therefore, by definition of subcoloring, the personal neighbor of c_{jk} has the other color.

We now are going to show that $\mathcal{C} \in$ NAE3SAT if and only if G has a 2-subcoloring.

Let b be a truth assignment for $\mathcal{C}$ in which every clause has at least one true and at least one false valued literal. We construct a 2-subcoloring of G as follows. If $b(v_i) = \mathsf{true}$ then all v_{ij} are red otherwise all v_{ij} are blue, and extend to a 2-subcoloring of $G(v_i)$. This is possible as we have seen by Lemma 3. Next, color c_{jk} differently from its personal neighbor $v \in \{v_{ij}, \overline{v_{ij}}\}$ (which is already colored by the previous step). By assumption on the assignment b, not all of c_{j1}, c_{j2}, c_{j3} can have the same color and due to Lemma 2 we may extend this coloring to a 2-subcoloring of $G(C_j)$. By construction of G, the obtained coloring is a 2-subcoloring of G.

Suppose that G has a 2-subcoloring in red and blue. We define the truth assignment b for $\mathcal{C}$ as follows: $b(v_i) = $ **true** if v_{ij} is red for some j; otherwise $b(v_i) = $ **false**. By Lemma 3, this assignment is well-defined. Due to Lemma 2, for each j, two of c_{j1}, c_{j2}, c_{j3} have different colors. By Lemma 4, the personal neighbors of these two vertices have different colors. Therefore, C_j has at least one true and at least on false literal by the truth assignment b.

2.2   Planar Graphs of Maximum Degree 4

In this section, we construct a triangle-free planar graph G' from the graph G obtained in the previous section, such that G is 2-subcolorable if and only if G' is 2-subcolorable.

Call an edge in G a *cv-edge* if it connects a vertex c_{jk} with its personal neighbor $v \in \{v_{ij}, \overline{v_{ij}}\}$. Note that G can be embedded in the plane, in polynomial time, such that every edge is a straight line and every edge crossing occurs only by two *cv*-edges.

Recall that in any 2-subcoloring of G, the endvertices of every *cv*-edge belong to different color classes (see Lemma 4). This makes the use of the "crossover" technique in proving NP-completeness of PLANAR GRAPH 3-COLORABILITY, described for example in [11], possible.

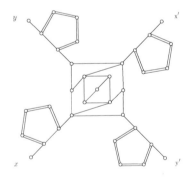

Fig. 4. The "crossover" graph H

The "crossover" in our construction is the graph H shown in Figure 4 and has the following properties (consult Lemma 1):

(1) In any 2-subcoloring of H, x and x' belong to the same color class, and y and y' belong to the same color class.
(2) H has a 2-subcoloring such that x and y belong to the same color class, and a 2-subcoloring such that x and y belong to different color classes.

The construction of the planar graph G' from G using crossover H is very similar to the construction for PLANAR GRAPH 3-COLORABILITY.

- To each cv-edge that is crossed by other cv-edge, add new vertices between each pair of adjacent crossings and one more vertex between c_{jk} and its nearest crossing.
- Replace each crossing by a copy of H, identifying the outlets x and x' with the nearest two vertices on one of the cv-edge involved, and y and y' with the nearest two vertices on the other cv-edge.

Observe, that the graph G' has a 2-subcoloring if and only if G does.

2.3   The Hardness of k-Subcoloring for $k \geq 3$

In this section we show that, for each fixed $k \geq 2$ the k-SUBCOLORABILITY is a NP-complete problem on graphs with maximum degree at most k^2.

Lemma 5. *If φ is a k-subcoloring of a graph G and H is an induced subgraph of G then the restriction of φ on H is a k-subcoloring of H.*

Note that Lemma 5 does not hold for subgraphs in general.

Lemma 6. *For every $k \geq 2$, the complete k-partite graph $K_{k,\dots,k,k+1}$ consisting of $k-1$ (small) partitions with k vertices and one (big) partition of $k+1$ vertices has exactly one (up to permutation) k-subcoloring; this k-subcoloring is also its unique k-coloring. The graph $K_{k,\dots,k,k+1}$ cannot be subcolored with less than k colors.*

We show a reduction from $(k-1)$-SUBCOLORABILITY to k-SUBCOLORABILITY. Let G be the graph for which the existence of a k-subcoloring is questioned and let $V(G) = \{v_1, \dots, v_n\}$. We define G' as follows:

- Take n copies $H_1, \dots, H_n$ of the $K_{k,\dots,k,k+1}$;
- in each H_i, label four distinct vertices of the big class with $x_1^i, x_2^i, x_3^i, x_4^i$, and label one vertex in each of the $k-1$ small classes with y_j^i, $1 \leq j \leq k-1$;
- add edges $(x_1^i, y_j^{i-1}), (x_2^i, y_j^{i-1})$ for all $1 \leq j \leq k-1, 2 \leq i \leq n$;
- add edges $(x_3^i, v_i), (x_4^i, v_i)$ for all $1 \leq i \leq n$.

We claim that G is $k-1$-subcolorable if and only if G' is k-subcolorable:

(1) Assume G can be subcolored with $k-1$ colors. Then subcolor, in each H_i, the $k-1$ small classes with these $k-1$ colors (each class gets one color), and take one new color for the big class. This is a k-subcoloring for G'.

(2) Consider a k-subcoloring of G'. Then, by Lemma 5, the restriction of this subcoloring on each H_i is a k-subcoloring. By Lemma 6, each H_i gets all k colors and each class in H_i is monochromatic. Moreover, the big classes of all H_is have the same color.

We show that, for $1 \leq i < n$, x_1^i and x_1^{i+1} have the same color. Assume not, then the color of x_1^{i+1} must occur in a small class of H_i, say y_1^i has this color.

But then $x_1^{i+1}, x_2^{i+1}, y_1^i$ induce a monochromatic path P_3 in G', contradicting the definition of k-subcoloring.

No vertex in G can have the color occurring in the big classes of the H_is. Therefore, the restriction of the k-subcoloring of G' on G is a $k-1$-subcoloring of G.

Lemma 7. *For $k \geq 3$, $\Delta(G') = \max\{\Delta(G) + 2, k^2\}$:*

Proof. Observe that $\Delta(G') = \max\{\Delta(G) + 2, d_{G'}(x_1^2), d_{G'}(y_1^2)\}$. By the construction, $d_{G'}(x_1^2) = (k-1)k + (k-1) = k^2 - 1$ and also $d_{G'}(y_1^2) = (k-2)k + (k+1) + 2 = k^2 - k + 3$, hence, for $k \geq 3$, the Lemma follows.

Proof of Theorem 2 The case $k = 2$ is proven by Theorem 1. The statement for $k \geq 3$, follows from the construction and Lemma 7 and by noting that if G has maximum degree at most $(k-1)^2$ then the graph G' constructed from G as above has maximum degree at most k^2.

3   Graphs with Bounded Treewidth

The notion of treewidth was introduced by Robertson and Seymour in [16] via tree decompositions.

Let $G = (V, E)$ be a graph. The *tree decomposition* of G is a pair $(T, \mathcal{X} = \{X_i : i \in V(T)\})$, where T is a tree, each X_i is a subset of V and the following is satisfied

1. For each edge $(u, v) \in E$ exists $i \in V(T)$ such that $u, v \in X_i$.
2. For any $v \in V(G)$ the indices of sets X_i containing v induce a nonempty connected subtree of T.

The width of a tree decomposition $(T, \mathcal{X})$ is $\max_{i \in V(T)}\{|X_i|\} - 1$ and the treewidth of G is the minimum width among all possible tree decompositions. We denote treewidth by $tw(G)$. If the treewidth of G is bounded by a constant c, a tree decomposition of width $\leq c$ of G can be constructed in linear time ([3]).

For fixed k, k-SUBCOLORABILITY can be expressed in monadic second order logic, which is a language to describe graph properties, using the following constructions: quantifications over vertices, edges, sets of edges, sets of vertices, membership tests, adjacency tests and logic operations. By the results of Courcelle ([8]) it is known that each problem that can be stated in monadic second order logic can be solved in linear time on graphs with bounded treewidth. Unfortunately, the multiplicative constant grows very fast, essentially it is a tower of 2's whose height is the number of quantifier-alternations of the monadic second order logic formula. In our case the height is at least linear in k.

We present a decision algorithm that for fixed k tests whether the subchromatic number $\chi_s(G) \leq k$ for graphs G with bounded treewidth. The algorithm involves dynamic programming to find a proper k-subcoloring. For simplicity we restrict ourself to nice tree decomposition.

The *nice tree decomposition* of G ([13]) is a tree decomposition such that: T is a rooted binary tree and for each $i \in V(T)$ at least one of the following cases applies:

- i is a leaf and $|X_i| = 1$, then we call i a *leaf node*.
- i has one son j and $X_i = X_j \cup \{v\}$ for some $v \in V(G) \setminus X_j$, then we call i an *introduce node*.
- i has one son j and $X_i = X_j \setminus \{v\}$ where $v \in X_j$, then we call i a *forget node*.
- i has two sons j, j', and $X_i = X_j = X_{j'}$ then we call i a *join node*.

Any tree decomposition can be transformed in linear time into a nice tree decomposition of the same width [13].

Denote by G_i the subgraph of G induced by vertices of X_i and G'_i the subgraph of G induced by vertices of $\bigcup_{j \succ i} X_j$, where $j \succ i$ means that j is a descendant of i i.e. i lies on the path from j to the root of T.

Let ϕ be a k-subcoloring of G_i, then the *color clique* of ϕ is any inclusion-maximal set $K \subseteq V(G_i) = X_i$, such that all vertices of K have the same color under ϕ and are mutually adjacent in G. In other words, a color clique is any clique that belongs to a color class of ϕ.

We introduce a new data structure $(\mathsf{Tab}_i)_{i \in V(T)}$ such that each entry Tab_i consists of several pairs $(\phi, grow_{i,\phi})$, where each ϕ is a feasible k-subcoloring of G_i, that might be extended to a subcoloring of G'_i and $grow_{i,\phi}$ is a function assigning to each color clique K of ϕ a boolean variable which helps us to properly define a new k-subcoloring in the inductive step. Note that a single ϕ may occur in some entry of Tab_i several times with different functions $grow_{i,\phi}$. However, as G has bounded tree width, the number of all pairs $(\phi, grow_{i,\phi})$ is bounded by a constant. The evaluation of Tab_i goes as follows:

1. If i is a *leaf node* then Tab_i contains all k possible k-subcolorings ϕ of $G_i = (\{v\}, \emptyset)$ and for the only color clique $K = \{v\}$ and all ϕ set $grow_{i,\phi}(K) =$ true.

2. Let i be a *forget node* with the son j, and Tab_j has been already computed. Then let Tab_i contain all entries from Tab_j, restricted to set X_i. If K was a color clique of some k-subcoloring $(\phi', grow_{j,\phi'}) \in \mathsf{Tab}_j$, and ϕ is the restricted k-subcoloring, then set $grow_{i,\phi}(K \setminus \{v\}) =$ false only if K contains the vertex $v = X_j \setminus X_i$. For all other color cliques L of ϕ let $grow_{i,\phi}(L) = grow_{j,\phi'}(L)$. Remove duplicated entries in Tab_i, if some exists.

3. Let i be an *introduce node* with the son j, $v \in V(G)$ is the added vertex and Tab_j is already known. Then for every pair $(\phi', grow_{j,\phi'}) \in \mathsf{Tab}_j$ and every k-subcoloring ϕ of G_i, such that ϕ restricted onto X_j is equal to ϕ', find a color clique K of ϕ containing v. If $K = \{v\}$ or $grow_{j,\phi'}(K \setminus \{v\}) =$ true, then add into Tab_i entry $(\phi, grow_{i,\phi})$ where $grow_{i,\phi}(K) =$ true and set $grow_{i,\phi}(L) = grow_{j,\phi'}(L)$ for all other color cliques $L \neq K$ of ϕ.

4. Let i be a *join node* with sons j and j' and ϕ a k-subcoloring of X_i. Then for all possible combinations of $(\phi, grow_{j,\phi}) \in \mathsf{Tab}_j$ and $(\phi, grow_{j',\phi}) \in \mathsf{Tab}_{j'}$ add the entry $(\phi, grow_{j,\phi} \wedge grow_{j',\phi})$ into Tab_i if and only if for each color

clique K of ϕ the value of $grow_{j,\phi}(K) \vee grow_{j',\phi}(K)$ is true. Again, if some
entries are present more times, store only one.

5. Compute the values of $\mathtt{Tab}_i$ for all vertices i in the tree, as described in steps
1–4. The graph G allows a k-subcoloring if and only if the table entry $\mathtt{Tab}_r$
is nonempty for the root r.

To see that the algorithm is correct we think that steps 2, 3 and 4 deserve
further explanation.

In step 2 we remember in the function $grow$ the fact, that a certain color
clique K has already lost a vertex v, and future extension of K by v' would
cause that the color class will contain induced P_3, since the edge (v, v') does not
belong to G. Therefore in step 3 we try extend only those color cliques which
might be extended. The same argument is used in step 4, since it is impossible
to identify two color cliques when both of them already forget a vertex. Note
that various functions $grow_{i,\phi}$ for a single k-subcoloring ϕ may appear during
executing steps 2 and 4.

This discussion concludes the proof of the first part of Theorem 3. For a graph
G with tree-decomposition of width c the decision of k-SUBCOLORABILITY can
be performed as fast as the evaluation of the table $(\mathtt{Tab}_i)_{i \in V(T)}$, that is in time
$O(n2^c k^{c+1})$. This expression is linear in n.

Note that finding the minimum k such that G allows a k-subcoloring can be
done in time $O(n^{c+2})$ by running at most n tests for all $k < n$.

4   Cographs

Cographs belong to the class of graphs with bounded cliquewidth. Due to the
results of Courcelle [9] problems that are expressible in monadic second order
logic are linear time solvable in graph with bounded cliquewidth, but the constants involved are equally bad as in the case of graphs with bounded treewidth,
as discussed in beginning of Section 3. as in the case for graphs with bounded
treewidth. Hence there is a linear time algorithm to decide k-SUBCOLORABILITY,
k fixed, for graphs of bounded cliquewidth. In the full version we also show that
when k is part of the input, the k-SUBCOLORABILITY is polynomial solvable for
graphs of bounded cliquewidth.

In this section we present a $O(n^3)$ algorithm to compute the subchromatic
number of cographs (graphs of bounded cliquewidth at most 2). In particular,
k-SUBCOLORABILITY (k arbitrary) is efficiently solvable for cographs.

Cographs [6] are inductively defined as follows.

– Every single vertex graph is a cograph.
– If G_1 and G_2 are two (disjoint) cographs, then the disjoint union $G_1 \cup G_2$
  and the join $G_1 + G_2$ are cographs.

The join graph $G_1 + G_2$ is obtained from $G_1 \cup G_2$ by adding all edges between
vertices in G_1 and vertices in G_2.

With each cograph G we can associate a rooted binary labeled tree T_G, called
the cotree of G. The leaves of the cotree T_G represent vertices of G. Internal nodes

of T_G have a label $\cup$ or $+$. If $G = G_1 \cup G_2$, respectively, $G = G_1 + G_2$, then the root of T_G carries label $\cup$, respectively, label $+$, and its two sons are root nodes of the subtrees representing G_1 and G_2, respectively.

There is a linear time algorithm for recognizing whether a given graph G is a cograph and, if so, for constructing a cotree T_G of G (see [7]).

Our algorithm for determining the subchromatic number of a cograph relies on the following notion. A subcoloring ϕ of a graph G is of *type* (i, j) if ϕ has exactly i color classes each of which induces a clique (called *small classes*) and exactly j remaining classes (called *big classes*). If ϕ is of type (i, j), we also call ϕ an (i, j)-subcoloring.

Consider an (i, j)-subcoloring ϕ of a graph G which arose by disjoint union $\cup$ or by join $+$ of two graphs G_1 and G_2. In the following, for each $t = 1, 2$ we denote by ϕ_t be the restriction of ϕ on G_t and assume that ϕ_t is of type (i_t, j_t).

Lemma 8. *If $G = G_1 + G_2$ then for any (i, j)-subcoloring ϕ of G it holds:*
$i \geq \max\{i_1, i_2\}$ *and* $j = j_1 + j_2$.

Proof. The first equality follows from the fact, that any small class of ϕ may consist of at most two small color classes, one in ϕ_1 and one in ϕ_2. The second equality express the fact that big color class of ϕ is big in exactly one of ϕ_1 or ϕ_2.

Lemma 9. *If ϕ_1 and ϕ_2 are subcolorings of type (i_1, j_1) and (i_2, j_2), then a $(\max\{i_1, i_2\}, j_1 + j_2)$-subcoloring of $G = G_1 + G_2$ can be obtained from ϕ_1 and ϕ_2.*

Proof. A $(\max\{i_1, i_2\}, j_1 + j_2)$-subcoloring of G can be obtained by a combination $\min\{i_1, i_2\}$ small classes of G_1 and G_2.

Lemma 10. *If $G = G_1 \cup G_2$ then $j \geq \max\{j_1, j_2\}$, $i + j \geq i_t + j_t$ $(t = 1, 2)$, and $i + 2j \geq i_1 + j_1 + i_2 + j_2$.*

Proof. Let

- r denote the number of the big classes C of ϕ such that, for each $t = 1, 2$, $C \cap G_t$ is a big class in ϕ_t,
- r_t denote the number of big classes C of ϕ such that, for $t \neq t'$, $C \cap G_t$ is a small class in ϕ_t but $C \cap G_{t'}$ is a big class in $\phi_{t'}$,
- q_t denote the number of big classes of ϕ belonging only to ϕ_t,
- s denote the remaining big classes of ϕ, which are small both in G_1 and G_2.
- l_t denote the number of the small classes of ϕ belonging to ϕ_t.

The first statement of the Lemma follows directly from:

$$i = l_1 + l_2, \qquad j = r + r_1 + r_2 + q_1 + q_2 + s,$$
$$i_1 = r_1 + l_1 + s, \qquad j_1 = r + r_2 + q_1,$$
$$i_2 = r_2 + l_2 + s, \qquad j_2 = r + r_1 + q_2.$$

Moreover, $i_1 + j_1 = i + j - l_2 - q_2$ and $i_2 + j_2 = i + j - l_1 - q_1$, hence the second statement holds. The third statement then follows from $i_1 + j_1 + i_2 + j_2 = 2(i + j) - l_1 - l_2 - q_1 - q_2 = i + 2j - q_1 - q_2$.

In view of Lemmas 8 and 9, we are interested in (i, j)-subcolorings with small number j of big classes. A way to obtain such a subcoloring of $G = G_1 \cup G_2$ from ϕ_1 and ϕ_2 is as follows: We first merge the $\min\{j_1, j_2\}$ pairs of big classes of ϕ_1 and ϕ_2, and then combine as many as possible of the $|j_1 - j_2|$ remaining big classes together with some small classes into a new color class of ϕ. The number of remaining small classes of ϕ_1 is $k_1 := i_1 - \min\{i_1, \max\{0, j_2 - j_1\}\}$. Similarly ϕ_2 contains $k_2 := i_2 - \min\{i_2, \max\{0, j_1 - j_2\}\}$ remaining small classes.

Note that $k_1 = i_1$ (if $j_1 \geq j_2$) or $k_2 = i_2$ otherwise. Finally we combine k, $0 \leq k \leq \min\{k_1, k_2\}$, small classes of ϕ_1 with k small classes of ϕ_2 and get a $(k_1 + k_2 - 2k, \max\{j_1, j_2\} + k)$-subcoloring ϕ' of $G = G_1 \cup G_2$.

This and Lemmas 8 and 9 suggest the following algorithm for determining $\chi_s(G)$, assuming that the cograph G and its cotree T_G with root r are given.

For each node x of T_G the algorithm stores in a table Tab_x the type (i, j) of all possible (i, j)-subcolorings of the induced subgraph G_x of G corresponding to x that are relevant for computing $\chi_s(G)$. The tables are computed in bottom-up order:

(1) For each leave x of T_G, put $(1, 0)$ into Tab_x.
(2) If x has label $+$ and sons x_1, x_2, then for all combinations of entries $(i_1, j_1) \in \mathsf{Tab}_{x_1}$ and $(i_2, j_2) \in \mathsf{Tab}_{x_2}$ put into Tab_x the entry (i, j), where $i = \max\{i_1, i_2\}$ and $j = j_1 + j_2$.
(3) If x has label $\cup$ and sons x_1, x_2, then for all combinations of entries $(i_1, j_1) \in \mathsf{Tab}_{x_1}$ and $(i_2, j_2) \in \mathsf{Tab}_{x_2}$ perform the following computation:
    (3.1) Set $k_1 := i_1 - \min\{i_1, \max\{0, j_2 - j_1\}\}$ and
        $k_2 := i_2 - \min\{i_2, \max\{0, j_1 - j_2\}\}$.
    (3.2) For each k varying from 0 to $\min\{k_1, k_2\}$ put into Tab_x the entry (i, j), where $i = k_1 + k_2 - 2k$ and $j = \max\{j_1, j_2\} + k$.
(4) Return $\chi_s(G) = \min\{i + j \ : \ (i, j) \in \mathsf{Tab}_r\}$.

Note that the number of all entries (i, j) stored in each Tab_x is bounded by n^2. Moreover, as discussed by Lemmas 8 and 9 and after Lemma 10, if $(i, j) \in \mathsf{Tab}_x$ then there exists a subcoloring of G_x of type (i, j).

We say that $(i', j') \preceq (i, j)$ if simultaneously holds that $j' \leq j$ and $i' + j' \leq i + j$. It is clear that from a subcoloring of type (i', j') any subcoloring of type (i, j) with $(i', j') \preceq (i, j)$ can be derived by adding extra colors or claiming some small color classes as big.

The following lemma shows the correctness of the algorithm (proof omitted):

Lemma 11. *For every node x of T_G and every subcoloring ϕ of G_x of type (i, j), there exists a pair $(i', j') \in \mathsf{Tab}_x$ such that $(i', j') \preceq (i, j)$.*

The direct application of Lemma 11 concludes the proof of the second part of Theorem 3 in time $O(nk^3)$.

Acknowledgments

The third author (VBL) thanks Hoàng-Oanh Le for her interesting comments. He also thanks the second author (KJ) for inviting him to visit his research group in Kiel.

References

1. D. Achlioptas, The complexity of G-free graph colourability, *Discrete Math*, 165/166 (1997), 31-38.
2. M.O. Albertson, R.E. Jamison, S.T. Hedetniemi, S.C. Locke, The subchromatic number of a graph, *Discrete Math.*, 74 (1989), 33-49.
3. H.L. Bodlaender, A linear time algorithm for finding tree-decompositions of small treewidth. *SIAM J. Computing*, 25 (1996), 1305-1317.
4. I. Broere, C.M. Mynhardt, Generalized colorings of outerplanar and planar graphs, *Proc. Graph Theory with Applications to Algorithms and Computer Science*, Kalamazoo, Mich., (1984), 151-161.
5. J.L. Brown, D.G. Corneil, On generalized graph colorings, *J. Graph Theory*, 11 (1987), 87-99.
6. D.G. Corneil, H. Lerchs, L. Stewart Burlingham, Complement reducible graphs. *Discrete Appl. Math.*, 3 (1981), 163-174.
7. D.G. Corneil, Y. Perl, L.K. Stewart, A linear recognition algorithm for cographs, *SIAM J. Computing*, 14 (1985), 926-934.
8. B. Courcelle, Graph rewriting: an algebraic and logical approach, *Handbook of Theoretical Computer Science*, volume B, (1990), 192-242.
9. B. Courcelle, The monadic second-order logic of graphs I: Recognizable sets of finite graphs, *Information and Computation*, 85 (1991), 12-75.
10. P. Erdős, Bipartite subgraphs of graphs, *Math. Lapok*, 18 (1967), 283-288.
11. M. Garey, D.S. Johnson, *Computers and Intractability: A Guide to the Theory of NP-completeness*, W.H. Freeman, San Fransisco, 1979.
12. M.R. Garey, D.S. Johnson, L. Stockmayer, Some simplified NP-complete graph problems, *Theoretical Computer Science*, 1 (1976), 237-267.
13. T. Kloks, *Treewidth — Computations and Approximations*, Lecture Notes in Computer Science no. 842, Springer-Verlag, 1994.
14. F. Maffray, M. Preissman, On the NP-completeness of k-colorability problem of triangle-free graphs, *Discrete Math.*, 162 (1996), 313-317.
15. C.M. Mynhardt, I. Broere, Generalized colorings of graphs, *Proc. Graph Theory with Applications to Algorithms and Computer Science*, Kalamazoo, Mich., (1984), 583-594.
16. N. Robertson and P.D. Seymour, Graph minors. II. Algorithmic aspects of treewidth, *Journal of Algorithms*, 7 (1986), 309-322.
17. T.J. Schaefer, The complexity of the satisfability problem, *Proc. 10th Ann. ACM Symp. on theory of computing* (1978), 216-226.

Approximation of Pathwidth
of Outerplanar Graphs

Fedor V. Fomin[*,1] and Hans L. Bodlaender[**,2]

[1] Faculty of Mathematics and Mechanics, St.Petersburg State University,
Bibliotechnaya sq.2, St.Petersburg, 198904, Russia
fomin@gamma.math.spbu.ru
[2] Department of Computer Science, Utrecht University,
P. O. Box 80.089, 3508 TB Utrecht, the Netherlands
hansb@cs.uu.nl

Abstract. There exists a polynomial time algorithm to compute the pathwidth of outerplanar graphs [3], but the large exponent makes this algorithm impractical. In this paper, we give an algorithm, that given a biconnected outerplanar graph G, finds a path decomposition of G of pathwidth at most twice the pathwidth of G plus one. To obtain the result, several relations between the pathwidth of a biconnected outerplanar graph and its dual are established.

1 Introduction

Much research has been done to compute the pathwidth of graphs. The notion of pathwidth first appeared in the theory on graph minors by Robertson and Seymour [8], and is equivalent to several other graph theoretic notions, e.g., vertex separation number, interval thickness, node search number. See [2,1] for overviews.

In this paper, we consider the problem to approximate the pathwidth of outerplanar graphs. In [3], it was shown that the pathwidth of graphs with bounded treewidth can be computed in polynomial time. As outerplanar graphs have treewidth two, we know that pathwidth is polynomial time computable for outerplanar graphs. However, the exponent in the running time of this algorithm is rather large — already one step in the algorithm requires to work with sets of size $O(n^{11})$. In this paper, we give a linear algorithm, which approximates the pathwidth of a 2-connected outerplanar graph with multiplicative factor two. Our algorithm is based on structural results on the relation between the

* The work of this author was done in part while he was at the Centro de Modelamiento Matemático, Universidad de Chile and UMR 2071-CNRS, supported by FONDAP and while he was a visiting postdoc at DIMATIA-ITI partially supported by GAČR 201/99/0242 and by the Ministry of Education of the Czech Republic as project LN00A056.
** The research of this author was partially supported by EC contract IST-1999-14186: Project ALCOM-FT (Algorithms and Complexity - Future Technologies).

A. Brandstädt and V.B. Le (Eds.): WG 2001, LNCS 2204, pp. 166–176, 2001.
© Springer-Verlag Berlin Heidelberg 2001

pathwidth of a 2-connected outerplanar graph and its dual, which are interesting in their own right. This 'dual' relation combined with the results of Ellis et al. [4] that the pathwidth of trees can be computed in linear time are the main ingredients of our algorithm. Also, we show how to construct the corresponding path decomposition in $O(n \log n)$ time.

In [5], Govindran et al. give an $O(n \log n)$ time algorithm for approximating the pathwidth of an outerplanar graph with a multiplicative factor of three. We improve upon this paper for biconnected outerplanar graphs.

2   Definitions and Notations

We use the following notations: $G = (V, E)$ is an undirected and finite graph with vertex set V and the edge set E which is assumed to be without self-loops or parallel edges. A *plane graph* is a particular drawing of a planar graph in the plane without crossings. An *outerplane graph* is a planar embedding of an outerplanar graph with every vertex on the exterior face. Edges of an outerplane graph that are not on the boundary of the exterior face are called *internal*. If $G = (V, E)$ is a plane graph then $G^* = (V^*, E^*)$ denotes its geometric dual. The *weak dual* of a plane graph G is the graph obtained from the dual G^* by deleting the vertex corresponding to the exterior face of G.

Observation 1 *The weak dual of an outerplane graph is a forest.*

Observation 2 *The weak dual of a 2-connected outerplane graph is a tree.*

The notion of pathwidth was introduced by Robertson and Seymour [8]. (See [2] and [7] for surveys.) A *path decomposition* of a graph $G = (V, E)$ is a sequence $(X_1, X_2, \ldots, X_r)$ of subsets of V, such that

- $\bigcup_{1 \le i \le r} X_i = V$.
- For all $\{v, w\} \in E$, there is an i, $1 \le i \le r$, with $v, w \in X_i$.
- For all $1 \le i_0 < i_1 < i_2 \le r$, $X_{i_0} \cap X_{i_2} \subseteq X_{i_1}$.

The *width* of path decomposition $(X_1, X_2, \ldots, X_r)$ is $\max_{1 \le i \le r} |X_i| - 1$. The *pathwidth* of a graph is the minimum width over its path decompositions.

The notion of treewidth is strongly related to the notion of pathwidth. In this paper, we use a variant of this notion, which we call *semi treewidth*.

A *semi tree decomposition* of a graph $G = (V, E)$ is a pair $(\{X_i \mid i \in I\}, T = (I, F))$, $\{X_i \mid i \in I\}$ a collection of subsets of V and $T = (I, F)$ a tree, such that

- $\bigcup_{i \in I} X_i = V$.
- For all $\{v, w\} \in E$, there is an $i \in I$ with $v, w \in X_i$, or there are nodes $i_0, i_1 \in I$ with $v \in X_{i_0}$, $w \in X_{i_1}$, and i_0 and i_1 are adjacent in T.
- For all $i_0, i_1, i_2 \in I$, if i_1 is on the path in T from i_0 to i_2, then $X_{i_0} \cap X_{i_2} \subseteq X_{i_1}$.

The *width* of a semi tree decomposition $(\{X_i \mid i \in I\}, T = (I, F))$ is $\max_{i \in I} |X_i| - 1$, and the *semi treewidth* of a graph is the minimum width over its semi tree decompositions.

The definition of tree decomposition is obtained by replacing the second condition in the definition above by

- For all $\{v, w\} \in E$, there is an $i \in I$ with $v, w \in X_i$.

The treewidth of a graph is the minimum width over its tree decompositions. Notice that *path decomposition* of a graph G can be defined as a tree decomposition with a tree T being path.

3   Pathwidth of Outerplane Graphs

The main purpose of this section is to prove the following theorem.

Theorem 1. *Let G be a 2-connected outerplane graph without loops and multiple edges, and let G^* be the dual of G. Then:*

$$\mathrm{pw}(G^*) \leq \mathrm{pw}(G) \leq 2\,\mathrm{pw}(G^*) + 2.$$

Proof. By Lemma 2, $\mathrm{pw}(G^*) \leq \mathrm{pw}(G)$. By Lemma 3, there is a planar inner triangulation H of G such that $\mathrm{pw}(H^*) \leq \mathrm{pw}(G^*) + 1$. Notice that $\mathrm{pw}(G) \leq \mathrm{pw}(H)$. Applying Lemma 7 for H^* we have that $\mathrm{pw}(H) \leq 2\,\mathrm{pw}(H^*)$. So we have:

$$\mathrm{pw}(G^*) \leq \mathrm{pw}(G) \leq \mathrm{pw}(H) \leq 2\,\mathrm{pw}(H^*) \leq 2\,\mathrm{pw}(G^*) + 2.$$

$\square$

The lemmas needed for the proof above will follow in the remainder of this section. We need the following fact about pathwidth of trees.

Theorem 2 (Ellis et al. [4]).

1. *Every tree T of pathwidth $k + 1$, $k \geq 1$, has a vertex u such that the forest $T \setminus \{u\}$ has at least three connected components of pathwidth $\geq k$;*
2. *For any tree T, $\mathrm{pw}(T) \leq k + 1$, $k \geq 1$, if and only if there is a path P such that every connected component of the forest $T \setminus V(P)$ has pathwidth $\leq k$.*

Lemma 1. *Let $G^* = (V^*, E)$ be the dual and let $T = (V_T, E_T)$ be the weak dual of a 2-connected outerplane graph $G = (V, E)$. Then $\mathrm{pw}(T) + 1 = \mathrm{pw}(G^*)$.*

Proof. Let $v \in V^*$ be the vertex corresponding to the exterior face of G, i.e., $T = G^* \setminus \{v\}$.

Let us prove that $\mathrm{pw}(G^*) \geq \mathrm{pw}(T) + 1$. Suppose that $\mathrm{pw}(T) = k$. Then by Theorem 2 there is a vertex $u \in V_T$ such that at least three branches at u have pathwidth $\geq k - 1$. Let T_1, T_2, T_3 be these branches.

Let $(X_1, X_2, \ldots, X_r)$ be a path decomposition of G^*. In this path decomposition there are sets X_{i_j} containing at least k vertices of T_j, $j \in \{1, 2, 3\}$, as the pathwidth of each T_j is at least $k - 1$. Without loss of generality, we suppose that $i_1 < i_2 < i_3$. Choose $a \in X_{i_1} \setminus X_{i_2}$ and $b \in X_{i_3} \setminus X_{i_2}$. Then X_{i_2} is an (a, b)-separator. But there are at least two vertex disjoint (a, b)-paths in G^* containing no vertex of T_2 (one passing through u and one through v.) Hence $\mathrm{pw}(G^*) \geq (k + 2) - 1 = k + 1$.

The proof of $\mathrm{pw}(G^*) \leq \mathrm{pw}(T) + 1$ is obvious because $T = G^* - v$.   □

Lemma 2. *Let G be a 2-connected outerplane graph without loops and multiple edges. Then $\mathrm{pw}(G^*) \leq \mathrm{pw}(G)$.*

Proof. Let $T = (V_T, E_T)$ be the weak dual of G. By Lemma 1, $\mathrm{pw}(T) = \mathrm{pw}(G^*) - 1$. We now will show that $\mathrm{pw}(T) \leq \mathrm{pw}(G) - 1$, using induction to the pathwidth of T. In the case $\mathrm{pw}(T) = 1$ the result clearly holds.

Suppose now that for every 2-connected outerplane graph having a weak dual of pathwidth $\leq k$, the lemma is correct.

Let G be a 2-connected outerplane graph and T be its weak dual with $\mathrm{pw}(T) = k + 1$. Then by Theorem 2 there is a vertex $u \in V_T$ such that the graph $T \setminus \{v\}$ has at least three components T_1, T_2, T_3 of pathwidth $\geq k$. Let H_1, H_2, H_3 be the subgraphs of G having T_1, T_2, T_3 as weak duals. These graphs are 2-connected and outerplane and by induction hypothesis $\mathrm{pw}(H_i) \geq k + 1$, $i \in \{1, 2, 3\}$.

Because G is a 2-connected outerplane graph without loops and multiple edges, the face of G corresponding to vertex u is bounded by a cycle C of length ≥ 3. See Figure 1 for a schematic diagram. Notice that for every $i \in \{1, 2, 3\}$ the subgraph H_i has some vertices of C and there is a path p_i in C from H_j to H_k avoiding H_i, $i \neq j \neq k$. Let $(X_1, X_2, \ldots, X_k)$ be a path decomposition of G^*. In this path decomposition there are sets X_{i_j} containing at least $k + 2$ vertices of H_j, $j \in \{1, 2, 3\}$. We may assume, without loss of generality, that $i_1 < i_2 < i_3$. Then X_{i_2} separates X_{i_1} and X_{i_3}. Therefore X_{i_2} contains a vertex of p_2 and $|X_{i_2}| \geq k + 2 + 1 = k + 3$.   □

A *triangulation* of a plane graph G is a maximal plane supergraph, *i.e.* a plane supergraph of G with every face (including the exterior face) a triangle.

An *inner triangulation* of a plane graph G is a plane supergraph of G, such that all interior faces are a triangle, i.e., we allow the exterior face not to be a triangle, and no edges are added to the outerface of G. Notice that every inner triangulation is a maximal outerplane graph.

Lemma 3. *Let G be a 2-connected outerplane graph without loops and multiple edges. Then there exists an inner triangulation H of G such that $\mathrm{pw}(G^*) \geq \mathrm{pw}(H^*) - 1$.*

Proof. Consider the following *face split* operation on plane graphs: take an interior face that is not a triangle, and add an edge between two non-adjacent vertices of the face. Clearly, when we repeat the face split operation until it is

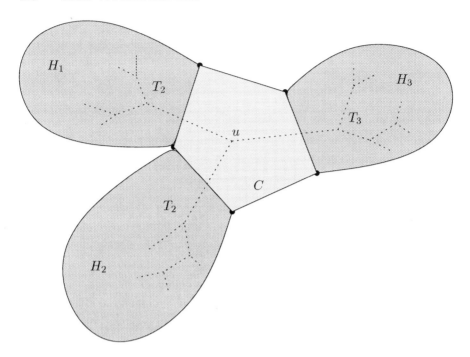

Fig. 1. Illustration to the proof of Lemma 2

no longer possible, we end up with an inner triangulation of the original graph. It remains to be shown that we can do this such that the pathwidth of the dual graphs do not change.

The face split operation on G corresponds to the following operation on the dual G^*, or, similarly, to the weak dual T. Take a vertex v of degree at least 4 in G^*. Let $N[v]$ be the set of the vertices that are adjacent to v. Assume a clockwise ordering of the neighbors of v, $v_1, v_2, \ldots, v_s$. (The ordering is forced by the embedding, up to a cyclic shift.) Now, partition the set $N[v]$ into two sets M and N, where M consists of the vertices in some consecutive part of the ordering, i.e., $M = \{v_i, v_{i+1}, \ldots, v_j\}$, $1 \leq i < j \leq s$. M and N have size at least two. Now, transform G^* as follows: delete v with all its incident edges, add new vertices u and w with an edge $\{u, w\}$, and make u adjacent to all vertices of M and w adjacent to all vertices of N. Notice that the degree of u in new graph is $|N| + 1$ and the degree of w is $|M| + 1$. We say that the result of this transformation is obtained from G^* by a *vertex splitting* of v. A graph H^* is said to be a *vertex split* of G^* if H^* is obtained from G^* by a sequence of vertex splittings.

Notice that if $G = (V, E)$ is an outerplane graph and $F^* = (V_F^*, E_F^*)$ is a vertex split of $G^* = (V^*, E^*)$ then $F = (V_F, E_F)$ is outerplane and $V_F = V$, $E_F \supseteq E$.

Let T be a weak dual tree of G. We wish to prove first that there is a split T_S of T such that $\mathrm{pw}(T) \geq \mathrm{pw}(T_S) - 1$ and $\Delta(T_S) = 3$.

By Theorem 2, every tree of pathwidth $\leq k+1$ has a path P such that every branch at this path (connected component of $T \setminus V(P)$) has pathwidth at most k. Using this fact one can find the split T_S easily. In fact, choose the path P as in Theorem 2 and split (if necessarily) the vertices of P. Since the new path has the same branches, such splittings do not increase the pathwidth. Then we can split the branches of P recursively, unless such a branch consists of a single vertex. Splitting a branch that is a single vertex means that the vertex is replaced by a path: this increases the pathwidth of the branch by one.

If we do the corresponding face splits to G, then we obtain the desired inner triangulation H of G. In fact, by Lemma 1, $\mathrm{pw}(G^*) = \mathrm{pw}(T) + 1$ and $\mathrm{pw}(H^*) = \mathrm{pw}(T_S) + 1$. Therefore $\mathrm{pw}(G^*) = \mathrm{pw}(T) + 1 \geq \mathrm{pw}(T_S) = \mathrm{pw}(H^*) - 1$.   □

Note that a graph remains outerplanar when we apply an inner triangulation.

Lemma 4. *Let G_0 be a graph, and let G_1 be obtained from G_0 by removing all vertices of degree two whose neighbors are adjacent. Then $\mathrm{pw}(G_1) \leq \mathrm{pw}(G_0) \leq \mathrm{pw}(G_1) + 1$.*

Proof. As G_1 is a subgraph of G_0, clearly, $\mathrm{pw}(G_1) \leq \mathrm{pw}(G_0)$.

Suppose we have a path decomposition $(X_1, X_2, \ldots, X_r)$ of G_1. For each vertex v in G_0 with degree two whose neighbors w, x are adjacent, find a set X_i with $w, x \in X_i$ and add after X_i in sequence of the path decomposition a set $X_i \cup \{v\}$. This gives a path decomposition of G_0 whose width is at most one larger than the given path decomposition of G_1.   □

Lemma 5. *Let H be a 2-connected inner triangulated outerplane graph. Let $T = (V_T, E_T)$ be the weak dual of H. Let H^- be the graph, obtained by removing all vertices of degree two from H. Then there is a semi tree decomposition $(\{X_i \mid i \in V_T\}, T)$ of H^- with width 1.*

Proof. Choose an arbitrary leaf node v_0 from T, and view T as a rooted tree with root v_0. We take $X_{v_0} = \emptyset$. For all nodes $w \neq v_0$, consider the edge from w to its parent in T. This edge is dual (crosses in the diagram) an edge, say $\{y, z\}$ from H. Then take $X_w = \{y, z\}$.

We claim that $(\{X_i \mid i \in V_T\}, T)$ defined in this way is a semi tree decomposition of H^-. We will verify the second and third condition of semi tree decomposition; the first then follows directly from the second.

Consider an edge $\{y, z\}$ from H^-. If $\{y, z\}$ is an internal edge, then it crosses an edge say $\{v, w\}$ from T. Suppose v is the parent of w in T. Then, by construction, $\{y, z\} = X_w$. Now, suppose $\{y, z\}$ is not an internal edge. Suppose $\{y, z\}$ is adjacent to an internal face, represented by a vertex v from T. As the graph is triangulated, this face must also be a triangle. The face must be adjacent to two other internal faces in H otherwise y or z would have degree two in H, and hence $\{y, z\}$ would not be an edge in H^-. Say these faces are represented by nodes w and x. As the root of T is a node of degree one in T, either w or x is a child of v in T. Suppose w is a child of v in T. Then $y \in X_v$ and $z \in X_w$ or $z \in X_v$ and $y \in X_w$. (See diagram 2.)

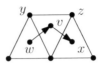

Fig. 2. Illustration to the proof of Lemma 5

Finally, for a node z in H^-, note that the edges from T that cross an edge with z as endpoint form a path in T. By the choice of a leaf node as root of T, this is a directed path and hence the collection of sets X_v that z induces a path in T.   □

Lemma 6. *Let (X, T) be a semi tree decomposition of a graph $G = (V, E)$ with width k. Then,*

$$\mathrm{pw}(G) \leq (k + 1)(\mathrm{pw}(T) + 1) - 1.$$

Proof. Let $(Y_1, Y_2, \ldots, Y_r)$ be a path decomposition of T of pathwidth $\mathrm{pw}(T)$. Consider the sequence $(Z_1, Z_2, \ldots, Z_r)$ with $Z_i = \bigcup_{j \in Y_i} X_i$.

We now will verify that this sequence is a path decomposition of G. First, suppose $i_1 < i_2 < i_3$, and $v \in Z_{i_1} \cap Z_{i_3}$. Hence, $v \in X_{j_1}$ for some $j_1 \in Y_{i_1}$, and $v \in X_{j_3}$ for some $j_3 \in Y_{i_3}$. By the properties of path decomposition, we have that Y_{i_2} must contain a node from the path from j_1 to j_3 in T; call this node j_2. By the definition of semi tree decomposition, $v \in X_{j_2}$, and hence $v \in Z_{i_2}$.

Next, consider an edge $\{v, w\} \in E$. If there is an edge $\{i_1, i_2\}$ in T with $v \in X_{i_1}$ and $w \in X_{i_2}$, then there must be an j, $1 \leq j \leq r$ with $i_1, i_2 \in Y_j$. Hence, $v, w \in Z_j$. The case that there is an $i \in I$ with $v, w \in X_i$ is easy.

One directly sees that the width of this path decomposition is at most $(k + 1)(\mathrm{pw}(T) + 1) - 1$.   □

Lemma 7. *Let H be a 2-connected inner triangulated outerplane graph with dual H^*. Then $\mathrm{pw}(H) \leq 2 \cdot \mathrm{pw}(H^*)$.*

Proof. Let H^- be the graph obtained by removing the nodes of degree two from H. Note that, as H is triangulated, the neighbors of every vertex of degree two in H are adjacent. By Lemma 4, $\mathrm{pw}(H) \leq \mathrm{pw}(H^-) + 1$. Let T be the weak dual of H. By Lemma 5, there is a semi tree decomposition (X, T) of H^- of width 1, hence by Lemma 6, $\mathrm{pw}(H^-) \leq 2 \cdot \mathrm{pw}(T) + 1$. Now

$$pw(H) \leq \mathrm{pw}(H^-) + 1 \leq 2 \cdot \mathrm{pw}(T) + 2 = 2 \cdot \mathrm{pw}(H^*)$$

□

4 An Approximation Algorithm for Biconnected Outerplanar Graphs

In this section, we give an algorithm, that given a 2-connected outerplanar graph $G = (V, E)$ with n vertices finds a path decomposition of G that has a width that is at most two times the pathwidth of G plus 2. The algorithm uses $O(n \log n)$ time, and follows the structure of the proof, given in the previous section. Note that if we just want to have a bound on the pathwidth of G, then the algorithm can run in linear time.

Step 1: Remove loops and parallel edges. If we allow our input graph to have self-loops (edges of the form $\{v, v\}$), or parallel edges, then we just remove such edges: this does not change the pathwidth of the graph.

Step 2: Compute an outerplane embedding of G. It is well known that given a graph G^*, one can test in linear time if G^* is outerplanar, and if so, find an embedding of G^* with all vertices on the exterior face. See e.g. [6,10].

Step 3: Compute the dual graph G^* of G. It is well known, that, given a plane embedding of a planar graph, one can find its dual in linear time.

Step 4: Compute the pathwidth of the dual graph G^*. The pathwidth of the dual graph G^* can be computed in linear time, in the following way. First, we take the tree T that is obtained from G^* by removing the vertex v representing the exterior face, i.e., T is the weak dual of G. Using the algorithm of Ellis et al. [4], we can compute in linear time the pathwidth of T, and compute a path decomposition $(X_1, \ldots, X_r)$ of T of optimal width in $O(n \log n)$ time. The pathwidth of G^* is one larger, by Lemma 1, and the path decomposition $(X_1 \cup \{v\}, X_2 \cup \{v\}, \ldots, X_r \cup \{v\})$ is a path decomposition of G^* of optimal width. By Euler formula the number of vertices in G^* is $O(n)$ and the pathwidth of G^* can be computed in $O(n)$ time.

If we are satisfied with a *bound* on the pathwidth of G that is at most two times plus 2 larger than the exact bound, then, by Theorem 1 we are done, as $2 \operatorname{pw}(G^*) + 2$ is such a bound. However, if we want a *path decomposition* of G whose width at most two times plus 2 than the pathwidth of G, then more work has to be done.

Step 5: Compute an inner triangulation H of G, such that $\operatorname{pw}(H^*) \le \operatorname{pw}(G^*) + 1$. In this step, we make the proof of Lemma 3 constructive, i.e., we compute an inner triangulation H of G, such that the pathwidth of G^* equals the pathwidth of H^*, where H^* is the dual of H.

Let T be again the weak dual of G. Suppose the pathwidth of T is k.

Now, first we find a path P in T, such that all branches at this path have pathwidth at most k. This can be done in linear time, using a minor modification of the algorithm from Ellis et al. [4]. When we have the path P, the split can easily be applied. After this, we continue recursively on the branches. As the pathwidth of a tree with t vertices is $O(\log t)$, every edge of T is involved in $O(\log t)$ recursive steps, and the total time of this step becomes $O(t \log t) = O(n \log n)$.

Step 6: Compute a path decomposition for the weak dual of H. Compute the weak dual T of the inner triangulation H, computed in the previous step, and use the linear time algorithm of Ellis et. al. [4] to compute an optimal path decomposition of T.

Step 7: Compute the semi tree decomposition of H^-. Let H^- be the graph obtained by removing the vertices of degree two from H. Following construction described in the proof of Lemma 5, we can make a semi tree decomposition (X, T) of width 1, with T the weak dual of H. It is not hard to see that this step can be done in linear time.

Step 8: Make a path decomposition of H^-. The proof of Lemma 6 can be made constructive in a straightforward way; we thus obtain a path decomposition of H^- of width $2 \operatorname{pw}(T) + 3 = 2 \operatorname{pw}(H^*) + 1$ of H^-.

Step 9: Make a path decomposition of H. Finally, we have to add back the vertices with degree two, while increasing the pathwidth with at most one, as in the proof of Lemma 4. This can be done in linear time by using the following method:

Suppose that the vertices in H are numbered $v_1, \ldots, v_n$. We use also a boolean array Z with indexes $\{1, \ldots, n\}$. Initially, all $Z[i]$ are false.

1. Suppose that S is the set of vertices of degree two. Make a set P of (ordered) pairs of vertices, with for every $v \in S$, its two neighbors as two ordered pairs in P, i.e., if v's neighbors are x and y, then both (x, y) and (y, x) belong to P. Maintain pointers from and to a vertex $v \in S$ and its corresponding pair in P.
2. Radix sort P, i.e., P is sorted with respect to the lexicographic ordering. This can be done in linear time, using the standard radix sort algorithm.
3. Now, we can build an array, with the ith entry pointing to a list of vertices v for which (v_i, v) belongs to P. This can be done in linear time, using the sorted list of the previous step.
4. Next, visit the bags in the path decomposition of H^- one by one. For each bag X_i, we do the following:
   (a) for every $v_j \in X_i$, make $Z[j]$ true.
   (b) for every $v_j \in X_i \setminus X_{i-1}$ and for every $v_{j'}$ with $(v_j, v_{j'}) \in P$, check if $Z[j']$ is true. If it is true, then we found the bag for the neighbors of a vertex v, namely the one corresponding to the pair $(v_j, v_{j'})$. We remove this pair and its reversed copy $(v_{j'}, v_j)$ from the array of step 3, and add a pointer from v to bag X_i.
   (c) for every $v_j \in X_i$, make $Z[j]$ false, i.e., now all values in Z are again false.

5. Now, for every vertex v of degree two in H, we can follow the pointer from v to a bag X_i. Add after X_i a bag which contains $X_i \cup \{v\}$. Note that in this way, we never create bags that are more than one larger than a bag in H^-.

When (v, w) and (w, v) belong to P, then consider the first bag that contains both v and w. In this bag, we either will consider v or w in step 4b, and hence the vertex of degree 2 associated with this pair will be pointing towards this bag.

The time is linear in the size of the path decomposition plus the size of H. The pathwidth of a tree with k vertices is $O(\log k)$ and by construction the number of vertices in a path decomposition is $O(n)$. Therefore this step can be performed in $O(n \log n)$ time.

Step 10: Obtain a path decomposition of G. The path decomposition of H obtained in Step 9 is also the path decomposition of G and the width of this decomposition is at most $2 \, \mathrm{pw}(G) + 2$.

5 Concluding Remarks

One of the most interesting question about pathwidth of outerplanar graphs we left open is the existence of fast practical exact algorithms or algorithms approximating pathwidth of outerplanar graphs with an additive factor. One of the possible ways in obtaining such algorithms is the proof of the following conjecture.

Conjecture 1. There is a constant c such that for any 2-connected outerplanar graph G without loops and multiple edges $\mathrm{pw}(G^*) \leq \mathrm{pw}(G) \leq \mathrm{pw}(G^*) + c$.

Moreover, we suggest that

Conjecture 2. For any 2-connected planar graph G without loops and multiple edges $\mathrm{pw}(G^*) - 1 \leq \mathrm{pw}(G) \leq \mathrm{pw}(G^*) + 1$.

Path decompositions of trees and of outerplanar graphs with n vertices can have $\Theta(n)$ bags X_i of size $\Theta(\log n)$, thus, when using a straightforward representation one may need already $\Omega(n \log n)$ time just for writing the output. However, more compact representations of path decompositions exists, e.g., mark for each vertex the first and last bag it belongs to, or one can use the equivalent notion of vertex separations. These representations have size linear in the number of vertices. As Skodinis [9] has shown that (with such representations) one can find an optimal path decomposition of a given tree in linear time, we conjecture that the algorithm of Section 4 can be made to run in linear time, but there are several unresolved matters in this, and we leave this as an open problem. As a side remark, we note that the algorithm of Govindran et al. [5] can be made to run in linear time, using Skodinis' algorithm and a corresponding representation of the path decompositions.

References

1. H. L. BODLAENDER, *Treewidth: Algorithmic techniques and results*, in Proceedings 22nd International Symposium on Mathematical Foundations of Computer Science, MFCS'97, Lecture Notes in Computer Science, volume 1295, I. Privara and P. Ruzicka, eds., Berlin, 1997, Springer-Verlag, pp. 19–36.

2. ——, *A partial k-arboretum of graphs with bounded treewidth*, Theor. Comp. Sc., 209 (1998), pp. 1–45.

3. H. L. BODLAENDER AND T. KLOKS, *Efficient and constructive algorithms for the pathwidth and treewidth of graphs*, J. Algorithms, 21 (1996), pp. 358–402.

4. J. A. ELLIS, I. H. SUDBOROUGH, AND J. TURNER, *The vertex separation and search number of a graph*, Information and Computation, 113 (1994), pp. 50–79.

5. R. GOVINDRAN, M. A. LANGSTON, AND X. YAN, *Approximating the pathwidth of outerplanar graphs*, Inform. Proc. Letters, 68 (1998), pp. 17–23.

6. S. L. MITCHELL, *Linear algorithms to recognize outerplanar and maximal outerplanar graphs*, Inform. Proc. Letters, 9 (1979), pp. 229–232.

7. R. H. MÖHRING, *Graph problems related to gate matrix layout and PLA folding*, in Computational Graph Theory, Comuting Suppl. 7, E. Mayr, H. Noltemeier, and M. Sysło, eds., Springer Verlag, 1990, pp. 17–51.

8. N. ROBERTSON AND P. D. SEYMOUR, *Graph minors. I. Excluding a forest*, J. Comb. Theory Series B, 35 (1983), pp. 39–61.

9. K. SKODINIS, *Computing optimal linear layouts of trees in linear time*, in Proceedings 8th Annual European Symposium on Algorithms, ESA'00, Lecture Notes in Computer Science, volume 1879, M. Paterson, ed., Berlin, 2000, Springer-Verlag, pp. 403-414.

10. M. M. SYSŁO, *Characterisations of outerplanar graphs*, Disc. Math., 26 (1979), pp. 47–53.

On the Monotonicity of Games Generated by Symmetric Submodular Functions

Fedor V. Fomin[*,1] and Dimitrios M. Thilikos[**,2]

[1] Faculty of Mathematics and Mechanics, St.Petersburg State University,
Bibliotechnaya sq.2, St.Petersburg, 198904, Russia
fomin@gamma.math.spbu.ru
[2] Departament de Llenguatges i Sistemes Informàtics,
Universitat Politècnica de Catalunya, Campus Nord – Mòdul C5,
c/Jordi Girona Salgado 1-3,
08034 Barcelona, Spain
sedthilk@lsi.upc.es

Abstract. Submodular functions have appeared to be a key tool for proving the monotonicity of several graph searching games. In this paper we provide a general game theoretic framework able to unify old and new monotonicity results in a unique min-max theorem. Our theorem, provides a game theoretic analogue to a wide number of graph theoretic parameters such as linear-width and cutwidth.

1 Introduction

A considerable part of graph theory is oriented to the following general problem: Given a non-trivial graph property P, find a complete characterization of the graphs that do not satisfy P. In general, such a characterization is achieved by describing some "forbidding" structure whose existence in G obstructs P from being satisfied. As a first example we mention the Kuratowski theorem asserting that a graph is non-planar iff it topologically contains an obstructing structure of two forbidden graphs. Many graph theoretic parameters have been characterized by their obstructing analogue and this characterization is typically achieved by a so called min-max theorem (for a good source of min-max theorems, see [5]). Several examples of such characterizations emerge from the work of Robertson and Seymour on their Graph Minors series. As a sample, we mention the characterization of treewidth via screens [21], of branchwidth via tangles [20],

---

[*] The work of this author was done in part while he was at the Centro de Modelamiento Matemático, Universidad de Chile and UMR 2071-CNRS, supported by FONDAP and while he was a visiting postdoc at DIMATIA-ITI partially supported by GAČR 201/99/0242 and by the Ministry of Education of the Czech Republic as project LN00A056.

[**] The work of the second author was supported by the EU project ALCOM-FT (IST-99-14186), by the Spanish CYCIT TIC-2000-1970-CE, and by the Ministry of Education and Culture of Spain, Grant number MEC-DGES SB98 0K148809.

A. Brandstädt and V.B. Le (Eds.): WG 2001, LNCS 2204, pp. 177–188, 2001.
© Springer-Verlag Berlin Heidelberg 2001

of pathwidth via blockages [3], and of carving-width via antipodalities [22] (see also [15,11]).

In many cases, it appears that min-max theorems are useful on the study of graph searching games. Roughly, a graph searching game has as objective to capture an omniscient fugitive residing on the vertices or/and the edges of a graph by systematically moving a specific number on searchers on it. Such a search strategy can in general allow "recontamination" in the sense that it can let the fugitive occupies parts of G that have already been searched. If this is not the case the strategy is called *monotone*. The "monotonicity question" for a graph searching game asks whether "allowing recontamination" makes the capturing of the fugitive easier than in the case where the player restricts his/her attention *only* to monotone strategies.

It frequently appears that the monotonicity question can be answered with the use of a suitable min-max theorem. The reason is that, in the majority of the cases, the minimum number of searchers required to capture the fugitive is equivalent to some known graph theoretic parameter. The existence of a min-max theorem for this parameter implies the existence of a forbidden structure that, in turn, indicates an escape strategy for the fugitive no matter whether the searching strategy is monotone or not. That way, the min-max theorems for treewidth and pathwidth, developed in [21] and [3], were the cornerstones for proving the monotonicity of the corresponding graph searching variants (see also [7]).

Our paper is motivated and constitutes an extension of the ideas in the proofs of the monotonicity of the agile fugitive search games examined in [13,14,18,23,2], [4,3,16] as well as the proofs of the min-max theorems in [22] and [20]. Our main observation is that the kernel argument of all these proofs is based on the fact that, in any game variant, the cost of the search can be expressed by a connectivity function that is a nonnegative-valued function on the set of subsets of a set M that is invariant over complement and satisfies the submodular property.

In this paper we show how a graph parameter can be generated by a connectivity function and we prove a general min-max theorem for it. Moreover, we define for any such parameter a one-player conquest game and we use our min-max theorem in order to prove its monotonicity. Our game framework is general in the sense that it provides a big variety of games depending on the choice of the connectivity function α that generates it. In particular, for suitable choices of α, it provides obstruction characterizations, game counterparts, and monotonicity proofs for the parameters of linear-width, cutwidth and their extensions. Finally, our general min-max theorem implies in a uniform way the monotonicity proofs of all the agile fugitive search games developed so far in [13,18,23,4,16].

To illustrate the main motivation of our research let us give a simple example of an "expansion" game. Suppose that we have a set of countries subject to join some organization. At every moment of time we can either add a bounded number of countries to the union or expel an arbitrary number of countries. Adding countries and keeping them in the union needs some resources, say for guarding its border from the countries outside the union. The question is how to add all the countries using the minimal amount of resources.

There is a number of interesting questions concerning this game. One of them is the "monotonicity question": can it be useful at some step to expel countries from the union? Another is the "min-max question": what kind of structure in a map provides necessary and sufficient conditions for obstructing the intended expansion? In this paper we provide the answers to both questions for a more general version of expansion game.

The paper is organized as follows. The main min-max theorem is proved in Section 2. In Section 3 we present our general game, and in Section 4 we present the consequences of the min-max theorem on its variants. In Section 5 we end up with further examples, remarks, and open problems.

2   A Min-max Theorem for Connectivity Functions

Given a finite set M, a function α mapping the subsets of M to integers is called a *connectivity function* for M [20] if the following two conditions are satisfied.

$$\forall_{A \subseteq M} \; \alpha(A) = \alpha(\overline{A}) \text{ (we denote } \overline{A} = M - A), \tag{1}$$

$$\forall_{A,B \subseteq M} \; \alpha(A \cup B) + \alpha(A \cap B) \leq \alpha(A) + \alpha(B). \tag{2}$$

Notice that for any $X \subseteq M$, $\alpha(X) \geq \alpha(\varnothing)$ because $2\alpha(X) = \alpha(X) + \alpha(M - X) \geq \alpha(\varnothing) + \alpha(M) = 2\alpha(\varnothing)$. It is more convenient for our purposes to think that $\alpha(\varnothing) = 0$ (replacing $\alpha'(X) = \alpha(X) - \alpha(\varnothing)$ if it is not the case).

For a given a finite set M, let $\mathcal{O}$ be a set of subsets of M and let α be a connectivity function for M. A sequence $\mathcal{A} = (A_1, \ldots, A_r)$ is a (k, m)-*expansion* in $\mathcal{O}$ (for M with respect to a connectivity function α) if

$$\forall_{i,1 \leq i \leq r} \; A_i \subseteq M \text{ and } \alpha(A_i) \leq k, \tag{3}$$

$$A_1, \overline{A_r} \in \mathcal{O}, \tag{4}$$

$$\forall_{i,1 \leq i \leq r-1} \; |A_{i+1} - A_i| \leq m. \tag{5}$$

If, additionally,

$$\forall_{i,1 \leq i \leq r} \; A_i \subseteq A_{i+1} \tag{6}$$

then we say that $\mathcal{A}$ is *monotone*. Finally, a (k, m)-expansion in $\{\varnothing\}$, i.e. expansion with $A_1 = \varnothing$ and $A_r = M$, is referred to as *complete*.

For facilitating the notation, in this section we will assume that all the expansions are defined with respect to a fixed connectivity function α. The following lemma uses the ideas of Bienstock-Seymour's monotonicity proof for graph crusades [4].

Lemma 1. *If there is a (k, m)-expansion in $\mathcal{O}$ for M then there is a monotone (k, m)-expansion in $\mathcal{O}$ for M.*

Proof. The proof of this lemma is omitted due to space restriction. It can be found in the full version of the paper [10]. □

Lemma 2. *Let $\mathcal{A} = (A_1, A_2, \ldots, A_r)$ be a monotone (k, m)-expansion for M. Then $\overline{\mathcal{A}} = (\overline{A}_r, \overline{A}_{r-1}, \ldots, \overline{A}_1)$ is also a monotone (k, m)-expansion for M.*

Proof. Since $\overline{\overline{A}}_1 = A_1$ and $\alpha(\overline{A}_i) = \alpha(A_i)$, we have that $\overline{\mathcal{A}}$ satisfy (3) and (4). For $i \in \{1, \ldots, r-1\}$, $A_{i+1} \supseteq A_i$ implies that $\overline{A}_i - \overline{A}_{i+1} = A_{i+1} - A_i$ and $\overline{A}_{i+1} \subseteq \overline{A}_i$. Therefore, (5) and (6) are also satisfied. □

For any integer k, we define a (k, m)-*obstacle* for M as the set $\mathcal{O}$ such that

(o1) Each $A \in \mathcal{O}$ is a subset of M with $\alpha(A) \leq k$.
(o2) If $A \in \mathcal{O}$, $B \subseteq A$ and $\alpha(B) \leq k$ then $B \in \mathcal{O}$;
(o3) If $A, B, C \subseteq M$, $A \cap B = \varnothing$, $\alpha(A) \leq k$, $\alpha(B) \leq k$, $|C| \leq m$, and $A \cup B \cup C = M$ then $(A \in \mathcal{O} \wedge B \notin \mathcal{O})$ or $(A \notin \mathcal{O} \wedge B \in \mathcal{O})$.

The aim of this section is to prove that the existence of a (k, m)-obstacle for M, obstructs the existence of a complete and monotone (k, m)-expansion for M and vice versa. In particular, we will prove the following min-max theorem.

Theorem 1. *Let M be a finite set and α a connectivity function on M. Then the following are equivalent.*

(i) *There exists no (k, m)-obstacle for M.*
(ii) *There exists a complete (k, m)-expansion for M.*
(iii) *There exists a complete and monotone (k, m)-expansion for M.*

For the proof of Theorem 1, we need first to prove Lemma 3 below. A set $\mathcal{O}$ is *partial (k, m)-obstacle* for M if it satisfies (o1) and (o2) and if there is no (k, m)-expansion for it. The next lemma is an analog of the blockage theorem (2.5) in [3].

Lemma 3. *Every partial (k, m)-obstacle for M is a subset of a (k, m)-obstacle for M.*

Proof. Let $\mathcal{O}$ be a partial (k, m)-obstacle for M. We assume that every $\mathcal{O}'$, where $|\mathcal{O}'| > |\mathcal{O}|$ and satisfying (o1) and (o2), either is not a partial (k, m)-obstacle for M, or is a subset of a (k, m)-obstacle for M. As base of the induction we can consider a set $\mathcal{O}_0$ containing any subset A of M, where $\alpha(A) \leq k$. Indeed the set $\mathcal{O}_0$ is not a partial (k, m)-obstacle. In fact, $\varnothing, M \in \mathcal{O}_0$ because $\alpha(\varnothing) = \alpha(M) = 0 \leq k$ and $(\varnothing)$ is a (k, m)-expansion in $\mathcal{O}_0$.

If $\mathcal{O}$ satisfies (o1), (o2) and (o3) then $\mathcal{O}$ is a (k, m)-obstacle for M and the lemma holds. Suppose that $\mathcal{O}$ does not satisfy (o3). Then there exist $A_1, A_2, C \subseteq M$ such that

- $\alpha(A_1) \leq k$, $\alpha(A_2) \leq k$, $|C| \leq m$;
- $A_1 \cup A_2 \cup C = M$ and $A_1 \cap A_2 = \varnothing$;

- $A_1, A_2 \in \mathcal{O}$ or $A_1, A_2 \notin \mathcal{O}$.

For our purpose it is more convenient to work with complements of A_1 and A_2. Let $T_1 = \overline{A}_1$ and $T_2 = \overline{A}_2$. Notice that

$$\alpha(T_1) \le k \text{ and } \alpha(T_2) \le k, \tag{7}$$

$$T_1 \cup T_2 = M, \tag{8}$$

$$|T_1 \cap T_2| \le m. \tag{9}$$

We claim that

$$T_1, T_2 \notin \mathcal{O}. \tag{10}$$

Indeed, notice first that, from (8), $A_1 \cap A_2 = \varnothing$ and thus $T_1 \supseteq A_2$. By (o2), if $T_1 \in \mathcal{O}$ then $A_2 \in \mathcal{O}$. Since $A_2 \in \mathcal{O}$, we have that $A_1 \in \mathcal{O}$. Then (A_1) is a (k, m)-expansion in $\mathcal{O}$, which is impossible. The proof of $T_2 \notin \mathcal{O}$ is similar.

We now choose T_1, T_2 satisfying (7)–(10) such that $|T_1|$ is minimal. We claim that

$$\text{If } X \subseteq T_1 \text{ and } \alpha(X) \le k \text{ then either } X \in \mathcal{O}, \text{ or } X = T_1, \text{ or } \overline{T}_1 \in \mathcal{O}. \tag{11}$$

We assume that $X \notin \mathcal{O}$ and $X \ne T_1$ and we will first prove that $\overline{X} \in \mathcal{O}$. For this, we will show that if $\overline{X} \notin \mathcal{O}$, then $X, \overline{X}$ satisfy (7)–(10), contradicting the choice of T_1, T_2. Indeed, (7) follows from $\alpha(X) \le k$ and (1); (8) is obvious and (9) follows as $X \cap \overline{X} = \varnothing$. We now conclude the proof of (11) noticing that $\overline{T}_1 \subset \overline{X}$ and (o2) gives $\overline{T}_1 \in \mathcal{O}$ as required.

For $i = 1, 2$ we define $\mathcal{O}_i$ as the set of all $X \subseteq T_i$ with $\alpha(X) \le k$. Notice that, for $i = 1, 2$, $T_i \notin \mathcal{O}$ implies $|\mathcal{O} \cup \mathcal{O}_i| > |\mathcal{O}|$. Therefore, the result follows from the induction hypothesis if we show that for some $i = 1, 2$, $\mathcal{O} \cup \mathcal{O}_i$ is a partial (k, m)-obstacle for M. Clearly, (o1) and (o2) are satisfied for both $\mathcal{O} \cup \mathcal{O}_i, i = 1, 2$ and therefore, it remains to prove that, for some $i = 1, 2$, there is no (k, m)-expansion in $\mathcal{O} \cup \mathcal{O}_i$. Assume, towards a contradiction, that, for $i = 1, 2$ there are (k, m)-expansions in $\mathcal{O} \cup \mathcal{O}_1$ and $\mathcal{O} \cup \mathcal{O}_2$ and, by Lemma 1, we can also assume that they are monotone.

Let us prove that

$$\text{There is a monotone } (k, m)\text{-expansion } \mathcal{X} = (X_1, X_2, \ldots, X_r) \text{ in } \mathcal{O}_1 \tag{12}$$
$$\text{with } X_1 = T_1 \text{ and } \overline{X}_r \in \mathcal{O}.$$

If $\overline{T}_1 \in \mathcal{O}$ then $\mathcal{X} = (T_1)$ is a (k, m)-expansion in $\mathcal{O}_1$ satisfying (12).

Suppose that $\overline{T}_1 \notin \mathcal{O}$. Let $\mathcal{X} = (X_1, X_2, \ldots, X_r)$ be a monotone (k, m)-expansion in $\mathcal{O} \cup \mathcal{O}_1$. Then $X_1, \overline{X}_r \in \mathcal{O}_1$. By Lemma 2, $\overline{\mathcal{X}} = (\overline{X}_r, \overline{X}_{r-1}, \ldots, \overline{X}_1)$ is also monotone (k, m)-expansion in $\mathcal{O} \cup \mathcal{O}_1$.

One of the sets X_1 and $\overline{X}_r$ is not in $\mathcal{O}$ because there is no (k, m)-expansion in $\mathcal{O}$. W.l.o.g we can assume that $X_1 \notin \mathcal{O}$. (Otherwise we can consider $\overline{\mathcal{X}}$.) As $X_1 \in \mathcal{O} \cup \mathcal{O}_1$, we have that $X_1 \in \mathcal{O}_1$ and, from the definition of $\mathcal{O}_1$, we conclude that $X_1 \subseteq T_1$. By (11), (because $X_1 \notin \mathcal{O}$ and $\overline{T}_1 \notin \mathcal{O}$) $X_1 = T_1$.

From the monotonicity of $\mathcal{X}$ it follows that $T_1 = X_1 \subseteq X_r$. We claim that either $\overline{X}_r \notin \mathcal{O}_1$ or $\overline{X}_r = \varnothing$. Indeed, if $\overline{X}_r \in \mathcal{O}_1$, then the definition of $\mathcal{O}_1$ implies that $\overline{X}_r \subseteq T_1$. Moreover, $T_1 \subseteq X_r$ implies that $\overline{X}_r \subseteq \overline{T}_1$ and $\overline{X}_r = \varnothing$ follows.

Notice now that (12) holds trivially if $\overline{X}_r = \varnothing \in \mathcal{O}$. Moreover, if $\overline{X}_r \notin \mathcal{O}_1$, then $\overline{X}_r \in \mathcal{O} \cup \mathcal{O}_1$ implies that $\overline{X}_r \in \mathcal{O}$ and this concludes the proof of (12).

Let $\mathcal{Y} = (Y_1, Y_2, \ldots, Y_s)$ be a monotone (k, m)-expansion in $\mathcal{O}_2$. Since there is no (k, m)-expansion in $\mathcal{O}$ we have that either Y_1 or $\overline{Y}_s$ is not in $\mathcal{O}$. W.l.o.g. we can assume that $Y_1 \notin \mathcal{O}$ (otherwise we replace $\mathcal{Y}$ by $\overline{\mathcal{Y}}$). Then $Y_1 \subseteq T_2$, yielding $|Y_1 - \overline{X}_1| \leq |T_2 - \overline{T}_1| \leq m$.

We now claim that $\overline{Y}_r \notin \mathcal{O}$. Indeed, if this is not the case, then

$$(\overline{X}_r, \overline{X}_{r-1}, \ldots, \overline{X}_1, Y_1, Y_2, \ldots, Y_s)$$

is a (k, m)-expansion in $\mathcal{O}$, contradicting to the fact that there are no (k, m)-expansions in $\mathcal{O}$.

Recall that $\overline{Y}_s \in \mathcal{O} \cup \mathcal{O}_2$ and, as $\overline{Y}_s \notin \mathcal{O}$, we get that $\overline{Y}_s \in \mathcal{O}_2$, thus $\overline{Y}_s \subseteq T_2$. But $Y_s \supseteq \overline{T}_2$ combined with (9), implies $|X_1 - Y_s| = |T_1 - Y_s| \leq |T_1 - \overline{T}_2| = |T_1 \cap T_2| \leq m$. As a consequence,

$$(\overline{X}_r, \overline{X}_{r-1}, \ldots, \overline{X}_1, Y_1, Y_2, \ldots, Y_s, X_1, X_2, \ldots, X_r)$$

is a (k, m)-expansion in $\mathcal{O}$. This is again a contradiction and the lemma is proved.
□

We can now proceed with the proof of Theorem 1.

Proof (Proof of Theorem 1). The fact that (ii) $\Rightarrow$ (iii) follows directly from Lemma 1.

Next, we will prove that $(iii) \Rightarrow (i)$. Suppose, in contrary that $\mathcal{O}$ is a (k, m)-obstacle for M and $\mathcal{A} = (A_1, \ldots, A_r)$ a complete and monotone (k, m)-expansion for M. Recall that from (3) we have that $\alpha(A_1) \leq k$ and from (o2), $\varnothing = A_1 \in \mathcal{O}$. (o3) now implies $A_r = M \notin \mathcal{O}$. Let i be the smallest $i, 0 \leq i < r$ such that $A_{i+1} \notin \mathcal{O}$. Let $A_{i+1} - A_i = C$ and (5) gives $|C| \leq m$. Notice that $A_i \cup C \cup \overline{A}_{i+1} = M$. Moreover, $A_i \cap \overline{A}_{i+1} = \varnothing$ and therefore (o3) implies that $A_i \notin \mathcal{O}$, a contradiction.

It remains now to prove $(i) \Rightarrow (ii)$. It is enough to show that there exists some (k, m)-expansion in $\{\varnothing\}$. Indeed, if this is not the case and since $\{\varnothing\}$ satisfies (o1) and (o2), we get that $\{\varnothing\}$ is a partial (k, m)-obstacle for M. From Lemma 3, $\{\varnothing\}$ is the subset of some (k, m)-obstacle for M, a contradiction.   □

3   A General Framework for Conquest Games

In this section we will introduce a general framework of a game on graphs, hypergraphs and sets based on the notion of (k, m)-expansions.

In particular, we assume that α is a connectivity function for some set M. The elements of M represent countries to be conquested. In the beginning, all the elements of M are considered unoccupied. The conquest proceeds in steps. Each step can either be the occupation of at most m new elements of M or the retreat from some already occupied elements of M. We define a *move* as a pair $P = (S, \omega)$, where S is a set of elements of M and ω equals 0 when these elements are removed from the set of conquested elements (retreat-move) and 1 when these elements are added to the set of conquested elements (attack-move). A *conquest strategy* for M is a sequence $\mathcal{E} = (P_1, \ldots, P_r)$ of moves. The *aggressivity* of a conquest strategy is $\max\{|S| \mid (S, 1) \in \mathcal{E}\}$, i.e. the maximal number of elements occupied at some step. Given a conquest strategy $\mathcal{E}$, we recursively define the *occupation sequence* of $\mathcal{E}$ as the sequence $(T_0, \ldots, T_r)$ where $T_0 = \varnothing$, $T_i = T_{i-1} \cup S$ if $P_i = (S, 1)$, and $T_i = T_{i-1} - S$ if $P_i = (S, 0)$. The α-*cost* (or simply *cost* when there is no doubt about α) of a conquest strategy $\mathcal{E}$ is defined as $\max_{0 \le i \le r} \alpha(T_i)$. We call a conquest strategy *monotone* if it does not contain any retreat move. We call a conquest strategy *successful* if T_r is the set of all the elements of M.

Lemma 4. *For any set M and any connectivity function α, there exists a successful (monotone) conquest strategy with α-cost k and aggressivity m if and only if there exists a complete (monotone) (k, m)-expansion for M with respect to α.*

Proof. Let $(A_1, \ldots, A_r)$ be a complete (monotone) (k, m)-expansion for M with respect to α. We construct a sequence of moves $\mathcal{E} = (P_1', P_1, \ldots, P_{r-1}', P_{r-1})$ such that for every $1 \le i \le r - 1$,

$$(P_i', P_i) = \begin{cases} ((A_i - A_{i+1}, 0), (A_{i+1} - A_i, 1)), & \text{if } \alpha(A_{i-1} \cap A_i) \le k; \\ ((A_{i+1} - A_i, 1), (A_i - A_{i+1}, 0)), & \text{if } \alpha(A_{i-1} \cup A_i) \le k. \end{cases}$$

Notice that, (2) and the fact that $\alpha(A_i) + \alpha(A_{i+1}) \le 2k$ imply that either $\alpha(A_{i-1} \cap A_i) \le k$ or $\alpha(A_{i-1} \cup A_i) \le k$. Therefore, $\mathcal{E}$ is well defined. It now follows directly from the definitions that $\mathcal{E}$ is a successful (monotone) conquest strategy with α-cost k and aggressivity m. Given now a successful (monotone) conquest strategy for G with α-cost k and aggressivity m, it is easy to see that its occupation sequence is a complete (monotone) (k, m)-expansion for $V(G)$ with respect to the function α. $\square$

Given a set M and a connectivity function α, we define a (k, m)-*ordered partition* of M as a linear ordering $(B_1, \ldots, B_r)$ where $\{B_1, \ldots, B_r\}$ is a partition of M, $\forall_{i, 1 \le i \le r} |B_i| \le m$, and $\forall_{i, 1 \le i \le r} \alpha(B_1 \cup \cdots \cup B_i) \le k$. For any set M, we define the (α, m)-*width* of M as the minimum k for which there exists a (k, m)-ordered partition of G.

We can now conclude this section with the following.

Theorem 2. *Let M be a set, α be a connectivity function for M, and k, m two integers. The following assertions are equivalent.*

1. *There exists a complete (k, m)-expansion for M with respect to the function α.*
2. *There exists a complete monotone (k, m)-expansion for M.*
3. *There exists no (k, m)-obstacle for M with respect to the function α.*
4. *The (α, m)-width of M is at most k.*

5. *There exists a successful conquest strategy for G with aggressivity m and α-cost k.*
6. *There exists a successful monotone conquest strategy for G with aggressivity m and α-cost k.*

Proof. The equivalence of (1), (2), and (3) follows directly from Theorem 1. The equivalence of (1) and (5) as well as the equivalence of (2) and (6) follows from Lemma 4. It remains to prove that (4) and (6) are equivalent. Let $(V_1, \ldots, V_r)$ be a (k, m)-ordered partition of $V(G)$. We construct $\mathcal{E} = ((S_1, 1), \ldots, (S_r, 1))$ where for $i = 1, \ldots, r$, $S_i = V_1$ and observe that $\mathcal{E}$ is a complete monotone (k, m)-expansion for $V(G)$, as required. Finally, if $\mathcal{E} = ((S_1, 1), \ldots, (S_r, 1))$ is a successful monotone conquest strategy, then it is enough to notice that $(S_1, \ldots, S_r)$ is a (k, m)-ordered partition of $V(G)$. □

4   Applications

In this section we will give some examples of games on graphs where the conquest game framework introduced in Section 2 can provide a min-max theorem.

4.1   Definitions

We give first general definitions of some well known width-type parameters for graphs.

The cutwidth of graphs has been extensively considered and emerged as a tool for the study of VLSI layouts (See [1,17] for further references). We give below a natural generalization of its definition.

Let H be a graph with the vertex set $V(H)$ and the edge set $E(H)$. For $X \subseteq V(H)$ let $\alpha_0(X)$ be the number of edges incident to vertices in X and $V(H) - X$. Let $\phi = (B_1, B_2, \ldots, B_n)$, $\max_{1 \leq i \leq n} |B_i| \leq m$, be an ordered partition of $V(H)$. For $1 \leq i \leq n$ we put $V_i = \cup_{j=1}^{i} B_j$ and

$$m\text{-cw}(H, \phi) = \max_{i \in \{1, \ldots, n\}} \alpha_0(V_i)$$

The *m-cutwidth* of H is $\min\{m\text{-cw}(H, \phi): \phi \text{ is an ordering of } V(H)\}$. So, in our terms, the m-cutwidth of a graph H is equivalent to its (α_0, m)-*width*. If in the definition above we set $m = 1$, we have the definition of *cutwidth*.

The notion of linear-width for graphs was introduced by Thomas [26]. Let G be an undirected and finite graph with vertex set $V(G)$ and edge set $E(G)$. For $X \subseteq E(G)$, let $\alpha_1(X)$ be the number of vertices incident to edges in X and $E(G) - X$. Let $\sigma = (e_1, e_2, \ldots, e_m)$ be an ordering of $E(G)$. For $i \in \{1, \ldots, m\}$ we put $E_i = \cup_{j=1}^{i} e_j$. We define

$$\text{lw}(G, \sigma) := \max_{i \in \{1, \ldots, m\}} \alpha_1(E_i),$$

and the *linear-width* of G is $\min\{\text{lw}(G, \sigma): \sigma \text{ is an ordering of } E(G)\}$. In a similar way, one can define the notion of linear-width for hypergraphs. In our terms the linear-width of G is the $(\alpha_1, 1)$-*width* of G. Certainly, it is possible to extend the linear-width to m-linear-width considering ordered partitions of edges in the fashion we did for m-cutwidth. This would define a parameter equivalent to (α_1, m)-*width*. Notice that both definitions given in this subsection can be generalized to hypergraphs and can have various extensions based on a weight assignment function for the vertices or the (hyper)edges of G. Any of these versions are equivalent to some (α, m)-width for suitable choices of M, α and m.

4.2   A Game for Cutwidth

We consider first the case where M is the vertex set of a graph G and α maps any of its subsets S to the number of edges with endpoints in both S and $V(G) - S$. That way, (α, m)-width is the m-cutwidth of G and Theorem 1 provides an obstruction characterization and a min-max theorem for m-cutwidth (which in case $m = 1$ is simply the cutwidth of G). For a more intuitive approach, we can consider the vertices of G as the countries of the world where existence of "common borders" between two countries implies the existence of an edge between the corresponding edges. The aggressivity of the game indicates how many countries are permitted to be occupied after each attack. The cost function α may indicate the resources the player can use in order to keep the occupied positions. For a more realistic scenario, we can assign a cost on each of the edges indicating the length of the corresponding border portion or simply the cost of guarding it. It is easy to check that, in any case, the cost function is a connectivity function and, therefore, Theorem 1 indicates that the player *can restrict* his/her attention only to monotone conquest strategies (those that do no contain retreats) as this does not imply any deterioration of his/her ability to take over the world.

4.3   Linear-Width and Search Games on Graphs

Recall that, using the terminology of the previous sections, the linear width of a graph G is equivalent to its $(\alpha_1, 1)$-width. Theorem 1 provides an obstruction characterization and a conquest game for linear-width. In what follows, we will show how this conquest game can modelize the three variants of the graph searching game.

A *mixed searching game* is defined in terms of a graph representing a system of tunnels where an agile and omniscient fugitive with unbounded speed is hidden (alternatively, we can formulate the same problem considering that the tunnels are contaminated by some poisonous gas). The object of the game is to *clear* all edges, using one or more *searchers*. An edge of the graph is cleared if any of the following occurs.

A: *both of its endpoints are occupied by a searcher*,
B: *a searcher slides along it*, i.e., a searcher is moved from one endpoint of the edge to the other endpoint.

A search is a sequence containing some of the following moves: (i) placing a new searcher on v, (ii): deleting a searcher from v, (iii): sliding a searcher on v along $\{v, u\}$ and placing it on u.

The object of a mixed search is to clear all edges using a search. The search number of a search is the maximum number of searchers on the graph during any move. The mixed search number, $\mathrm{ms}(G)$, of a graph G is the minimum search number over all the possible searches of it. A move causes *recontamination* of an edge if it causes the appearance of a path from an uncleared edge to this edge not containing any searchers on its vertices or its edges. (Recontaminated edges must be cleared again.) A search without recontamination is called *monotone*.

The *node (edge) search number*, $\mathrm{ns}(G)$ $(\mathrm{es}(G))$ is defined similarly to the mixed search number with the difference that an edge can be cleared only if **A** (**B**) happens.

The following is a combination of results in [4] and [25].

Theorem 3. *For any graph G the following hold:*

- *If G^p is the graph occurring from G after subdividing its pendant edges, then $\mathrm{ms}(G) = \mathrm{lw}(G^p)$. (We call pendant any edge with an endpoint of degree 1.)*
- *If G^e is the graph occurring from G after subdividing each of its edges, then $\mathrm{es}(G) = \mathrm{lw}(G^e)$.*
- *If G^n is the graph occurring from G after replacing every edge in G with two edges in parallel, then $\mathrm{ns}(G) = \mathrm{lw}(G^n)$.*

We mention that the mixed search number is equivalent to the parameter of proper-pathwidth defined by Takahashi, Ueno, and Kajitani in [23,24]. It is also known that the node search number is equivalent to the pathwidth, the interval thickness, and the vertex separation number (see [8,12,13,14,19]).

Theorem 3 gives a way to transform any searching game to a conquest game for linear-width. Therefore, the obstruction characterization for linear-width provided by Theorem 1, can serve as an obstruction characterization for any variant of the search parameters. Applying Theorem 1 and Theorem 3, we can directly derive the monotonicity results of [2,14,16] for the three variants of the search game. In particular, we have the following.

Theorem 4. *Let G be a graph and k, m two positive integers. The following assertions are equivalent.*

1. *There exists a k-expansion for $E(G^n)/E(G^e)/E(G^p)$ with respect to the function α_1.*
2. *There exists a monotone k-expansion for $E(G^n)/E(G^e)/E(G^p)$ with respect to the function α_1.*
3. *There exists no (k, m)-obstacle for $E(G^n)/E(G^e)/E(G^p)$ with respect to the function α_1.*
4. *There exists a node/edge/mixed search for G using k searchers.*
5. *There exists a monotone node/edge/mixed search for G using k searchers.*

5   Conclusions

It appears that Theorem 2 gives a framework to prove the monotonicity for any search/conquest game based on some connectivity function α. Several interesting versions of the conquest game can be generated like that. In particular, we can set α to be any connectivity function mapping vertex sets of $V(G)$ to non negative integers. As an example, we can define $\alpha : V(G) \to \{0, \dots, |V(G)|\}$ such that for $A \subseteq V(G)$, $\alpha(A)$ is the number of endpoints of the edges in $\{\{v, u\} \mid v \in A, u \in V(G) - A\}$, i.e. the number of countries lying on both sides of the frontier of the occupied area defined by A. By observing that this version of α is indeed a connectivity function, we can directly apply Theorem 1 and derive a min-max theorem as well as the monotonicity property of the corresponding game.

As a last application of Theorem 1, we define a conquest game on a plane graph where now the regions represent countries and the cost of the strategy is the maximum number of vertices incident to its frontier. This game seems similar to the "dual" version of the game of Subsection 4.3 when restricted to planar graphs. The only difference is that now the cost is determined by the vertices of the frontier and not by the edges. In both cases, the cost function is a connectivity function and the monotonicity property is a consequence of Theorem 1. We conjecture that the two games are equivalent, in the sense that when they have the same aggressivity, the existence of an optimal complete conquest strategy for the one implies the existence of an optimal complete conquest strategy for the other.

Another way of obtaining game-theoretical approaches to width type parameters is the study of conquest games with an "average" criteria of optimality. For example, instead of the α-cost of a conquest strategy $\mathcal{E}$ one can define the α-sum-cost of a conquest strategy $\mathcal{E}$ as $\sum_{0 \le i \le r} \alpha(T_i)$. Similar to Lemma 1, it is possible to prove monotonicity results for expansion games with minimal sum-cost. This idea can be used to obtain the monotonicity result for the game related to the sum bandwidth problem [6] (or optimal linear arrangement). (See also [9] for similar approach to the interval completion problem.) Nevertheless, to find an analog of Theorem 1, if any exists, appears to be an interesting and hard open problem.

In conclusion, we find interesting the question whether there are monotone search or conquest games where the cost is not expressed by a connectivity function.

References

1. S. L. Bezrukov, J. D. Chavez, L. H. Harper, M. Röttger, and U.-P. Schroeder, *The congestion of n-cube layout on a rectangular grid*, Discrete Math., 213 (2000), pp. 13–19. Selected topics in discrete mathematics (Warsaw, 1996).
2. D. Bienstock, *Graph searching, path-width, tree-width and related problems (a survey)*, DIMACS Ser. in Discrete Mathematics and Theoretical Computer Science, 5 (1991), pp. 33–49.

3. D. BIENSTOCK, N. ROBERTSON, P. D. SEYMOUR, AND R. THOMAS, *Quickly excluding a forest*, J. Comb. Theory Series B, 52 (1991), pp. 274–283.
4. D. BIENSTOCK AND P. SEYMOUR, *Monotonicity in graph searching*, J. Algorithms, 12 (1991), pp. 239–245.
5. B. BOLLOBAS, *Extremal Graph Theory*, Academic Press, London, 1978.
6. P. Z. CHINN, J. CHVÁTALOVÁ, A. K. DEWDNEY, AND N. E. GIBBS, *The bandwidth problem for graphs and matrices — a survey*, J. Graph Theory, 6 (1982), pp. 223–254.
7. N. D. DENDRIS, L. M. KIROUSIS, AND D. M. THILIKOS, *Fugitive-search games on graphs and related parameters*, Theor. Comp. Sc., 172 (1997), pp. 233–254.
8. J. A. ELLIS, I. H. SUDBOROUGH, AND J. TURNER, *The vertex separation and search number of a graph*, Information and Computation, 113 (1994), pp. 50–79.
9. F. V. FOMIN AND P. A. GOLOVACH, *Interval completion and graph searching*, SIAM J. Discrete Math., 13 (2000), pp. 454–464.
10. F. V FOMIN, D. M. THILIKOS, *On the Monotonicity of Games Generated by Symmetric Submodular Functions*. Technical Report 2001-010, Institute for Theoretical Computer Science, Charles University, Prague, Czech Republic, 2001.
11. E. C. FREUDER, *A sufficient condition of backtrack-free search*, J. ACM, 29 (1982), pp. 24–32.
12. N. G. KINNERSLEY, *The vertex separation number of a graph equals its path width*, Inform. Proc. Letters, 42 (1992), pp. 345–350.
13. L. M. KIROUSIS AND C. H. PAPADIMITRIOU, *Interval graphs and searching*, Disc. Math., 55 (1985), pp. 181–184.
14. ——, *Searching and pebbling*, Theor. Comp. Sc., 47 (1986), pp. 205–218.
15. L. M. KIROUSIS AND D. M. THILIKOS, *The linkage of a graph*, SIAM J. Comput., 25 (1996), pp. 626–647.
16. A. S. LAPAUGH, *Recontamination does not help to search a graph*, J. ACM, 40 (1993), pp. 224–245.
17. F. S. MAKEDON AND I. H. SUDBOROUGH, *On minimizing width in linear layouts*, Discrete Appl. Math., 23 (1989), pp. 243–265.
18. N. MEGIDDO, S. L. HAKIMI, M. R. GAREY, D. S. JOHNSON, AND C. H. PAPADIMITRIOU, *The complexity of searching a graph*, J. ACM, 35 (1988), pp. 18–44.
19. R. H. MÖHRING, *Graph problems related to gate matrix layout and PLA folding*, in Computational Graph Theory, Comuting Suppl. 7, E. Mayr, H. Noltemeier, and M. Sysło, eds., Springer Verlag, 1990, pp. 17–51.
20. N. ROBERTSON AND P. D. SEYMOUR, *Graph minors. X. Obstructions to tree-decomposition*, J. Comb. Theory Series B, 52 (1991), pp. 153–190.
21. P. D. SEYMOUR AND R. THOMAS, *Graph searching and a min-max theorem for tree-width*, J. Combin. Theory Ser. B, 58 (1993), pp. 22–33.
22. P. D. SEYMOUR AND R. THOMAS, *Call routing and the ratcatcher*, Combinatorica, 14 (1994), pp. 217–241.
23. A. TAKAHASHI, S. UENO, AND Y. KAJITANI, *Mixed-searching and proper-path-width*, Theoretical Computer Science, 137 (1995), pp. 253–268.
24. ——, *Minimal forbidden minors for the family of graphs with proper-path-width at most two*, IEICE Trans. Fundamentals, E78-A (1995), pp. 1828–1839.
25. D. M. THILIKOS, *Algorithms and obstructions for linear-width and related search parameters*, Discrete Applied Mathematics, 105 (2000), pp. 239–271.
26. R. THOMAS, *Tree decompositions of graphs. Lecture notes*, 1996. Georgia Institut of Technology, Atlanta, Georgia, 30332, USA.

Multiple Hotlink Assignment

Sven Fuhrmann[1], Sven Oliver Krumke[2], and Hans-Christoph Wirth[1]

[1] University of Würzburg
Department of Computer Science, Am Hubland, 97074 Würzburg, Germany
{fuhrmann, wirth}@informatik.uni-wuerzburg.de
[2] Konrad-Zuse-Zentrum für Informantionstechnik, Berlin (ZIB)
Takustraße 7, 14195 Berlin-Dahlem, Germany
krumke@zib.de

Abstract. The input for the hotlink assignment problem consists of a node weighted directed acyclic graph with a designated root node r. The goal is to minimize the weighted shortest path length rooted at r by adding a restricted number of outgoing arcs *(hotlinks)* to each node. The (h, k)-hotlink assignment problem is defined on k-regular complete trees, and at most h hotlinks can be assigned to each node. We contribute algorithms for the $(1, k)$, $(2, k)$, and $(k-1, k)$ hotlink assignment problem.

1 Introduction and Preliminaries

Fast access to large data bases is a main challenge in information technology. Probably one of the largest and most distributed data bases is the world wide web (WWW). A natural representation of the web is a directed graph, where nodes represent web pages and arcs represent links from one page to another. Given a root page and node weights which represent the access frequency of the corresponding pages, one tries to minimize the expected number of steps which are needed to get from the root page to the information nodes. This immediately corresponds to the well known shortest path tree problem, which motivates the restriction of the input graphs to the class of rooted trees.

The HOTLINK ASSIGNMENT problem is a *network upgrade* problem: We are allowed to add new arcs to the graph in order to insert shortcuts and decrease the expected path length.

The problem has been introduced in [1]. The authors investigate the case where the underlying graph is a complete binary tree and only the leaves have positive access weight. We generalize the results to complete k-regular trees.

Before defining the problem we briefly recall some definitions. A tree *rooted at node r* is a tree in a directed graph, where all nodes can be reached from the root node r. The *distance* $d(v, w)$ between two nodes v and w is the number of arcs on the directed path from v to w. The *out-degree* (or simply *degree*) of a node is the number of arcs leaving that node. A node of degree zero is called a *leaf*. If all internal nodes have equal degree k, the tree is called *k-regular*. The distance $d(r, v)$ from the root r is called the *depth* of node v. The *height* of the tree is the maximal depth. If all leaves have equal depth, the tree is called

A. Brandstädt and V.B. Le (Eds.): WG 2001, LNCS 2204, pp. 189–200, 2001.
© Springer-Verlag Berlin Heidelberg 2001

complete. Let (v, s) and (s, g) be a directed path of length 2 in the tree. Then g is the *son* of s and the *grandson* of v, while $p(g) := s$ is the *father* of g. A node $v' \neq v$ is a *brother* of node v if $p(v') = p(v)$.

Definition 1 ((h, k)-Hotlink Assignment problem). *An input instance of* HOTLINK ASSIGNMENT *consists of a k-regular complete tree $G = (V, R)$ rooted at a node $r \in V$ where $L \subset V$ is the set of leaves. Additionally there is a node weight function $w \colon L \to \mathbb{R}_0^+$ with $\sum_{v \in L} w(v) = 1$.*

The problem allows to construct a graph G' by inserting at most h arcs per node v, which are starting at v and ending at a node in the subtree hanging from v. The goal is to find an assignment such that the expected path length

$$E[G', w] := \sum_{v \in L} w(v) \cdot d_{G'}(r, v)$$

is minimized.

For convenience, the weight function w is extended to the total node set V by assigning each inner node the sum of the weight of all leaves in the subtree hanging from it. This can be achieved in linear time. In the sequel we use the vertex' name also to denote its weight.

Throughout the paper we denote by H the height of the tree. Observe that $E[G, w] = H$ for the initial graph.

Related Work

The $(1, 2)$-HOTLINK ASSIGNMENT has been investigated in [1]. The authors consider several weight distributions on the leaves. In particular, they present an algorithm with expected path length $3/4 \cdot H + 1/4$ for arbitrary and uniform distribution, and argue that this is optimal for uniform distribution.

The $(1, 2)$-HOTLINK ASSIGNMENT has been proven to be NP-hard even if restricted to general acyclic graphs [1]. A slight modification of the construction can be used to show the NP-hardness of the (h, k)-HOTLINK ASSIGNMENT problem for any fixed $h \in \mathbb{N}$. We omit the proof due to lack of space.

There are variants of the problem where the total sum of additional hotlinks is bounded rather than the number of hotlinks per node. In this situation it is an optimal strategy to assign all hotlinks starting at the root of the tree. This problem variant is known as *bookmark assignment* (see [1] and the references therein).

The HOTLINK ASSIGNMENT problem is a special case of a *graph augmentation* problem. Those types of problems are specified by a graph $G = (V, E)$ and a subset $E_0 \subset E$ which induces the initial graph $G[E_0]$. The goal is to find a minimum cost set $E' \subset E$ of edges such that the augmented graph $G[E_0 \cup E']$ satisfies some requirements. Augmentation problems have first been considered in [4] where an algorithm for obtaining 2-connectivity in an unweighted graph has been given. A survey on connectivity augmentation problems can be found in [6]. Other requirements include a bound on the diameter [3].

Network improvement problems are another category of problems on graphs, where weights and costs on edges and nodes are subject to change rather than the structure of the graph itself. For the case of a shortest path tree, Berman provided an algorithm which improves the length of a shortest path tree by edge weight modifications [2].

2 $(k-1, k)$-Hotlink Assignment

We start with the investigation of $(k-1, k)$-HOTLINK ASSIGNMENT. As pointed out in the introduction, an instance of this problem consists of a complete k-regular tree. For each vertex we are allowed to add up to $k-1$ hotlinks.

2.1   The Algorithm

The algorithm is a generalization of the approach presented in [1] for $(1, 2)$-HOTLINK ASSIGNMENT. The main idea is a recursive assignment starting from the root. For each node the algorithm determines a set of up to $k-1$ grandsons of maximal weight and assigns hotlinks pointing to them. While descending some care must be taken in order to avoid useless hotlinks. The reason why each hotlink spans exactly two levels is discussed in Section 5.1.

We now describe the algorithm. See Figure 1 for an illustration and Figure 2 for a detailed description.

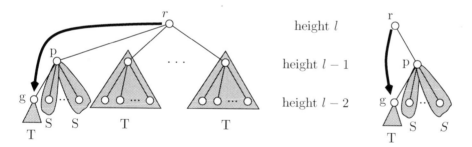

Fig. 1. Tree structure T (left) and S (right).

The algorithm starts with assigning an hotlink to root r according to a T-structure $T(l)$, where l is the height of the structure. It identifies a grandson g of r of maximal weight and adds the hotlink (r, g) to the graph.

Let $p := p(g)$ be the father of g. Obviously, a hotlink from p into the subtree rooted at g is useless, because for all affected paths the same gain can be performed by using the just added hotlink (r, g) instead. Hence the algorithm can ignore the subtree hanging from g in this situation. This is performed by assigning $k-1$ hotlinks to p according to S-structures $S(l-1)$. All remaining brothers of p are treated recursively according to a T-structure $T(l-1)$ as just described.

T-STRUCTURE

Input: Tree with root node r and tree height l
1 **if** $l \geq 2$ **then**
2   determine grandson g of r with maximal weight
3   insert hotlink (r, g)
4   call algorithm T-STRUCTURE$(g, l - 2)$
5   **for** each brother g' of g **do**
6   call algorithm S-STRUCTURE$(p(g), g', l - 1)$
7   **end for**
8   **for** each brother p' of $p(g)$ **do**
9   call algorithm T-STRUCTURE$(p', l - 1)$
10   **end for**
11 **end if**

S-STRUCTURE

Input: Tree with root node r and son p and tree height l
1 **if** $l \geq 2$ **then**
2   determine grandson g of r with maximal weight
3   insert hotlink (r, g)
4   call algorithm T-STRUCTURE$(g, l - 2)$
5   **for** each brother g' of g **do**
6   call algorithm S-STRUCTURE$(p, g', l - 1)$
7   **end for**
8 **end if**

Fig. 2. Algorithm for $(k - 1, k)$-HOTLINK ASSIGNMENT.

The assignment within an S-structure $S(l)$ with root r is as follows: The algorithm identifies a grandson g of r of maximal weight and adds the hotlink (r, g) to the graph. Similarly to the above case, the father $p(g)$ should be treated as a root of S-structures $S(l - 1)$ in order to avoid useless hotlinks.

The recursion can be terminated at height $l = 1$. We remark that the suggested algorithm uses the full power of assigning $k - 1$ hotlinks only at nodes rooting an S-structure.

2.2   Analysis of the Algorithm

We introduce the following notation for the nodes in tree $T(l)$: The sons of root r are denoted by a_i for $i = 1, \ldots, k$. The sons of node a_i are denoted by a_{ij} for $j = 1, \ldots, k$. Without loss of generality we can assume that the weight of a_{11} is a maximum of all grandsons. (Notice that from this assumption we can *not* deduce that weight $w(a_1)$ has maximal weight among the brothers of a_1.) In tree $S(l)$, we denote the grandsons of root r by $b_1, \ldots, b_k$, where b_1 is weight maximal.

Observe that if we consider average path lengths in a subtree, the weights of the leaves must be considered as being normalized:

$$\sum_{i=1}^{k}\sum_{j=1}^{k}a_{ij} = \sum_{i=1}^{k}a_i = 1 \quad \text{and} \quad \sum_{i=1}^{k}b_i = 1. \tag{1}$$

Denote by s_l and t_l the expected path length in a tree $S(l)$ and $T(l)$, respectively. Then we have $s_0 = t_0 = 0$ and $s_1 = t_1 = 1$. From the algorithm we can deduce the following recursion

$$\begin{aligned} t_l &= 1 + a_{11} \cdot t_{l-2} + (a_1 - a_{11}) \cdot s_{l-1} + (1 - a_1) \cdot t_{l-1} \\ s_l &= 1 + b_1 \cdot t_{l-2} + (1 - b_1) \cdot s_{l-1} \end{aligned} \tag{2}$$

which we will prove to have a solution of the form

$$\begin{aligned} t_l &\le \alpha \cdot l + \beta \\ s_l &\le \alpha \cdot l. \end{aligned} \tag{3}$$

We briefly describe how to guess suitable values for α and β. The conditions for s_l in (2) and (3), the fact $b_1 \ge 1/k$, and the assumption $\alpha \ge \beta$ yield $\beta \le (k+1)\alpha - k$. Then the condition for t_l is satisfied if

$$1 - \alpha \cdot (1 + a_{11}) + (a_{11} - a_1) \cdot \beta \le 0. \tag{4}$$

Clearly, $a_{11} - a_1 \le 0$. On the other hand, from (1) and the fact that $a_{11} \ge a_{ij}$ we can conclude that $a_{11} - a_1 \le (k^2 - k + 1)a_{11} - 1 =: d \cdot a_{11} - 1$. The latter estimation is sharper for $a_{11} < 1/d$.

Thus, with $a_{11} \ge 1/d$ and $a_{11} - a_1 \le 0$, equation (4) is satisfied if

$$\alpha \ge \frac{1}{1 + 1/d} = \frac{k^2 - k + 1}{k^2 - k + 2} \quad \text{and} \quad \beta \le (k+1) \cdot \alpha - k = \frac{(k-1)^2}{k^2 - k + 2}.$$

If we use equality as a definition for α and β, one can easily verify that (4) holds also for the remaining case $a_{11} < 1/d$.

Lemma 2. *Let* $\alpha := \frac{k^2 - k + 1}{k^2 - k + 2}$ *and* $\beta := \frac{(k-1)^2}{k^2 - k + 2}$. *Then recursion (2) yields*

$$t_l \le \alpha \cdot l + \beta \quad \text{and} \quad s_l \le \alpha \cdot l.$$

Proof. The proof uses induction on l. As observed above, we have $s_0 = t_0 = 0$ and $s_1 = t_1 = 1$ which show the claim for $l = 0$ and $l = 1$. By inspection, one observes that $t_2 = 2 - a_{11} \le 2 - 1/k^2$ and $s_2 = 2 - b_1 \le 2 - 1/k$ which both satisfy the claim for $l = 2$.

Consider $l = 3$. Observation $t_3 \le 3$ is sufficient for the claim if $k \ge 3$. Otherwise, if $k = 2$, we have $t_3 \le 5/2 = 3/4 \cdot 3 + 1/4$. For s_3 we observe $s_3 \le \left(3k^2 - k - (k-1)\right)/k^2$, and it is easy to verify that this satisfies the claim for $l = 3$. Notice that these bounds on s_l and t_l for $l \le 3$ are actually tight for the case of uniform distribution.

Thus the induction hypothesis is shown for $l \le 3$. The induction step is valid by the construction of α and β as described above. $\qquad\square$

Theorem 3. *There is an algorithm for problem* $(k-1, k)$-HOTLINK ASSIGN-MENT *with expected path length bounded by*

$$\frac{k^2 - k + 1}{k^2 - k + 2} \cdot H + \frac{(k-1)^2}{k^2 - k + 1}$$

where H is the height of the input tree. The running time is linear in the number of nodes.   □

As noted before, the algorithm for the $(k-1, k)$-HOTLINK ASSIGNMENT problem does not use the full power of assigning $k-1$ hotlinks to all nodes: the nodes which are a root of a T-structure have only one outgoing hotlink.

SECOND PHASE

Input: Tree structure from first phase

1 Let r be the current root
2 **if** r is the root of a T-structure **then**
3   Let $h \in \{0, 1\}$ be the number of hotlinks currently emanating from r
                                         { We can assign $k - 1 - h$ more links }
4   Let a_{11} be the target of the hotlink which emanated from r at end of first phase,
         $a_1 := p(a_{11})$
5   Choose $k - 1 - h$ nodes $\{a_i\}$ from brothers of a_1 such that the targets of hotlinks
         emanating from those nodes are weight maximal
6   Replace each of those hotlinks (a_i, x) by (r, x)
7 **end if**
8 Descend recursively to all sons of r

Fig. 3. Second phase of algorithm for $(k-1, k)$-HOTLINK ASSIGNMENT.

We claim that the results for $(k-1, k)$-HOTLINK ASSIGNMENT can be improved for $k > 3$ by applying a second algorithm afterwards (cf. Figure 3), which uses the following idea: Descending recursively in the tree, each time when a node r is found which has not already $k - 1$ outgoing hotlinks, then identify sons s of r where a hotlink (s, x) emanates from, and replace that hotlink by (r, x). If one ensures that during this phase no "overlapping" hotlinks are produced, then there is a nonnegative net improvement of the weighted path length. For the case of uniform probability distribution one can show by an analysis similar to the above that the upper bound on the path length can be decreased to

$$\frac{2 - 3k^2 + 2k^3}{2(k-1)k^2} \cdot H + O(1)$$

by the modification.

3 $(2, k)$-Hotlink Assignment

We now turn over to problem $(2, k)$-HOTLINK ASSIGNMENT. An instance of this problem consists of a complete k-regular tree, where at most 2 hotlinks per node can be assigned.

The idea of the algorithm is similar to the previous. We refer to Figure 4 for an illustration and to Figure 5 for a detailed description of the algorithm. For each node, we identify two grandsons g_1, g_2 with maximal weight and assign two hotlinks to them.

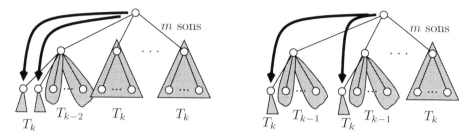

Fig. 4. Tree structure T_m.

T_m-STRUCTURE
Input: Tree with root node r of degree m and tree height l
1 **if** $l \geq 2$ and $m > 0$ **then**
2   determine two grandsons g_1, g_2 of r with maximal weight
3   insert hotlinks (r, g_1) and (r, g_2)
4   call T_k-STRUCTURE$(g_i, l-2)$ for both g_1 and g_2
5   **if** $p(g_1) = p(g_2)$ **then**
6   call T_{k-2}-STRUCTURE$(p(g_1), l-1)$
7   **else**
8   call T_{k-1}-STRUCTURE$(p(g_i), l-1)$ for both g_1 and g_2
9   **end if**
10   **for** all remaining sons v of r **do**
11   call algorithm T_k-STRUCTURE$(v, l-1)$
12   **end for**
13 **end if**

Fig. 5. Algorithm for $(2, k)$-HOTLINK ASSIGNMENT.

Let T_m be a tree structure which describes a tree whose root is of degree m. To avoid useless hotlinks, we proceed analogously to the strategy given in the previous section: Let $p_1 := p(g_1)$ and $p_2 := p(g_2)$. If $p_1 = p_2$, then we assign the father p_1 a tree structure T_{k-2}. All brothers of p_1 are assigned a tree structure T_k. If $p_1 \neq p_2$, then both fathers are assigned a tree structure T_{k-1}. Again, the remaining brothers are assigned a T_k-structure. Notice that in contrast to the strategy given in the previous section, the current strategy assigns two hotlinks to each node (except the lower levels).

Denote by t_l^m the expected path length in a tree $T_m(l)$. Consider the two hotlinks assigned to the root. If their targets share the same father, we have

$$t_l^m = 1 + a \cdot t_{l-2}^k + (b - a)t_{l-1}^{k-2} + (1 - b)t_{l-1}^k$$

$$\text{where } a := a_{11} + a_{12} \text{ and } b := \sum_{j=1}^{k} a_{1j} \tag{5}$$

Otherwise, if their targets have different fathers, then

$$t_l^m = 1 + a \cdot t_{l-2}^k + (b - a)t_{l-1}^{k-1} + (1 - b)t_{n-1}^k$$

$$\text{where } a := a_{11} + a_{21} \text{ and } b := \sum_{j=1}^{k}(a_{1j} + a_{2j}) \tag{6}$$

Again we conjecture $t_l^m \leq \alpha \cdot l + \beta$. To determine α and β we insert both (5) and (6) into the conjecture and get an overall estimation

$$t_l^m \leq \alpha l + \beta + (1 - (a + 1)\alpha).$$

To satisfy the conjecture, the rightmost term must vanish, i.e., bounded from above by zero. This happens if, using $a \geq 2/mk \geq 2/k^2$, we choose $\alpha \geq \frac{k^2}{k^2+2}$. This holds for arbitrary β.

Lemma 4. *For arbitrary $l \in \mathbb{N}$ and $m \in \{k - 2, k - 1, k\}$ we have*

$$t_l^m \leq \frac{k^2}{k^2 + 2} \cdot l + \frac{2}{k^2 + 2}.$$

Proof. The proof is by induction on l. For $l = 2$ one observes that

$$t_2^m = 2 - a \leq 2 - \frac{2}{mk} \leq 2 - \frac{2}{k^2}.$$

The right hand side is bounded by $\alpha \cdot 2 + \beta$ for arbitrary $\beta \geq \frac{2(k^2-2)}{k^2(k^2+2)}$. Notice that β has not been determined yet. For convenience, we choose the slightly weaker bound $\beta := \frac{2}{k^2+2}$ at this point.

Consider $l = 3$. We propose that the weights are uniformly distributed. By observation, for $k \geq 3$ we have

$$t_3^m \leq 3 - \frac{4k^2 - 4}{k^4} \leq 3 - \frac{4}{k^2 + 2} \leq \frac{k^2}{k^2 + 2} \cdot 3 + \frac{2}{k^2 + 2}.$$

For $k = 2$ the above estimation is valid only if the hotlinks point to different subtrees. If both hotlinks point into the same subtree, we must use the weaker estimation $t_3^m \leq 9/4$ which follows from the observation that the gain of the hotlinks sums up to at least $3/4$.

This shows the induction hypothesis for $l = 2$ and $l = 3$. From the construction of α as described above it follows that the induction step is also valid. □

Theorem 5. *There is an algorithm for problem $(2, k)$-HOTLINK ASSIGNMENT with expected path length bounded by*

$$\frac{k^2}{k^2 + 2} \cdot H + \frac{2}{k^2 + 2}$$

where H is the height of the input tree. The running time is linear in the number of nodes. □

4 $(1, k)$-Hotlink Assignment

The algorithm from Section 3 for the $(2, k)$-HOTLINK ASSIGNMENT problem can be slightly simplified to solve the $(1, k)$-HOTLINK ASSIGNMENT problem. Since there is only one hotlink emanating from each node, the distinction of cases (see Lines 5 to 9 in Figure 5) is no longer necessary. As a consequence, one can show the following result:

Theorem 6. *There is an algorithm for problem $(1, k)$-HOTLINK ASSIGNMENT with expected path length bounded by*

$$\frac{k^2}{k^2 + 1} \cdot H + \frac{1}{k^2 + 1}$$

where H is the height of the input tree. The running time is linear in the number of nodes. □

5   Conclusions

5.1   Optimal Hotlink Length

The two algorithms both use hotlinks which span two levels. We follow [1] to show that this is the optimal rule as long as the weight distribution is uniform. Consider an arbitrary tree with root r. If there is a hotlink added to the root which spans d levels, this hotlink reduces the path length by $d - 1$ for one node of weight k^{-d}. The same reduction applies for each additional hotlink (as long as the number of hotlinks does not exceed the number of grandsons) since the paths are independent.

Consequently the total gain of h hotlinks where each hotlink spans d levels is bounded from above by $h \cdot (d - 1) \cdot k^{-d}$. This factor is maximal for

$$d = 1 + \frac{1}{\ln k} \,.$$

This matches the result from [1] which states that the maximal gain is attained when $d = 2$ or $d = 3$ for the case $k = 2$. For $k \geq 3$, we have $1 < d < 2$ and can conclude that a hotlink spanning two levels is the optimal choice.

5.2   Lower Bounds for Uniform Distribution

We now generalize the results from [1] to prove a lower bound on the quality of any (h, k)-HOTLINK ASSIGNMENT strategy. We assume that the weight distribution is uniform. Consider a subtree of weight $w(r)$ hanging from a root r. As argumented above, the maximal gain of any assigned hotlink is attained when the hotlink spans two levels. As long as h does not exceed the number of available grandsons, the same arguments hold if h hotlinks are assigned to node r. Thus, the maximal gain which can be achieved by assigning hotlinks to node r is

$$h \cdot \frac{1}{k^2} \cdot w(r).$$

In order to derive an upper bound on the total gain in the tree, it suffices to sum the gain over all nodes. Observe that at level i we have k^i nodes each of weight $1/k^i$. Notice that nodes from the two lowest levels are not counted since any hotlink assigned to them does not improve the path length. Hence the total gain is bounded from above by

$$\sum_{i=0}^{H-2} k^i \cdot h \cdot \frac{1}{k^2} \cdot \frac{1}{k^i} = h \cdot \frac{H-1}{k^2}.$$

Recall that the weighted path length of the tree without hotlinks equals H. Thus we can conclude the following result:

Corollary 7. *There is no strategy for (h, k)-HOTLINK ASSIGNMENT with expected path length lower than*

$$H - h \cdot \frac{H-1}{k^2} = \frac{k^2 - h}{k^2} \cdot H + \frac{h}{k^2}$$

for each instance with uniform distribution.   □

From the above observations we can derive an algorithm which solves for any $h \leq k - 1$ the (h, k)-HOTLINK ASSIGNMENT problem on uniform distribution to optimality (i. e., the solution attains the lower bound). The algorithm starts with the root and descends recursively. For each node r, it indentifies the h leftmost sons $\{s_1, \ldots, s_h\}$ and assigns for each s_i a hotlink from r to the rightmost son of s_i. Since $h \leq k - 1$, this construction guarantees that all hotlinks are independent and contribute the maximal gain.

5.3   Summary of Results

We have presented algorithms for the $(1, k)$, the $(2, k)$ and the $(k-1, k)$-HOTLINK ASSIGNMENT problem, respectively. For all algorithms we provided upper bounds on the resulting path length. Moreover, we gave lower bounds (valid for uniform distribution only) which adjust the quality of the strategies. A summary can

Table 1. Results presented in this paper.

Problem	upper bound	lower bound (uniform dist.)
$(1,k)$-HA	$\left(1 - \frac{1}{k^2+1}\right) \cdot H + \frac{1}{k^2+1}$	$\left(1 - \frac{1}{k^2}\right) \cdot H + \frac{1}{k^2}$
$(2,k)$-HA	$\left(1 - \frac{2}{k^2+2}\right) \cdot H + \frac{2}{k^2+2}$	$\left(1 - \frac{2}{k^2}\right) \cdot H + \frac{2}{k^2}$
$(k-1,k)$-HA	$\left(1 - \frac{1}{k^2-k+2}\right) \cdot H + \frac{(k-1)^2}{k^2-k+2}$	$\left(1 - \frac{k-1}{k^2}\right) \cdot H + \frac{k-1}{k^2}$

be found in Table 1. It is not surprising that the factor of the linear term approaches 1 as $k \to \infty$ since the ratio of leaves which profit from the hotlinks tends to 0 in this situation.

Table 2 compares the results of the several strategies presented in this paper. Observe that, for $k = 2$, the known problem $(1,2)$-HOTLINK ASSIGNMENT is a special case of $(k-1,k)$-HOTLINK ASSIGNMENT, and that our analysis of the $(k-1,k)$-strategy matches the known results from [1] in that case. Furthermore, for $k = 3$, the strategies $(k-1,k)$ and $(2,k)$ both operate on 3-regular trees allowing two hotlinks per node. Here one can observe that the $(2,k)$-strategy performs better which may be a consequence of the fact that the other strategy does not always assign all possible hotlinks.

Table 2. Comparison of the strategies. The linear factor α of the terms $\alpha H + \beta$ is given also in decimal notation.

	$(1,k)$-HA bound	α	$(k-1,k)$-HA bound	α	$(2,k)$-HA bound	α
upper bound						
$k = 2$	$\frac{4}{5}H + \frac{1}{5}$	0.800	$\frac{3}{4}H + \frac{1}{4}$	0.750	$\frac{2}{3}H + \frac{1}{3}$	0.667
$k = 3$	$\frac{9}{10}H + \frac{1}{10}$	0.900	$\frac{7}{8}H + \frac{1}{2}$	0.875	$\frac{9}{11}H + \frac{2}{11}$	0.818
lower bound						
$k = 2$	$\frac{3}{4}H + \frac{1}{4}$	0.750	$\frac{3}{4}H + \frac{1}{4}$	0.750	$\frac{1}{2}H + \frac{1}{2}$	0.500
$k = 3$	$\frac{8}{9}H + \frac{1}{9}$	0.889	$\frac{7}{9}H + \frac{2}{9}$	0.778	$\frac{7}{9}H + \frac{2}{9}$	0.778

References

1. P. Bose, J. Czyzowicz, L. Gasieniec, E. Kranakis, D. Krizanc, A. Pelc, and M. V. Martin · *Strategies for hotlink assignment* · Proceedings of the 11th Annual International Symposium on Algorithms and Computation (ISAAC'00), Lecture Notes in Computer Science, vol. 1969, Springer, 2001, pp. 23–34.
2. O. Berman · *Improving the location of minisum facilities through network modification* · Annals of Operations Research **40** (1992), 1–16.

3. Y. Dodis and S. Khanna · *Designing networks with bounded pairwise distance* · Proceedings of the 31st Annual ACM Symposium on the Theory of Computing (STOC'99), 1999, pp. 750–759.

4. K. P. Eswaran and R. E. Tarjan · *Augmentation problems* · SIAM Journal on Computing **10** (1976), no. 2, 270–283.

5. D. S. Hochbaum (ed.) · *Approximation algorithms for NP-hard problems* · PWS Publishing Company, Boston, 1997.

6. S. Khuller · *Approximation algorithms for finding highly connected subgraphs* · in [5], 1997.

Small k-Dominating Sets in Planar Graphs with Applications

Cyril Gavoille[1], David Peleg[2], André Raspaud[1], and Eric Sopena[1]

[1] LaBRI, Université Bordeaux I, 351, cours de la Libération,
33405 Talence Cedex, France
{gavoille,raspaud,sopena}@labri.fr

[2] Department of Computer Science and Applied Mathematics, The Weizmann
Institute of Science, Rehovot, 76100 Israel
peleg@wisdom.weizmann.ac.il

Abstract. A subset of nodes S in a graph G is called k-*dominating* if, for every node u of the graph, the distance from u to S is at most k. We consider the parameter $\gamma_k(G)$ defined as the cardinality of the smallest k-dominating set of G. For planar graphs, we show that for every $\epsilon > 0$ and for every $k \geqslant (5/7 + \epsilon)D$, $\gamma_k(G) = O(1/\epsilon)$. For several subclasses of planar graphs of diameter D, we show that $\gamma_k(G)$ is bounded by a constant for $k \geqslant D/2$. We conjecture that the same result holds for every planar graph. This problem is motivated by the design of routing schemes with compact data structures.

1   Introduction

Given a graph $G = (V, E)$, denote by $d_G(u, v)$ the distance in G from u to v, and by $\Gamma_k(u) = \{v \in V \mid d_G(u, v) \leqslant k\}$ the ball of radius k centered at u (note that $\Gamma_k(u)$ is also defined for non integral k). We extend this notation to every subset $S \subseteq V$, letting $\Gamma_k(S) = \bigcup_{u \in S} \Gamma_k(u)$. We say that a subset of nodes $S \subseteq V$ is a k-*dominating set* for G (or, S k-dominates G, or S covers G) if $\Gamma_k(S) = V$. Let $\gamma_k(G)$ denote the cardinality of the smallest k-dominating set of G. Our goal is to bound $\gamma_k(G)$ for various planar graphs G and values of k.

To illustrate the properties of the parameter $\gamma_k(G)$, let us consider a tree T of diameter D. It is clear that if $k \geqslant D/2$, then $\gamma_k(T) = 1$; it suffices to consider the center of T. On the other hand, if $k < D/2$ then there exists some n-node tree T_0 of diameter D for which $\gamma_k(T_0) \geqslant 2(n-1)/D$. For instance, consider T_0 composed of a star $K_{1,p}$ with each edge subdivided into $q - 1$ nodes, where $n = pq + 1$, $D = 2q$ and $\gamma_{q-1}(T_0) = p$. Every two leaves of this tree are at distance $2q$, thus p nodes of T_0 are required to $(q - 1)$-dominate all the leaves.

More generally, for $k \geqslant D/2$, every n-node graph $G = (V, E)$ of diameter D satisfies $\gamma_k(G) < \sqrt{n(1 + \ln n)}$. To see this, we use a dual characterization of a k-dominating set as a set S which *hits* the collection $\{\Gamma_k(u) \mid u \in V\}$, i.e., such that $S \cap \Gamma_k(u) \neq \varnothing$ for every $u \in V$. Now, note that for $k \geqslant D/2$, $\Gamma_k(x) \cap \Gamma_k(y) \neq \varnothing$ for any two nodes x and y. Thus every set $\Gamma_k(x)$ k-dominates G. Hence, either there exists a node x such that $|\Gamma_k(x)| < \alpha$, where $\alpha = \sqrt{n(1 + \ln n)}$, and we

A. Brandstädt and V.B. Le (Eds.): WG 2001, LNCS 2204, pp. 201–216, 2001.
© Springer-Verlag Berlin Heidelberg 2001

are done, or $|\Gamma_k(x)| \geqslant \alpha$ for every node x. In the latter case, the claim follows by a result of Lovász [11] about cover sets, which states that there exists a k-dominating set of G of size $\beta < n(1 + \ln n)/\min_x |\Gamma_k(x)| \leqslant \sqrt{n(1 + \ln n)}$.

For planar graphs, it is known that the size of a dominating set (i.e., 1-dominating set) is bounded by 3 if $D = 2$, and by 10 if $D = 3$, cf. [12]. Thus, for $k \geqslant D/2$, and every planar graph G, $\gamma_k(G) \leqslant 3$ if $D = 2$, and $\gamma_k(G) \leqslant 10$ if $D = 3$.

A recent result concerns planar triangulations. A triangulation of the plane is a planar graph with an embedding on the plane such that each face, except maybe the outer face, is a triangle. In [2] it is shown that every planar triangulation G in which every internal node (i.e., that does not belong to the outer face) has degree at least 6, satisfies $\gamma_k(G) \leqslant 2$, again for $k \geqslant D/2$.

Section 2 presents our main results. For planar graphs, we show that for every $\epsilon > 0$ and for every integer $k \geqslant (5/7 + \epsilon)D$, $\gamma_k(G) = O(1/\epsilon)$. For outerplanar graphs of diameter D, we show that $\gamma_k(G) \leqslant 2$ for $k \geqslant D/2$. We conjecture that $\gamma_k(G)$ is bounded by a constant for every planar graph of diameter D for $k \geqslant D/2$.

In Section 3 we show study bounded treewidth and chordality graphs. More precisely, we show that if G has treewidth bounded by t, then $\gamma_k(G) \leqslant t + 1$ for $k \geqslant \lfloor D/2 \rfloor$, and $\gamma_k(G) = 1$ for $k \geqslant \lfloor D/2 \rfloor + \lfloor c/2 \rfloor$ if G has no chordless cycle of length greater than c.

The motivation for studying this parameter stems from the design of routing schemes with compact data structures. In particular, it is shown in Section 4 that for every graph G of diameter D and every $k \geqslant 0$, G has an interval routing scheme [14,17] with dilation at most $D + k$ and compactness at most $\frac{1}{2}\gamma_k(G) + 1$. (The dilation of a routing scheme is the length of the longest route, and its compactness is the maximum number of intervals over any arc.) Moreover, the upper bound on the compactness can be reduced to $\frac{1}{4}\gamma_k(G) + o(\gamma_k(G))$, if $\gamma_k(G)/\log n \to +\infty$. This result improves (by a multiplicative factor) and generalizes the compactness vs. dilation trade-offs of [10] and of [16]. Our result also implies that for every constant $\epsilon > 0$, every planar graph of diameter D has an interval routing scheme with dilation at most $\lceil (12/7 + \epsilon)D \rceil$ and constant compactness.

2   Plane Graphs

2.1   Preliminaries

Hereafter, we assume that $G = (V, E)$ is connected and of diameter D. The following basic property is important for the remainder of the paper.

Proposition 1. *Let $S \subset V$, let $k \geqslant \lfloor D/2 \rfloor$, and let $u, v \notin \Gamma_k(S)$. Then no shortest path between u and v goes through any node of S.*

Proof. Assume, towards contradiction, that there is some shortest path from u to v that contains a node $w \in S$. Then $d_G(u, v) = d_G(u, w) + d_G(w, v)$. Since

$u, v \notin \Gamma_k(S)$, necessarily $u, v \notin \Gamma_k(w)$. Thus $d_G(u, w) > k$, and more precisely, $d_G(u, w) \geqslant \lfloor k \rfloor + 1$ since the distance is integral. Similarly $d_G(v, w) \geqslant \lfloor k \rfloor + 1$. It implies that $d_G(u, v) \geqslant 2 \lfloor k \rfloor + 2 \geqslant 2 \lfloor D/2 \rfloor + 2 \geqslant D + 1$, in contradiction with the fact that D is the diameter of G. □

A subset $S \subset V(G)$ is a *separator* of G if $G \setminus S$ is composed of two or more connected components.

Proposition 1 yields the following immediate corollary, keeping in mind that every path joining two nodes of different connected components in $G \setminus S$, for some separator S, has to go through S.

Corollary 1. *Let $k \geqslant \lfloor D/2 \rfloor$, let S be a separator of G, and let $U = V \setminus \Gamma_k(S)$. Then there exists a unique connected component C of $G \setminus S$ such that $U \subseteq V(C)$.*

2.2   Outerplanar Graphs

Let us start by proving that for outerplanar graphs, $\gamma_k(G)$ is bounded by a constant for $k \geqslant D/2$. Recall that a graph H is a *minor* of a graph G if H can be obtained from G by a sequence of zero or more node deletions, edge deletions or edge contractions. We say that G is H-*free* if it does not contain H as a minor. Let us denote by $K_{p,q}$ the complete bipartite graph with p nodes in one partition and q nodes in the other. We prove the following.

Theorem 1. *If G is $K_{2,t+1}$-free then $\gamma_k(G) \leqslant t$ for every $k \geqslant D/2$, and this is tight for a subdivision of $K_{t,t}$.*

Proof. Let u and v be any two nodes in G with $d_G(u, v) = D$. The *level* of the node x is defined as its distance from u, denoted $l(x) = d_G(u, x)$. Let $V_i = \{x \in V \mid l(x) = i\}$ for every $0 \leqslant i \leqslant D$. Let M be the set of nodes $x \in V_k$ such that there exists a path $(x, x_1, x_2, \ldots, x_\ell, v)$ from x to v satisfying $l(x_i) > k$ for every $1 \leqslant i \leqslant \ell$.

We first claim that the set M has cardinality at most t. To prove this, we show that if M contains $t + 1$ distinct nodes then G contains $K_{2,t+1}$ as a minor, a contradiction. Indeed, consider the graph H composed of the $t + 1$ nodes of M and the $2(t + 1)$ paths linking M to u and v, and delete the edges and nodes of $G \setminus H$. Then it is easy to see that by contractions we obtain a copy of $K_{2,t+1}$.

Now let $P = (u, u_1, u_2, \ldots, u_p, v)$ be any path linking u and v and let q be the largest subscript such that $l(u_q) = k$. Clearly, $l(u_i) > k$ for every i in the range $q < i \leqslant p$, and thus $u_q \in M$. Hence every path linking u and v must go through the set M which, therefore, is a separator in G. (In particular, u and v belong to different connected components of $G \setminus M$).

We now prove (by contradiction again) that M k-dominates G. Suppose that there exists a node x in G such that $d_G(x, M) > k$. As observed above, M separates u from v, hence either u or v does not belong to the connected component of x in $G \setminus M$. In the former case we get

$$d_G(u, x) \geqslant d_G(u, M) + d_G(x, M) > 2k \geqslant D,$$

and in the latter case

$$d_G(v,x) \geqslant d_G(v,M) + d_G(x,M) > D - k + k = D \ .$$

Therefore, either $d_G(u,x)$ or $d_G(v,x)$ is strictly greater than D, a contradiction.

We thus get that M k-dominates G, hence $\gamma_k(G) \leqslant |M| \leqslant t$, completing the proof.   □

As every outerplanar graph is $K_{2,3}$-free, we get the following.

Corollary 2. *If G is an outerplanar graph of diameter D then $\gamma_k(G) \leqslant 2$, for every $k \geqslant D/2$.*

2.3   Planar Graphs

We assume from now on that G is a plane graph, that is, a planar graph with an embedding in $\mathbb{R}^2$. More precisely, the nodes are points of $\mathbb{R}^2$ and the edges are simple curves that can cross or meet each other on the nodes only. A connected subset of $\mathbb{R}^2$ is called a *region* of the plane. Given a subgraph H of G and a point $w \notin H$ (i.e., a point $w \in \mathbb{R}^2$ that does not belong to an edge or a node of H), we denote by $\mathrm{reg}(w,H)$ the unique region of $\mathbb{R}^2 \setminus H$ containing w. Note that if H is a tree, then $\mathbb{R}^2 \setminus H$ consists of one region only. If $\mathbb{R}^2 \setminus H$ is composed of exactly two non-empty regions, we denote by $\overline{\mathrm{reg}}(w,H)$ the complementing region, such that $\mathbb{R}^2 = \mathrm{reg}(w,H) \cup \overline{\mathrm{reg}}(w,H) \cup H$.

Given an integer $\psi \geqslant 3$, a ψ-*gon* w.r.t. G and k is a subgraph H of G defined by a sequence $(x_1, \ldots, x_\psi)$ of distinct nodes, and a sequence $(Q_1, \ldots, Q_\psi)$ of paths such that the following four conditions hold:

1. Q_i is a shortest path from x_i to $x_{(i \bmod \psi)+1}$, for every $i \in \{1, \ldots, \psi\}$;
2. $\mathbb{R}^2 \setminus H$ is composed of at most two non-empty regions;
3. $x_\psi \notin \Gamma_k(Q_1 \cup \cdots \cup Q_{\psi-2})$;
4. $d(x_i, x_{(i \bmod \psi)+1}) > k$, for every $i \in \{1, \ldots, \psi\}$.

The point x_ψ is referred to as the *extremity* of the ψ-gon. A ψ-gon H is called *weak* if conditions 3 and 4 do not necessarily hold. A weak 3-gon is hereafter called a *triangle*.

Roughly speaking, a ψ-gon consists of ψ shortest paths, each of length at least $k+1$, possibly sharing some edges. Fig. 1 presents a 4-gon H w.r.t. a plane graph G of diameter $D = 4$ and $k = 2$. Note that the x_i's may border to different regions induced by H.

Given a path P and two nodes u and v on P, we denote by $P[u,v]$ the subpath of P between u and v.

Let $Q_1, \ldots, Q_p$ be a set of shortest paths, and for each $i \in \{1, \ldots, p\}$ let $(z_1^i, \ldots, z_{t_i}^i)$ be the sequence of nodes of $P \cap Q_i$ encountered in this order along a walk on P from u to v (possibly $t_i = 0$ if $P \cap Q_i = \varnothing$). The *simplified path of P on $Q_1 \cup \cdots \cup Q_p$* is the path defined by

$$S = P[u, z_1^1] \cup Q_1[z_1^1, z_{t_1}^1] \cup P[z_{t_1}^1, z_1^2] \cup \cdots \cup Q_i[z_1^i, z_{t_i}^i] \cup P[z_{t_i}^i, z_1^{i+1}] \cup \cdots$$
$$\cup Q_p[z_1^p, z_{t_p}^p] \cup P[z_{t_p}^p, v] \ .$$

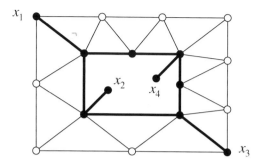

Fig. 1. A plane graph G with a 4-gon H for $k = 2$ (bold edges).

Roughly speaking, S uses shortcuts of P on Q_1, then shortcuts of P on Q_2, and so on. Note that if P is a shortest path, then S is also a shortest path between u and v. (If $t_i = 0$ for all i, then $S = P$.) Fig. 2 shows the simplified path S of P on $Q_1 \cup Q_2$. Note that S is unique, but other shortcuts are possible, for instance, $Q_1[u, z_2^1] \cup P[z_2^1, z_1^2] \cup Q_2[z_1^2, v]$.

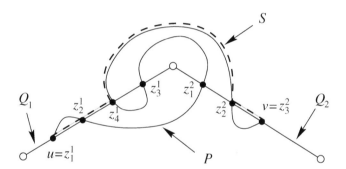

Fig. 2. The simplified path S of P on $Q_1 \cup Q_2$.

For notational convenience, we use $\Gamma_k(H)$ where H is a graph instead of $\Gamma_k(V(H))$.

The next lemma shows an important property of ψ-gons. The proofs of the next two lemmas can be founded in [9].

Lemma 1. *Let H be a p-gon of G with $p \leqslant 4$ and let $k \geqslant \lfloor D/2 \rfloor$. Then for every $w \notin \Gamma_k(H)$:*

1. *there exists a shortest path A from w to x_p which is entirely contained in* reg$(w, H) \cup H$;
2. *there exists a shortest path B from w to x_i, $i < p$, which is entirely contained in* reg$(w, H) \cup H$.

Moreover, A and B can be chosen such that the two graphs $H_1 = B \cup (Q_i \cup Q_{i+1} \cup \cdots \cup Q_{p-1}) \cup A$ and $H_2 = B \cup (Q_{i-1} \cup Q_{i-2} \cup \cdots \cup Q_1) \cup Q_p \cup A$ are, respectively, a $(p - i + 2)$-gon and an $(i + 2)$-gon of G.

Lemma 2. *Let Q be a path of q nodes. For every integer $\lambda \geqslant 0$, $\gamma_\lambda(Q) = \lceil q/(2\lambda + 1) \rceil$.*

2.4   Main Result

In this section we prove our main result, namely, that $\gamma_k(G) < 3/\epsilon + 6$ for every planar graph G and $k \geqslant (5/7 + \epsilon)D$ (for any $\epsilon > 0$).

Towards proving this result, we introduce and study the related notion of a *k-dominating subgraph*. Given a graph G, a connected subgraph H is called a k-dominating subgraph of G if $V(H)$ is a k-dominating set for G. Our main result is derived from the following technical theorem concerning k-dominating subgraphs, which may be of independent interest.

Theorem 2. *For every planar graph G of diameter D, and for every $k \geqslant \lfloor 5D/7 \rfloor$, there exists a k-dominating subgraph H of G such that H is composed of at most 6 shortest paths of G, and $|V(H)| \leqslant 6D - 1$.*

Proof. We present a constructive proof, implying a polynomial time algorithm for the construction of such a k-dominating subgraph H. Without loss of generality we assume hereafter that $k < D$, as the result clearly holds for $k = D$. The algorithm constructs H gradually. Letting $\delta(H)$ denote the *deviation* of a candidate subgraph H from being a k-dominating subgraph, i.e., $\delta(H) = |V \setminus \Gamma_k(H)|$, we attempt to progress by repeatedly decreasing $\delta(H)$.

Step 1: We start assuming that $k \geqslant \lfloor D/2 \rfloor$. Let x_1 and x_2 be two nodes such that $d_G(x_1, x_2) > k$, and let Q_1 be any shortest path connecting them.

If $V(Q_1)$ is k-dominating, then taking $H = Q_1$ we are done, as H is connected and $|V(H)| \leqslant D$.

Step 2: We construct a 3-gon H for G as follows. Let $x_3 \notin \Gamma_k(Q_1)$, and let Q_2 (resp., Q_3) be a shortest path between x_3 and x_1 (resp., between x_3 and x_2), chosen such that $H = Q_1 \cup Q_2 \cup Q_3$ induces at most two regions of $\mathbb{R}^2$. This can be done by ensuring that for every $t \in Q_2 \cap Q_3$, the subpaths of Q_2 and Q_3 from x_3 to t coincide, i.e., $Q_2[x_3, t] = Q_3[x_3, t]$, and analogously, that for every $t \in Q_2 \cap Q_1$, $Q_2[x_1, t] = Q_1[x_1, t]$, and that for every $t \in Q_3 \cap Q_1$, $Q_3[x_2, t] = Q_1[x_2, t]$. One can readily verify that H is a 3-gon of G.

Step 3: H is a 3-gon.

If $V(H)$ is a k-dominating set then we are done, as H is connected and $|V(H)| \leqslant 3D$.

Otherwise, let $w \notin \Gamma_k(H)$. By Lemma 1, G has a $(3 - i + 2)$-gon and a $(i + 2)$-gon. Since $1 \leqslant i < 3$, G has a 4-gon. Set H to be this 4-gon.

Step 4: H is a 4-gon.

If $V(H)$ is a k-dominating set, then we are done, as H is connected and $|V(H)| \leqslant 4D$.

Otherwise, let $m = \delta(H) > 0$, and let $w \notin \Gamma_k(H)$. Assume that H is defined by the nodes $(x_1, \ldots, x_4)$ and the shortest paths $(Q_1, \ldots, Q_4)$. By Lemma 1, let A be the shortest path between w and x_4 that is wholly contained in $\text{reg}(w, H) \cup H$, and let B be the second shortest path between w and x_i, $i < 4$, that is wholly contained in $\text{reg}(w, H) \cup H$. Let $H_1 = B \cup (Q_i \cup Q_{i+1} \cup \cdots \cup Q_3) \cup A$ and $H_2 = B \cup (Q_{i-1} \cup \cdots \cup Q_1) \cup Q_4 \cup A$ be the respective $(6-i)$-gon and $(i+2)$-gon whose existence is asserted by Lemma 1. Note that H_1 and H_2 are 3-, 4- or 5-gons. Figure 3 depicts H_1 when $i = 2$, in which case $H_1 = B \cup Q_2 \cup Q_3 \cup A$, namely, it is obtained by following the cycle w, x_2, x_3, x_4, w.

Let $H' = H_1 \cup H_2 = A \cup H \cup B$. If $V(H')$ is a k-dominating set, then we are done, as H' is connected and $|V(H')| \leqslant 6D - 1$. Indeed, in the worst-case H' is composed of 6 paths of length D that meet in 5 nodes $(x_1, \ldots, x_4$ and $w)$, thus has $5 + 6(D - 1) = 6D - 1$ nodes. So hereafter assume that $V(H')$ is not k-dominating. Let $m' = \delta(H') > 0$, and let $w' \notin \Gamma_k(H')$. Note that $m' \leqslant m - 1$ because $V(H')$ covers $\Gamma_k(H)$ and also w (as $V(H) \subset V(H')$).

H' defines at most three regions, namely, $\overline{\text{reg}}(w, H)$, and $\text{reg}(w, H)$, which in turn is split by H_1 into two sub-regions, $R_1 = \overline{\text{reg}}(z, H_1)$ and $R_2 = \text{reg}(w, H) \setminus (R_1 \cup H_1)$, where z is an arbitrary node of $V(H_1) \setminus V(H_2)$ (such a node must exist). See Fig. 3. Note that on Fig. 3 two incident paths may partially overlap: for instance $A[w, t] = B[w, t]$ for some t is possible. The proof anyway is not based on this particular example.

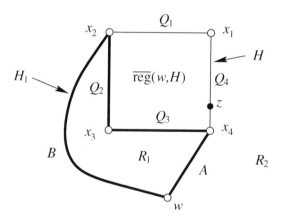

Fig. 3. The ψ-gon H_1 (in bold edges). A node $w' \notin \Gamma_k(H')$ has to belong to R_1 or R_2.

Corollary 1 implies that $w' \notin \overline{\text{reg}}(w, H)$, since w and w' must belong to the same region. So assume that $w' \in R_j$, for $j = 1$ or 2. Note that $R_j = \text{reg}(w', H_j)$. We now split the discussion into two cases.

Case 1: H_j is a 3-gon or a 4-gon.
In this case, we prove that $\Gamma_k(H_j) = \Gamma_k(H')$. Clearly, $\Gamma_k(H_j) \subseteq \Gamma_k(H')$. The converse is proved by contradiction. By Corollary 1, H_j covers all the points of $\overline{\text{reg}}(w', H_j)$, and thus H' covers those points as well since $H_j \subset H'$. Assume,

to the contrary, that there is some $w'' \in R_j$ such that $w'' \in \Gamma_k(H')$ but $w'' \notin \Gamma_k(H_j)$. This implies that there exists some $u \in V(H') \setminus V(H_j)$ such that $w'' \in \Gamma_k(u)$. Any shortest path from u to w'' must intersect H_j at some node v because H_j is a separator. Thus $w'' \in \Gamma_k(v)$, proving that $w'' \in \Gamma_k(H_j)$, contradiction. Thus $\Gamma_k(H_j) \supseteq \Gamma_k(H')$ as well.

It follows that the number of nodes uncovered by H_j is $m' \leqslant m-1$. Depending on whether H_j is a 3-gon or a 4-gon, we now set $H \leftarrow H_j$ and go to Step 3 or Step 4 and reiterate, and by induction the number of uncovered nodes $\delta(H)$ will decrease to 0.

Case 2: H_j is a 5-gon.

Note that in this case the shortest path B is between w and x_1 or x_3, w.l.o.g. x_1. (The other case is handled symmetrically by exchanging node names x_1 and x_3.)

Assume there is a path B' between w and x_2 that is wholly contained in $\text{reg}(w, H) \cup H$. Then, we can remove the path B constructed while applying Lemma 1, and rename B' into B. The ψ-gons H_1 and H_2 resulting from considering the new simplified paths A and B are necessarily 4-gons, and we can conclude by Case 1.

Similarly, if there is a path A' between w and x_3 that is wholly contained in $\text{reg}(w, H) \cup H$, then we can remove path A and rename A' into A. The 4-gons resulting from considering the new simplified paths A and B allow us to go again to Case 1.

We are thus left with the situation when none of the shortest paths from w to x_2 and from w to x_3 is wholly contained in $\text{reg}(w, H) \cup H$.

Let X be a shortest path from w to x_2, and let Y be a shortest path from w to x_3. (Without loss of generality we assume that X and Y are simplified on H). By assumption, X and Y are not wholly contained in $\text{reg}(w, H) \cup H$. Therefore, X must intersect Q_3 or Q_4, and Y must intersect Q_1 or Q_4. Let u be the farthest node from w such that $u \in X \cap (Q_3 \cup Q_4)$, and let v be the farthest node from w such that $v \in Y \cap (Q_1 \cup Q_4)$.

Case 2.1: $u \in Q_3$.

Similarly to the proof of Lemma 1, X and Y intersect at some point $\alpha \in \overline{\text{reg}}(w, H)$. (See Fig. 4.) Then $X[w, u]$ is wholly contained in $\text{reg}(w, H) \cup H$ and intersects Q_3 leading to x_3. As both X and Y induce shortest paths from w to α, $|X[w, u]| + |X[u, \alpha]| = |Y[w, \alpha]|$, or $|X[w, u]| = |Y[w, \alpha]| - |X[u, \alpha]|$. Also, $|Q_3[u, x_3]| \leqslant |X[u, \alpha]| + |Y[\alpha, x_3]|$. Hence $|X[w, u]| + |Q_3[u, x_3]| \leqslant |Y[w, \alpha]| + |Y[\alpha, x_3]| = |Y[w, x_3]|$, so $X[w, u] \cup Q_3[u, x_3]$ is a shortest path from w to x_3 wholly contained in $\text{reg}(w, H) \cup H$, contradicting the definition of Y. Thus Case 2.1 is not possible.

Case 2.2: $v \in Q_1$.

This case is symmetric to the previous one; the path $Y[w, v] \cup Q_1[v, x_2]$ is a shortest path wholly contained in $\text{reg}(w, H) \cup H$, contradicting the definition of X.

Case 2.3: $u, v \in Q_4$.

In this case we are left with a configuration similar to the one depicted on Fig. 5.

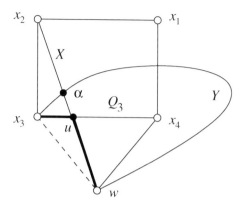

Fig. 4. Case 2.1.

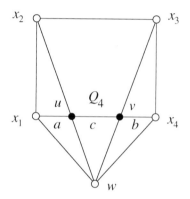

Fig. 5. The final configuration.

We now assume that $k \geqslant \lfloor 5D/7 \rfloor$. Let $a = d_G(x_1, u)$, $b = d_G(v, x_4)$ and $c = d_G(u, v)$. Without loss of generality we assume that $a \leqslant b$. Note that since $a + c + b \leqslant D$, we have $a \leqslant (D - c)/2$. Note also that we have $d_G(u, x_2) \leqslant D - \lfloor k \rfloor - 1$ and $d_G(v, x_3) \leqslant D - \lfloor k \rfloor - 1$ because $w \notin \Gamma_k(u)$ and $w \notin \Gamma_k(v)$.
    If $c \geqslant D/7$, then

$$d_G(x_1, x_2) \leqslant a + d_G(u, x_2) \leqslant \frac{1}{2}(D - c) + D - \lfloor k \rfloor - 1$$
$$\leqslant \frac{1}{2}\left(D - \frac{1}{7}D\right) + D - \frac{5}{7}D + 1 - 1 = \frac{5}{7}D - 1. \tag{1}$$

It follows that $d_G(x_1, x_2) \leqslant 5D/7 - 1 < k$: a contradiction since H is a 4-gon, forcing to have $d_G(x_1, x_2) > k$.
    If $c \leqslant D/7$, then

$$d_G(x_2, x_3) \leqslant d_G(x_2, u) + c + d_G(v, x_3) \leqslant 2D - 2\lfloor k \rfloor - 2 + \frac{1}{7}D$$

$$\leqslant 2D - 2\left(\frac{5}{7}D\right) - 2 + \frac{1}{7}D \;=\; \frac{5}{7}D - 2 \,. \tag{2}$$

It follows that $d_G(x_2, x_3) \leqslant 5D/7 - 2 < k$, leading to contradiction since H is a 4-gon, forcing to have $d_G(x_2, x_3) > k$.

Hence the Case 2.3 is impossible as well. The theorem follows.   □

We are now ready to state and prove our main result.

Theorem 3. *For every planar graph G of diameter D, for every $\epsilon > 0$, and for every integer $k \geqslant (5/7 + \epsilon)D$, we have $\gamma_k(G) < 3/\epsilon + 6$.*

Proof. The connected subgraph H constructed in Theorem 2 consists of the union of at most 6 shortest paths, each of length at most D, and is a k-dominating set for G for $k \geqslant \lfloor 5D/7 \rfloor$. Choosing a minimum cardinality λ-dominating set for each path, we obtain a $(k + \lambda)$-dominating set B for G. By Lemma 2, $|B| \leqslant 6 \lceil D/(2\lambda + 1) \rceil$.

Let us set $\lambda = \lceil \epsilon D \rceil$. B is therefore a k'-dominating set for G for every integer $k' \geqslant \lfloor 5D/7 \rfloor + \lceil \epsilon D \rceil \geqslant (5/7 + \epsilon)D$. We have, for $\epsilon > 0$,

$$|B| \leqslant 6\left\lceil \frac{D}{2\lambda + 1} \right\rceil \;<\; \frac{6D}{2\lceil \epsilon D \rceil + 1} + 6 \;<\; \frac{6D}{2\epsilon D + 1} + 6 \;<\; \frac{3}{\epsilon} + 6,$$

completing the proof of Theorem 3.   □

2.5   Conjecture and Worst-Case

We leave open the problem of bounding $\gamma_k(G)$ for $k \geqslant D/2$, and make the following conjecture.

Conjecture 1. There exists a constant c_0 such that for every planar graph G of diameter D, and for every $k \geqslant D/2$, $\gamma_k(G) \leqslant c_0$.

Here we show that $c_0 \geqslant 4$.

Theorem 4. *For every even diameter $D \geqslant 8$, there exists a planar graph G of diameter D for which $\gamma_{D/2}(G) = 4$.*

Proof. Let $D = 2t$ for $t \geqslant 4$. The graph denoted by G_t is composed of

(1) two nodes A and B called *poles*;
(2) D disjoint paths $P_1, \ldots, P_D$ called *meridians*, each of length D, joining A to B;
(3) a cycle of length $2D$, called *equator*, joining the middle nodes of each meridians and with exactly one node of degree two between two consecutive meridians.

(See Fig. 6 for $t = 4$).

Observe that for a cycle C_{2D} of length $2D$, $\gamma_{D/2}(C_{2D}) = 2$. Hence taking $S = \{A, B\}$, it is clear that all the nodes of the meridians are covered, but

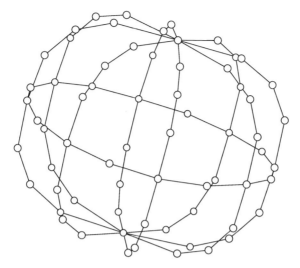

Fig. 6. The planar graph embedded on a sphere for $D = 8$ such that $\gamma_4(G) = 4$.

the degree 2 nodes of the equator are note covered. Thus to complete S into a dominating set, we have to add two opposite nodes u, v on the equator. The resulting set $S = \{A, B, u, v\}$ is then a $D/2$-dominating set of G_t. It follows that $\gamma_{D/2}(G_t) \leqslant 4$.

Now if we take two opposite nodes different from A and B belonging to two distinct meridians we will not cover all the nodes of the $2t$ meridians. In this case we will need more than 4 nodes ($t \geqslant 4$) to cover all the nodes of the meridians. This completes the proof of the lower bound.   □

3   Treewidth and Chordality Bounded Graphs

In this section we establish some additional results concerning the value of $\gamma_k(G)$ for various graph classes. We use the notion of tree-decomposition introduced by Robertson and Seymour in their work on graphs minors [13]. A *tree-decomposition* of a graph $G = (V, E)$ is a tree $T = (\tilde{V}, \tilde{E})$ whose nodes are subsets of V, such that

1. $\bigcup_{X \in \tilde{V}} X = V$;
2. for every edge $(u, v) \in E$, there exists a node set $X \in \tilde{V}$ such that $u, v \in X$; and
3. for every $X, Y, Z \in \tilde{V}$, if Y is on the path from X to Z in T, then $X \cap Z \subseteq Y$.

The *width* of the tree-decomposition T is $\omega(T) = \max_{X \in \tilde{V}}\{|X| - 1\}$. The *treewidth* of a graph G is $\min_T\{\omega(T)\}$, where the minimum is taken over all the tree-decompositions T of G. Note that trees have treewidth 1. We use the following property whose proof can be founded in [3, pp. 256].

Property 1. Let $\{X_1, X_2\}$ be any edge of a tree-decomposition T of G, and let T_1, T_2 be the components of $T \setminus \{X_1, X_2\}$, with $X_1 \in T_1$ and $X_2 \in T_2$. Then $I = X_1 \cap X_2$ separates $U_1 = \left(\bigcup_{A \in T_1} A\right) \setminus I$ from $U_2 = \left(\bigcup_{B \in T_2} B\right) \setminus I$ in G.

Theorem 5. *For every graph $G = (V, E)$ of diameter D and treewidth t, and for every $k \geqslant \lfloor D/2 \rfloor$, we have $\gamma_k(G) \leqslant t + 1$.*

Proof. (Outline) Let $T = (\tilde{V}, \tilde{E})$ be a width t tree-decomposition of G, i.e., such that $|X| \leqslant t + 1$ for every $X \in \tilde{V}$. We prove the theorem by showing that there exists some set $X^* \in \tilde{V}$ that k-dominates G. For $X \in \tilde{V}$, let $T(X)$ denote the subtree of T rooted at X, and let $\mathrm{Succ}(X) = \bigcup_{Y \in T(X)} Y$. Also let $\mathrm{Far}(X) = V \setminus \Gamma_k(X)$.

We construct a cycle-free path $X_1, X_2, \ldots, X_m$ in T step by step, such that X_1 is the root of T, $X_m = X^*$, and X_i does not k-dominate G for $i < m$. The initial step consists of selecting X_1 and checking whether it k-dominates G or not. Starting with a current node X_i of T, Step i consists of choosing the unique child X_{i+1} of X_i such that $\mathrm{Succ}(X_{i+1})$ contains $\mathrm{Far}(X_{i+1})$. It suffices to prove by induction on i that Step i is doable, namely, that there exists such a unique child X_{i+1}. Indeed, since the length of a cycle-free path in T is finite, we stop the process, in the worst-case, reaching a leaf of T. The full proof can be founded in [9]. □

It is known that graphs of genus g and diameter D have a treewidth bounded by $O(gD)$, cf. [4]. For planar graph one can tighten this bound. For that we need the following definition. A plane graph is a 1-*outerplanar embedding* if all its nodes lie on the outer face. For $r \geqslant 2$, a plane graph is an r-*outerplanar embedding* if deleting all the nodes of the outer face yields an $(r-1)$-outerplanar embedding of the remaining graph. A graph is r-outerplanar if it has a r-outerplanar embedding. We refer to 1-outerplanar graphs (such as trees and $K_{2,3}, K_4$-free graphs) simply as outerplanar graphs.

Theorem 6. *For every planar G of diameter D, and for every $k \geqslant \lfloor D/2 \rfloor$, we have $\gamma_k(G) \leqslant 3D + 3$.*

Proof. Assume that G is r-outerplanar for $r \geqslant 1$. We first assume that $r \geqslant 2$. Consider an r-outerplanar embedding of G. Let V_1 denote the outer layer of G, namely, the set of all the nodes lying on the outer face of G. Then, successively for $i = 2, \ldots, r$, let V_i denote the set of all the nodes lying on the outer face of $G \setminus (V_1 \cup \cdots \cup V_{i-1})$. For every i, consider some $u_i \in V_i$. Observe that for every $j > i$, we have $d_G(u_i, u_j) \geqslant j - i$. Indeed, if $d_G(u_i, u_j) < j - i$, then there is a path of at most $j - i - 1$ edges linking a node of V_i to a node of V_j, which is impossible for a plane embedding of G (there is no edge between a node of V_i and a node of V_{i+2}). Thus $d_G(u_1, u_r) \geqslant r - 1$, implying that $D \geqslant r - 1$. Obviously, the latter result holds for $r = 1$.

Now, in [1], it is shown that r-outerplanar graphs have treewidth at most $3r - 1$. The claim follows by Theorem 5. □

A graph is *c-chordal* if every induced cycle is of length at most c. Triangulated graphs are precisely 3-chordal graphs. Using a result of [7] that shows that c-chordal graphs have a tree-decomposition $T = (\tilde{V}, \tilde{E})$ such that for every $X \in \tilde{V}$ and all $u, v \in X$, $d_G(u, v) \leqslant c/2$, one can show (full proof is given in [9]).

Theorem 7. *For every c-chordal graph G of diameter D, and for every $k \geqslant \lfloor D/2 \rfloor + \lfloor c/2 \rfloor$, we have $\gamma_k(G) = 1$.*

4   Application to Compact Routing

Let us now switch to discussing an application of our results in the area of communication networks. A point-to-point communication network is modeled as a graph $G = (V, E)$, where the set of nodes represent the processors of the network and every pair of two opposite arcs represents a bidirectional communication link. A *routing scheme* R is a distributed algorithm whose role is to deliver messages between nodes of the network. The routing scheme consists of certain *distributed data structures* in the network, and a *delivery protocol*, which can be invoked in any node u with two parameters: a *routing label* $\mathcal{L}(v)$ of the destination node v, and the message's information field. The message is delivered to v via a sequence of transmissions determined uniquely by the distributed data structure, i.e., at each intermediate node along the route, the routing scheme decides the next neighbor to which the message should be forwarded. The length of the route traversed by a message from u to v in the graph G according to the routing scheme R is denoted by $\rho_R(u, v)$. The *dilation* of a routing scheme R is the maximal route length of a path traversed by a message, formally $\max_{u \neq v} \{\rho_R(u, v)\}$.

An *interval routing scheme* R on G is a routing scheme consisting of a pair $(\mathcal{L}, \mathcal{I})$, generated in the preprocessing step, where $\mathcal{L}$ is a *node-labeling*, $\mathcal{L} : V \to \{1, \ldots, n\}$, and $\mathcal{I}$ is an *arc-labeling*, $\mathcal{I} : E \to 2^{\mathcal{L}(V)}$, that satisfy the following condition. For any node u, the collection of sets that label all the outgoing arcs of u forms a partition of the name range (possibly excluding u itself[1]). Formally, for every $u \in V$, denoting by E_u the set of arcs incident to u,

1. $\bigcup_{e \in E_u} \mathcal{I}(e) = \{1, \ldots, n\} \setminus \{\mathcal{L}(u)\}$,
2. $\mathcal{I}(e_1) \cap \mathcal{I}(e_2) = \varnothing$ for every two distinct arcs $e_1, e_2 \in E_u$.

The delivery protocol is defined as follows. Given a destination node v, set the first header to be $h = \mathcal{L}(v)$. Also, for every node u, receiving a message with header h, first check if $\mathcal{L}(u) = h$, and deliver the message locally if equality holds (meaning that u is the message's final destination). If not, then forward the message with the same header h on the output port corresponding to the unique arc (u, w) such that $h \in \mathcal{I}(u, v)$, namely, the arc whose labeling set contains the destination label.

---

[1] A labeling excluding u from its arc-labels is termed *strict*. Although non-strict labeling may produce more compact schemes, in this paper we restrict our attention to strict labeling only.

Note that any interval routing scheme can be implemented by classical routing tables. The main difference is in the coding of the table.

Given an integer n and a subset $I \subseteq \{1, \ldots, n\}$, define the *compactness of I* w.r.t. n, denoted $c_n(I)$, as the smallest integer k such that I can be represented by the union of k intervals $[a, b]$ of consecutive integers from $\{1, \ldots, n\}$, with n and 1 being considered as consecutive (cyclically). The *compactness* of an interval routing scheme $R = (\mathcal{L}, \mathcal{I})$ on G is the maximum, over all arcs $e \in E$, of the compactness $c_n(\mathcal{I}(e))$ of the set $\mathcal{I}(e)$ labeling e. The *total compactness* is the sum $\sum_{e \in E} c_n(\mathcal{I}(e))$.

Because interval routing schemes can be implemented by routing tables, we deal with the compactness of a routing table. Intuitively, smaller compactness and degrees imply smaller routing tables. The interval routing strategy presented in [14], based on routing on a minimum spanning tree, has compactness 1 (for every graph of diameter D) and dilation at most $2D$. It is known that there are worst-case graphs for which every routing table of compactness 1 has dilation at least $2D - 3$ [15]. It is also known that every graph has routing table of compactness $\sqrt{n \ln n} + O(1)$ with dilation at most $\lceil 1.5D \rceil$ [10], whereas there are worst-case graphs for which every routing table of compactness k has dilation at least $\lfloor 1.5D \rfloor - 1$ for every $k = \Omega(\frac{n}{D \log(n/D)})$ [5]. Other results can be found on the recent survey [6].

Here we show a trade-off between dilation and compactness.

Theorem 8. *Let G be a graph of diameter D with n nodes and m edges. Let $k \geqslant 0$, and let $t = \gamma_k(G)$. Then, G has a routing table with dilation at most $D + k$, compactness at most $t/2 + 1$, and total compactness at most $tn + 2m$. Moreover, if $t / \log n \to +\infty$, then the compactness can be reduced to $t/4 + o(t)$.*

Proof. Let $\{x_1, \ldots, x_t\}$ be a k-dominating set in G of minimal cardinality $t = \gamma_k(G)$. Construct a partition of the graph nodes into t pairwise disjoint connected regions $R_i \subseteq \Gamma_k(x_i)$ around each x_i, such that each region R_i consists of nodes closest to x_i (breaking tie arbitrarily) and the union of the regions covers all the nodes. Obviously the radius of each region is at most k.

For every x_i, construct a shortest path spanning tree T_i for G. Let $\hat{T}_i$ be the restriction of T_i to the region R_i, i.e., $\hat{T}_i = T_i \cap R_i$.

Partition the range of integers $[1, n]$ into t contiguous intervals I_i of size $|R_i|$, for $i \in \{1, \ldots, t\}$. For each region R_i, assign each node $v \in R_i$ a distinct label $\mathcal{L}(v)$ from the range I_i, in DFS order, starting at x_i. Use these labels to define an interval routing scheme on $\hat{T}_i$ as in [14]. The compactness is 2 (instead of 1) because a cyclic interval in I_i has to be represented by two sub-intervals of I_i. For every two nodes $u, w \in R_i$, this scheme yields a shortest route on $\hat{T}_i$ (albeit perhaps not shortest in G).

In addition, for every $1 \leqslant i \leqslant t$ and for every node $w \notin R_i$, add the interval I_i to the arc connecting w to its parent in T_i.

As the regions are disjoint, it is easy to verify that for every node v, all the intervals assigned to the arcs of v are pairwise disjoint. Overall, at most $t/2 + 1$ intervals were assigned to each arc because every subset of at most t intervals has no more than $t/2$ non contiguous intervals (one more sub-interval is required for the own region of v). In fact, exactly $t + \deg(v)$ intervals altogether ($\deg(v) + 1$ intervals for R_i and $t - 1$ for all the others) are assigned to the arcs of a node v of degree $\deg(v)$. Thus the total compactness is $tn + 2m$.

Observe that the resulting route between nodes belonging to the same region has length at most $2k \leqslant D + k$. As for routing from a node $u \in R_i$ to a node $w \in R_j$, $j \neq i$, note that the first segment of the route, until reaching R_j, proceeds along a shortest path from u to x_j, and once entering R_j, the remaining segment of the route follows a shortest path. Hence the total length of the route cannot exceed $d_G(u, x_j) + d_G(x_j, w) \leqslant D + k$.

Actually, the bound on the compactness can be slightly reduced to $t/4 + o(t)$, for t large enough, using the result of [8]. This result shows that, for every t, there exists a permutation π of $\{1, \ldots, t\}$ such that the compactness of every set F of a family $\mathcal{F}$ of subsets of $\{1, \ldots, t\}$, with $|\mathcal{F}| < \exp(t/2)/t$, satisfies:

$$c_t(\pi(F)) < \frac{t}{4} + \frac{1}{4}\sqrt{2t \ln(|\mathcal{F}| \cdot n)},$$

where $\pi(F)$ denotes the set $\{\pi(x) \mid x \in F\}$. Let $\mathcal{I}(e)$ be the set of labels assigned to arc e by the previous interval routing scheme. Set $\mathcal{F} = \{F(e) \mid e \in E(G)\}$, where $F(e) = \{i \mid I_i \subseteq \mathcal{I}(e)\}$ corresponds to the set of region's indices contained in $\mathcal{I}(e)$. Note that the range of labels of each region (except for v's) is wholly contained in $\mathcal{I}(e)$. Since $\bigcup_{j=1}^{t} I_j = [1, n]$, we have that $c_n(\mathcal{I}(e)) \leqslant c_t(F(e)) + 1$. Observe that the relative order of the ranges can be chosen arbitrary: all the labels of some I_j can be chosen to be larger or smaller than all the labels of any other range. So, indices of I_j's ranges can be permuted by some permutation π of $[1, t]$ in order to decrease $c_t(\pi(F(e)))$. By [8], if $|\mathcal{F}| < \exp(o(t))/t$, then there exists π_0 such that $c_t(\pi_0(F(e))) < t/4 + o(t)$, for every $F(e) \in \mathcal{F}$. Note that $|\mathcal{F}| = 2m$. So, the condition $|\mathcal{F}| < \exp(o(t))/t$ can be replaced by $t/\log n \to +\infty$, as $m = O(n^2)$. □

This result can be viewed as a generalization of [10], originally proved for $k = \lceil D/2 \rceil$. In that paper it was shown that every graph G satisfies $\gamma_k(G) \leqslant \lceil \sqrt{n \ln n} \rceil$, and the compactness derived therein was $\lceil \sqrt{n \ln n} \rceil + 1$. Thus Theorem 8 improves the compactness bound obtained in [10] by an asymptotic factor of 4. The same kind of construction has been also implicitly used in [16]. Specifically, it is shown in [16] that every graph G satisfies $\gamma_k(G) \leqslant \sqrt{n}$ for $k \geqslant 2D/3$. Thus, every graph has routing table of compactness at most $\sqrt{n}/4 + o(\sqrt{n})$ for dilation $\lceil 5D/3 \rceil$. (The original compactness result of [16] was $\lceil \sqrt{n} \rceil + 1$.)

Combining Theorem 3 and Theorem 8, we obtain the following.

Corollary 3. *Let $\epsilon > 0$ be an arbitrary constant. Every planar graph has a routing table with dilation at most $\lceil (12/7 + \epsilon)D \rceil$, and constant compactness.*

Note that Conjecture 1 for $c_0 = 4$ implies that every planar graph of diameter D enjoys a routing table with dilation at most $\lceil 1.5D \rceil$ and compactness at most 3.

References

1. Hans Leo Bodlaender. A partial k-arboretum of graphs with bounded treewidth. *Theoretical Computer Science*, 209:1–45, 1998.
2. Victor Chepoi and Yann Vaxes. On covering bridged plane triangulations with balls. Manuscript submitted, 2000.
3. Reinhard Diestel. *Graph Theory (second edition)*, volume 173 of Graduate Texts in Mathematics. Springer, February 2000.
4. David Eppstein. Diameter and treewidth in minor-closed graph families. *Algorithmica*, 27:275–291, 2000.
5. Cyril Gavoille. On the dilation of interval routing. *The Computer Journal*, 43(1):1–7, 2000.
6. Cyril Gavoille. A survey on interval routing. *Theoretical Computer Science*, 245(2):217–253, 2000.
7. Cyril Gavoille, Michal Katz, Nir A. Katz, Christophe Paul, and David Peleg. Approximate distance labeling schemes. In 9th *Annual European Symposium on Algorithms (ESA)*, volume Lectures Notes in Computer Science. Springer, August 2001. To appear.
8. Cyril Gavoille and David Peleg. The compactness of interval routing. *SIAM Journal on Discrete Mathematics*, 12(4):459–473, October 1999.
9. Cyril Gavoille, David Peleg, André Raspaud, and Eric Sopena. Small k-dominating sets in planar graphs with applications. Research Report RR-1258-01, LaBRI, University of Bordeaux, 351, cours de la Libération, 33405 Talence Cedex, France, May 2001.
10. Rastislav Královič, Peter Ružička, and Daniel Štefankovič. The complexity of shortest path and dilation bounded interval routing. *Theoretical Computer Science*, 234(1-2):85–107, 2000.
11. Laszlo Lovász. On the ratio of optimal integral and fractional covers. *Discrete Mathematics*, 13:383–390, 1975.
12. G. MacGillivray and K. Seyffarth. Domination numbers of planar graphs. *Journal of Graph Theory*, 22(3):213–229, 1996.
13. Neil Robertson and Paul D. Seymour. Graph minors. II. Algorithmic aspects of tree-width. *Journal of Algorithms*, 7:309–322, 1986.
14. Nicola Santoro and Ramez Khatib. Labelling and implicit routing in networks. *The Computer Journal*, 28(1):5–8, February 1985.
15. Savio S. H. Tse and Francis C. M. Lau. An optimal lower bound for interval routing in general networks. In 4th *International Colloquium on Structural Information & Communication Complexity (SIROCCO)*, pages 112–124. Carleton Scientific, July 1997.
16. Savio S. H. Tse and Francis C. M. Lau. Some results on the space requirement of interval routing. In 6th *International Colloquium on Structural Information & Communication Complexity (SIROCCO)*, pages 264–279. Carleton Scientific, July 1999.
17. Jan van Leeuwen and Richard B. Tan. Interval routing. *The Computer Journal*, 30(4):298–307, 1987.

Lower Bounds for Approximation Algorithms for the Steiner Tree Problem

Clemens Gröpl, Stefan Hougardy, Till Nierhoff, and Hans Jürgen Prömel

Humboldt-Universität zu Berlin
Institut für Informatik
10099 Berlin, Germany
{groepl,hougardy,nierhoff,proemel}@informatik.hu-berlin.de

Abstract. The Steiner tree problem asks for a shortest subgraph connecting a given set of terminals in a graph. It is known to be APX-complete, which means that no polynomial time approximation scheme can exist for this problem, unless P=NP. Currently, the best approximation algorithm for the Steiner tree problem has a performance ratio of 1.55, whereas the corresponding lower bound is smaller than 1.01. In this paper, we provide for several Steiner tree approximation algorithms lower bounds on their performance ratio that are much larger. For two algorithms that solve the Steiner tree problem on quasi-bipartite instances, we even prove lower bounds that match the upper bounds. Quasi-bipartite instances are of special interest, as currently all known lower bound reductions for the Steiner tree problem in graphs produce such instances.

1   Introduction

Given a graph $G = (V, E)$, a length function on its edges, and a set $R \subseteq V$ of *terminals*, a *Steiner tree* is a connected subgraph of G spanning all vertices in R. The Steiner tree problem in graphs is to find a shortest Steiner tree. This problem is a classical NP-hard problem [10] and even worse it is also known to be APX-complete, i.e., there exists some constant $c > 1$ such that no polynomial time approximation algorithm for this problem can have a performance ratio smaller than c, unless P=NP. (The performance ratio of an approximation algorithm is the largest ratio between the length of a solution found by the algorithm and the optimal length.)

The Steiner tree problem appears in many different applications, as for example in VLSI-design, in the design of telecommunication networks, and in the reconstruction of phylogenetic trees in biology. Therefore it is an important task to find good approximation algorithms for this problem and, eventually, to prove that no better approximation algorithms can be possible, unless P=NP.

Currently, the best approximation algorithm for the Steiner tree problem is due to Robins and Zelikovsky [7] and has a performance ratio of $1 + \frac{1}{2} \ln 3 < 1.550$. On the other hand, the largest known lower bound for the approximability of the Steiner tree problem has a value not exceeding 1.01 [4,12].

This large gap between lower and upper bounds suggests that there might be still some room for improvements on both sides. There exist several approximation algorithms for the Steiner tree problem in graphs that achieve a performance ratio less than 2 [6]. But

A. Brandstädt and V.B. Le (Eds.): WG 2001, LNCS 2204, pp. 217–228, 2001.
© Springer-Verlag Berlin Heidelberg 2001

Table 1. Summary of results

	approximation algorithm	known upper bound	our lower bound
general graphs	Relative Greedy Algorithm	1.694	1.333
	Loss Contracting Algorithm	1.550	1.200
quasi-bipartite graphs	Iterated 1-Steiner Heuristic	1.500	1.500
	Loss Contracting Algorithm	1.279	1.195
	Greedy-MSS	$1.21\overline{6}$	$1.21\overline{6}$

for none of these algorithms it is known whether their analysis is tight. Therefore it is not clear which of these algorithms actually achieves the best performance ratio.

An instance of the Steiner tree problem is *quasi-bipartite* if the set $V \setminus R$ is independent. Quasi-bipartite graphs are an important special case as the instances resulting from lower bound reductions [4,12] have this form. For these graphs there exist several approximation algorithms for which better performance ratios have been proved than in the general case.

In this paper, we provide for several Steiner tree approximation algorithms lower bounds on their performance ratio. For two algorithms that solve the Steiner tree problem on quasi-bipartite instances, we even prove lower bounds that match the upper bounds. Such lower bound results, even if they are not tight, allow to estimate the quality of a given performance analysis and can provide ideas on how to improve it. Moreover, these lower bounds give good examples of instances on which every Steiner tree approximation algorithm must perform well, and thus might yield to better approximation algorithms for the Steiner tree problem.

All algorithms considered in this paper are based on a general greedy framework which we describe in Section 2. In Sections 3 and 4 we present lower bounds for two approximation algorithms for the Steiner tree problem in general graphs. Lower bounds for approximation algorithms for the quasi-bipartite case are given in Sections 4, 5 and 6. For two of these algorithms we even provide tight lower bounds. Our results are summarized in Table 1.

2   Notations and a General Framework for Greedy Algorithms

Given a graph $G = (V, E)$, a length function on its edges, and a set $R \subseteq V$ of *terminals*, a *Steiner tree* is a connected subgraph of G spanning all vertices in R. We denote a Steiner minimum tree by SMT and its length by smt. A Steiner tree usually contains not only vertices from R but also from $V \setminus R$. These vertices are called *Steiner vertices*. Every Steiner tree can be split into so called *full components*. A full component is a Steiner tree for a subset of R in which every terminal is a leaf. The length of a full component is the sum of its edge lengths. Therefore, a Steiner tree is a collection of full components which is connected and covers R.

{ FC is a set of full components }
$i \leftarrow 0$
While an improving full component exists:
    Find $T_{i+1} \in FC$ that minimizes f_i
    $i \leftarrow i + 1$
$i_{max} \leftarrow i$
Output Steiner tree using $T_1, \ldots, T_{i_{max}}$

Fig. 1. A general framework for greedy algorithms.

Most of the known approximation algorithms for the Steiner tree problem in graphs fit into the general framework shown in Figure 1. Let FC be a set of full components. The algorithm chooses the $(i+1)$-st full component $T_{i+1} \in FC$ such that a certain function $f_i: FC \rightarrow \mathbb{R}_+$ is minimized. The algorithm stops when no further improvement is possible. The precise meaning of this condition depends on the function f_i. Finally a Steiner tree is output that uses the full components that have been selected in the while-loop.

3   Relative Greedy Algorithm

In this section we consider Zelikovsky's relative greedy algorithm [13], which achieves a performance ratio of less than 1.694. This algorithms fits into the general greedy framework for Steiner tree approximation algorithms as follows. Let $G = (V, E)$ be a graph with a weight function on the edges and $R \subseteq V$ be the set of terminals. Fix some constant $k \in \mathbb{N}$ and define FC to be the set of all full components with at most k terminals. The weight of a full component $X \in FC$ is simply its length and will be denoted by $|X|$. Note that FC can be computed in polynomial time.

From G one can easily derive the so called *terminal distance graph* G' which is a complete graph on the set R where the weight of an edge is the length of a shortest path in G connecting the two endpoints of this edge. If $T_1, \ldots T_i$ are some full components in FC then $MST(R/T_1 \ldots T_i)$ denotes a minimum spanning tree in the graph that is obtained from G' by contracting each of the full components $T_1, \ldots, T_i$. Its length is denoted by $mst(R/T_1 \ldots T_i)$.

Now the function f_i on a full component $X \in FC$ is defined as follows:

$$f_i(X) := \frac{|X|}{mst(R/T_1 \ldots T_i) - mst(R/T_1 \ldots T_i X)} .$$

The while-loop of the algorithm will be executed, as long as there exists some full component $X \in FC$ with $f_i(X) \leq 1$. Zelikovsky proved that on termination of the relative greedy algorithm the following inequality holds:

$$|T_1| + \ldots + |T_{i_{max}}| \leq (1 + \ln 2 + \varepsilon) \cdot smt .$$

If ε is chosen small enough this value is smaller than 1.694 .

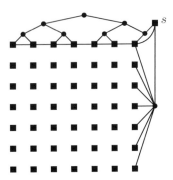

Fig. 2. Part of the instance F_3.

Theorem 1. *The performance ratio of the relative greedy algorithm is at least* $4/3$.

Proof. We will construct a family of instances F_i for the Steiner tree problem and prove that the ratio between the length of a Steiner minimum tree and the length of a Steiner tree that is found by the relative greedy algorithm tends to $4/3$ as $i \to \infty$.

The instance F_i has $(2^i - 1)(2^i - 1) + 1$ terminals which are arranged in a $(2^i - 1) \times (2^i - 1)$-grid with one extra terminal s. We assume that s belongs to every row and every column of the $(2^i - 1) \times (2^i - 1)$-grid, i.e., we can think of s as the last row and last column of a $2^i \times 2^i$-grid. For each row, we have a binary tree, called the *row tree*, which has depth i and the 2^i terminals of a row as its leaves. For each column we have a star, called the *column star*, which has the 2^i terminals of a column as its endpoints. No two row trees or column stars have a vertex in common, except for terminals. Thus, in total the instance F_i has $(2^i - 1)(2^i - 1) + (2^i - 1) = 2^i(2^i - 1)$ non-terminal vertices. Finally, we connect any two consecutive terminals in a row by an edge. In Figure 2 one row tree and one column star and the edges between any two consecutive terminals in one row of the instance F_3 are shown.

We now have to specify the weights for the edges of the instance F_i. For all row trees all edges within one level have the same weight. The edges of the first and second level of the binary tree, i.e., the six edges incident to the root of the binary tree and its two children, have weight $1/4$. The 2^i edges of the i-th level of the binary tree, i.e., the edges incident to the terminals have weight 1. All other edges in the binary tree have weight $1/2$. The total weight of all row trees is therefore

$$(2^i - 1)\left(2^i \cdot 1 + (2^i - 8) \cdot \frac{1}{2} + 6 \cdot \frac{1}{4}\right) = (2^i - 1)\left(3 \cdot 2^{i-1} - \frac{5}{2}\right).$$

Next we specify the weights for the edges of the column stars. All edges within one column star have the same weight, which depends on the column. The columns are numbered from left to right by $1, 2, \ldots, 2^i - 1$. We number a column star with a value that is obtained from the binary representation of the column number read from right to left. E.g., for $i = 3$ the binary representation of the coulmn numbers is 001, 010, 011, 100, 101, 110, 111 and these numbers read from right to left give 100, 010, 110, 001, 101, 011, 111. Thus the column stars are numbered $4, 2, 6, 1, 5, 3, 7$ from left to right.

Table 2. Small lower bound instances for the relative greedy algorithm.

i	lower bound	edge weights (levelwise)	lower bound	edge weights (levelwise)
3	$\frac{47}{38} \doteq 1.2368$	$1, \frac{1}{4}, \frac{1}{4}$	1.2895	$1, \frac{1}{4}, \frac{1}{4}$
4	$\frac{225}{172} \doteq 1.3081$	$1, \frac{1}{2}, \frac{1}{4}, \frac{1}{4}$	1.3151	$1, \frac{1}{2}, \frac{1}{4}, \frac{1}{4}$
5	$\frac{965}{728} \doteq 1.3255$	$1, \frac{1}{2}, \frac{1}{2}, \frac{1}{4}, \frac{1}{4}$	1.3290	$1, \frac{1}{2}, \frac{3}{8}, \frac{1}{4}, \frac{1}{4}$
6	$\frac{3981}{2992} \doteq 1.3305$	$1, \frac{1}{2}, \frac{1}{2}, \frac{1}{2}, \frac{1}{4}, \frac{1}{4}$		

All edges in the column stars with numbers 1 to $2^{i-1} + 2$ have the same weight. We define it as $(2 - 2^{1-i})(1 - 3 \cdot 2^{-i-1})$. The edges in all other column stars get the weight $2 - 2^{1-i}$. As each column star has 2^i edges, the total weight of all column stars is

$$2^i \cdot (2 - 2^{1-i}) \left((2^{i-1} + 2)(1 - 3 \cdot 2^{-i-1}) + (2^{i-1} - 3) \right) = 2 \cdot (2^i - 1) \cdot (2^i - \frac{7}{4} - \frac{3}{2^i}).$$

Finally we have to specify the edges connecting two consecutive terminals in a row. They are all defined as $(2 - 3 \cdot 2^{-i})(2 - 2^{1-i})$.

Note that the set of all row trees is a solution to the Steiner tree problem (in fact, it is easily seen that it is the optimum solution for the instances F_i). We claim that the relative greedy algorithm started with $k = 2^i$ will return the set of all column stars as solution. Therefore the performance ratio of the relative greedy algorithm is at least

$$\frac{2 \cdot (2^i - 1) \cdot (2^i - \frac{7}{4} - \frac{3}{2^i})}{(2^i - 1)(3 \cdot 2^{i-1} - \frac{5}{2})}.$$

This expression tends to $4/3$ as i goes to infinity, which proves the statement of the theorem.

To prove that the relative greedy algorithm indeed returns the set of column stars as solution we assume that it selects as the j-th full components the column star numbered with j, i.e., we assume that the ties between the weights of the column stars are broken exactly in this way. By adding a weight of $j \cdot \varepsilon$ to the j-th column star one could break all the ties, but this would make the calculations more complicated. It is now a straightforward but quite tedious calculation to show that the cost function of the relative greedy algorithm is minimized in the j-th step by the column star numberd with j. Details of this calculation are omitted in this version of the paper due to space restrictions. □

In column two of Table 2 the lower bounds that are obtained by the construction described in the proof of Theorem 1 are given. The third column contains the weights of the edges of the row trees. The construction given in the proof is not optimal as is indicated by columns four and five of this table. The lower bounds can be improved slightly by choosing better weights for the edges of the column stars and row trees. For $i = 3, 4, 5$ we computed such optimal weights and obtained the lower bounds given in column four of Table 2. We were not able to calculate a general formula for the lower bound of these instances, but it seems to be the case that the lower bounds also converge to $4/3$.

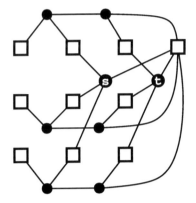

Fig. 3. A 1.2 lower bound instance for the loss contracting algorithm. All edges of the row trees have weight 5. The edges incident to s and t have weights 9 and 6, respectively.

4   Loss Contracting Algorithm

Currently the best approximation algorithm for the Steiner tree problem in graphs the loss contracting algorithm of Robins and Zelikovsky [7] which has a performance ratio of at most $1 + \frac{\ln 3}{2} < 1.550$. The loss contracting algorithm is very similar to the relative greedy algorithm, but instead of contracting a selected full component T entirely, it contracts only a subset $Loss(T)$ of its edges in order to achieve the same effect at a lower price. The *loss* of a full component T is a minimum length forest contained in T such that every Steiner vertex of T is connected to a terminal. Its length is denoted by $loss(T)$. The contracted parts are connected by a minimum spanning tree using edges from the terminal distance graph. We denote its length by $m(\cdot) := mst(R/Loss(\cdot))$. Clearly, the length of the Steiner tree corresponding to the full components $T_1, \ldots, T_i$ is $loss(T_1, \ldots, T_i) + m(T_1, \ldots, T_i)$.

The loss contracting algorithm fits into the framework of Figure 1. In each step, the loss contracting algorithm chooses a full component that minimizes the function

$$f_i(T) := \frac{loss(T)}{m(T_1 \ldots T_i) - m(T_1 \ldots T_i T)}.$$

It stops when there are no more improving full components, i. e., $\min_{T \in FC} f_i(T) \geq 1$.

Theorem 2. *The performance ratio of the loss contracting algorithm is at least 1.2.*

Proof. We use a construction similar to the one used in the proof of Theorem 1. Unfortunately, this construction breaks down for grids of size larger than 4×4. We explain the best weight assignment for a 4×4-grid as shown in Figure 3. The edges of the row trees have weight 5. The edges incident to s and t have weights 9 and 6, respectively. We denote the full components containing s and t by T_s and T_t. The optimal Steiner tree uses the row trees and has length 75.

Notice that the edges in $MST(R)$ connecting the vertices of the upper row of terminals have lengths 10, 15, and 10 from left to right. For each row tree, the *loss* is

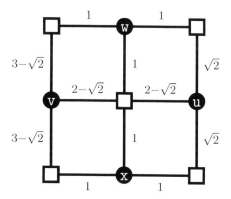

Fig. 4. A quasi-bipartite 1.195 lower bound instance for the loss contracting algorithm.

10 and f_0 is $10/\big(3 \cdot 35 - (2 \cdot 35 + 15)\big) = 0.5$. One can easily check that sub-trees of the row trees are worse. Since $f_0(T_s) = 9/\big(3 \cdot (35 - (20 + 9))\big) = 0.5$ and $f_0(T_t) = 6/\big(3 \cdot (35 - (25 + 6))\big) = 0.5$, the algorithm is allowed to choose T_s. (Choosing T_s could be forced by perturbing the edge weights.) In the second step, the value of f_1 for each (complete) row tree is $10/\big(3 \cdot 29 - (2 \cdot 29 + 15)\big) = 10/14 > 0.7$, and $f_1(T_t) = 6/\big(3 \cdot (29 - (10 + 9 + 6))\big) = 0.5$. Therefore T_t is chosen. Now the value of f_2 for each row tree is $10/\big(3 \cdot 25 - (2 \cdot 25 + 15)\big) = 1$, and the algorithm stops with a solution of length $loss(T_s, T_t) + m(T_s, T_t) = (9 + 6) + 3 \cdot (10 + 9 + 6) = 90$. We get a lower bound of $90/75 = 1.2$. □

Robins and Zelikovsky [7] also showed that the performance ratio of the loss con-tracting algorithm is at most 1.279 if the graph is quasi-bipartite. A quasi-bipartite lower bound instance for the loss contracting algorithm is shown in Figure 4. It is only slightly weaker than the one shown in Figure 3 and comes fairly close to the performance ratio proved in [7].

Theorem 3. *The performance ratio of the loss contracting algorithm in quasi-bipartite graphs is at least* $(5 - \sqrt{2})/3 > 1.195$.

Proof. The quasi-bipartite lower bound instance is shown in Figure 4. It is based on the observation that *one* full component chosen by the algorithm can 'destroy the benefit' of *two* full components of the optimal solution. The Steiner minimum tree consists of the full components T_w and T_x and has length 6. $MST(R)$ uses four diagonal edges and has length 8.

At the beginning, we have $f_0(T_u) = \big(2 - \sqrt{2}\big)/\big(8 - (2 \cdot 2 + 2 \cdot \sqrt{2})\big) = 0.5$ and $f_0(T_v) = \big(2 - \sqrt{2}\big)/\big(8 - (2 \cdot 2 + 2 \cdot (3 - \sqrt{2}))\big) = 1/\sqrt{2} > 0.5$, whereas $f_0(T_w) = f_0(T_x) = 1/\big(8 - (2 \cdot 2 + 2 \cdot 1)\big) = 0.5$. Therefore we may assume that the loss contracting algorithm decides for T_u. $MST(R/Loss(T_u))$ has the same form as $MST(R)$, but the two edges to the right have shrunk from 2 to $\sqrt{2}$. Therefore $m(T_u) = 4 + 2\sqrt{2}$.

In the next step, we have $f_1(T_v) = \big(2 - \sqrt{2}\big)/\big((4 + 2\sqrt{2}) - (2 \cdot (3 - \sqrt{2}) + 2 \cdot \sqrt{2})\big) = 1/\sqrt{2}$ and $f_1(T_w) = f_1(T_x) = 1/\big((4 + 2\sqrt{2}) - (1 + 1 + 2 + \sqrt{2})\big) = 1/\sqrt{2}$.

We may assume that the loss contracting algorithm chooses T_v. The edges on the left hand side of $MST(R/Loss(T_u, T_v))$ have shrunk from 2 to $3 - \sqrt{2}$. Now we have $m(T_u, T_v) = 2 \cdot (3 - \sqrt{2}) + 2 \cdot \sqrt{2} = 6$.

As $f_2(T_w) = f_2(T_x) = 1/(6 - (1 + 1 + (3 - \sqrt{2}) + \sqrt{2})) = 1$, no further improvement is possible and the algorithm will stop at this point without using T_w and T_x. The length of the Steiner tree found by the loss contracting algorithm is $|T_u| + |T_v| = 10 - 2\sqrt{2}$. The lower bound follows. □

5   Iterated 1-Steiner Heuristic

The iterated 1-Steiner heuristic is a simple greedy local search heuristic. Recall that the Steiner minimum tree for a set of required vertices R can be reconstructed from its set of Steiner vertices I as $SMT(R) = MST(R \cup I)$. Starting from a spanning tree for the terminal set, i.e. $I = \emptyset$, the iterated 1-Steiner heuristic adds in each step a single Steiner vertex to I if this will reduce the size of the MST. The algorithm stops if no such improvement is possible. A vertex s is accepted if

$$f_i(s) := mst(R \cup \{s_1, \ldots, s_i, s\}) - mst(R \cup \{s_1, \ldots, s_i\}) < 0 \,.$$

Since iterated 1-Steiner starts from a spanning tree in the terminal distance graph and $mst(R) \leq 2 \cdot smt(R)$, it is clear that its solution is never more than a factor of 2 away from the optimum. No better performance ratio has been proven. Minoux [11] showed how to accelerate iterated 1-Steiner in quasi-bipartite graphs, but did not improve the performance ratio.

Rajagopalan and Vazirani gave a $3/2 + \varepsilon$ approximation algorithm for the Steiner tree problem in quasi-bipartite graphs which is based on the primal-dual method. They pointed out that approximation algorithms for the Steiner tree problem in quasi-bipartite graphs are a significant step towards better approximation algorithms for general instances. By now, the best approximation ratio known for the quasi-bipartite case is 1.279 and belongs to the loss contracting algorithm of Robins and Zelikovsky [7]. In the same paper, they also showed that already the simple iterated 1-Steiner heuristic achieves an approximation ratio of $3/2$ in quasi-bipartite graphs. Here we give a family of instances for which this ratio is asymptotically attained. This shows that the analysis of iterated 1-Steiner in quasi-bipartite graphs in [7] is tight. Interestingly, their result does not rely on a greedy (locally optimal) choice of the next Steiner vertex which is included into the set of Steiner vertices I. It is not necessary to find a vertex s that minimizes $f_i(s)$; non-positivity suffices.

A slight improvement of the iterated 1-Steiner heuristic is to remove Steiner vertices which have degree one or two in $MST(R \cup I)$. (These may result from changes in the topology during the course of the algorithm.) Steiner vertices of degree one can only add to the length of the minimum spanning tree without connecting a terminal, and Steiner vertices of degree two can be removed safely because a shortcut is present in the terminal distance graph due to the triangle inequality.

Theorem 4. *The performance ratio of the iterated 1-Steiner heuristic in quasi-bipartite graphs is $3/2$ (even if Steiner vertices of degree 1 or 2 are removed).*

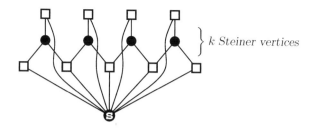

Fig. 5. A simple lower bound instance for iterated 1-Steiner.

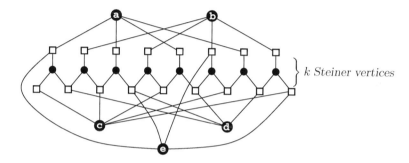

Fig. 6. A lower bound instance for iterated 1-Steiner with vertex removal.

Proof. A bad instance for the iterated 1-Steiner heuristic without vertex removal is shown in Figure 5. All edges have unit weight. Assume that the algorithm has already chosen the k Steiner vertices of degree 3 in the row (each of them reduced the value of $mst(R \cup I)$ by one when it was added to I). Then the iterated 1-Steiner heuristic will stop at this point without including s, because this would increase the length of the minimum spanning tree from $3k$ to $2k + 1 + k$. The Steiner minimum tree, however, consists of the star around s. We get a lower bound of $\frac{3k}{2k+1}$, which converges to $\frac{3}{2}$ as $k \to \infty$.

The instance from Figure 5 does not work for the iterated 1-Steiner heuristic with vertex removal, since it will remove all the Steiner vertices from I and keep only s for the final solution. Consider Figure 6 instead. Again we assume that the set I consists of the Steiner vertices of degree 3. We have split the vertex s from Figure 5 into four vertices a, b, c, and d, each of degree $\sim \frac{k}{2}$. Another vertex e of degree four is connected to one terminal from each of these stars around a, b, c, and d. The connection points are spread out in such a way that for each $u \in \{a, b, c, d, e\}$, no vertex in I is adjacent to more than one terminal from the star around u. This implies that no vertex of I can have degree one in $MST(R \cup I \cup \{u\})$ and consequently, $mst(R \cup I \cup \{u\}) = mst(R \cup I) + 1$. Therefore iterated 1-Steiner with vertex removal will stop at this point. The resulting lower bound is $\frac{3k}{2k+5} \to \frac{3}{2}$. □

6   Greedy-MSS

Recently, a new approximation algorithm for the Steiner tree problem in quasi-bipartite graphs was presented in [5]. The analysis of this algorithm uses a matroid-style exchange argument, and uses ideas from the analysis of the greedy algorithm for set cover. (In fact, it can even be extended to the minimum spanning set problem in polymatroids [3].)

The resulting algorithm Greedy-MSS [5] achieves a performance ratio of 1.217 for the Steiner tree problem in quasi-bipartite graphs where all edges incident to a Steiner vertex have the same weight. Such instances are called *uniformly quasi-bipartite*. Previously the best perfomance ratio for this class of instances was 1.279 and due to the loss contracting algorithm of [7]. (See also Section 4.)

In quasi-bipartite instances of the Steiner tree problem in graphs, every full component contains a unique Steiner vertex. Algorithm Greedy-MSS tries to minimize the length-per-connection ratio greedily, i. e., we have

$$f_i(T) := \frac{|T|}{c_i(T)}$$

in the general greedy framework (Fig. 1). Here $|T|$ denotes the length of the full component T and $c_i(T)$ is defined as the difference between the number of components before and after T is added to the subhypergraph $(R, \{T_1, \ldots, T_i\})$. The algorithm stops when $(R, \{T_1, \ldots, T_p\})$ is a spanning subhypergraph. While the number of full components around a particular Steiner vertex which might be added in step i can in general be exponential, it is easy to see that Greedy-MSS will always choose one that yields as many new connections as possible among them, and due to the uniformity condition all of these full components have the same weight. Therefore, the algorithm only has to evaluate one full component for each Steiner vertex in each phase. Greedy-MSS can be implemented to run in $O(\#V \#R)$ time.

Theorem 5. *The performance ratio of algorithm Greedy-MSS is* $73/60 = 1.21\overline{6}$.

Proof. The upper bound on the performance ratio was proved in [5]. Here we give an instance for which the performance ratio $73/60$ of Greedy-MSS is attained.

The worst case instance is shown in Figure 7. All edges have length 1. The terminals $\{r_{i,j} \mid 1 \leq i \leq 4,\ 1 \leq j \leq 12\}$ are arranged in form of a grid. (For notational convenience we will number the rows from bottom to top.) We also have a terminal r_0 which is shown in the top left corner. The set of potential Steiner vertices is $X \cup Y$. The set X consists of the vertices $\{x_1, \ldots, x_{12}\}$, where x_j is connected to $r_{1,j}, \ldots, r_{4,j}$ and r_0. A Steiner tree using the Steiner vertex set X has total length $5 \cdot 12 = 60$, which is in fact optimal. The set Y consists of vertices $y_{i,j}$ which are connected to r_0 and a subset of terminals in a row, as indicated by the dashed rectangles. Formally, we have $Y = \{y_{i,j} \mid 1 \leq i \leq 4,\ 1 \leq j \leq 12/i\}$, and $y_{i,j}$ is connected to $r_{i,i(j-1)+1}, \ldots, r_{i,ij}$ and r_0. Next we show that Greedy-MSS may end up with a Steiner tree using the Steiner vertex set Y. Since the total length of this Steiner tree is $3 \cdot 5 + 4 \cdot 4 + 6 \cdot 3 + 12 \cdot 2 = 73$, the theorem follows.

Note that in the unweighted quasi-bipartite case, $f_i(T) = |T|/(|T|-1)$ for every full component T that does not create a cycle. Therefore Greedy-MSS will always choose a

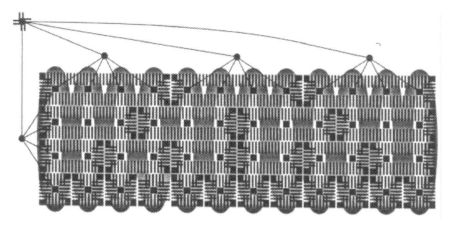

Fig. 7. A worst-case instance for Greedy-MSS.

full component that connects as many connected components as possible. Assume that the Steiner vertices $y_{4,1}, \ldots, y_{4,3}$ have already been chosen. Then the full components around the Steiner vertices in X have been 'nibbled off' such that the size of any largest full component that does not create a cycle has dropped down to 4. Therefore Greedy-MSS might continue choosing the Steiner vertices $y_{3,1}, \ldots, y_{3,4}$. Going forth in this way, we arrive at Y and stop since the graph is connected. $\qquad\Box$

A smaller worst case instance for Greedy-MSS was given in[5]. It was inspired by our proof of the performance ratio of Greedy-MSS, but requires edge weights.

References

1. S. Arora, C. Lund, R. Motwani, M. Sudan and M.Szegedy, *Proof verification and hardness of approximation problems*, Proceedings 33rd Annual Symposium on Foundations of Computer Science (1992), 14–23.
2. A. Borchers and D.-Z. Du, *The k-Steiner ratio in graphs*, SIAM J. Computing 26 (1997), 857–869.
3. G. Baudis, C. Gröpl, S. Hougardy, T. Nierhoff, H.J. Prömel, *Approximating minimum spanning sets in hypergraphs and polymatroids*, Technical report, Humboldt-Universität zu Berlin, Institut für Informatik, 2000.
4. M. Bern and P. Plassmann, *The Steiner problems with edge lengths 1 and 2*, Information Processing Letters 32 (1989), 171–176.
5. C. Gröpl, S. Hougardy, T. Nierhoff, H. J. Prömel, *Steiner trees in quasi-bipartite graphs*, Technical report, Humboldt-Universität zu Berlin, Institut für Informatik, 2000.
6. C. Gröpl, S. Hougardy, T. Nierhoff, H.J. Prömel, *Approximation algorithms for the Steiner tree problem in graphs*, technical report, Humboldt-Universität zu Berlin, 2000. To appear in: D.-Z. Du, X. Cheng, Steiner Trees in Industries, Kluwer Academic Publishers, Dordrecht.
7. G. Robins, A. Zelikovsky, *Improved Steiner tree approximation in graphs*, In Proc. Symposium on Discrete Algorithms, 770–779, 2000.
8. J. Håstad, *Some optimal inapproximability results*, manuscript, 1996.

9. J. Håstad, *On bounded occurence constraint satisfaction*, Information Processing Letters 74 (2000) 1–6.

10. R.M. Karp, *Reducibility among combinatorial problems*, In: Complexity of Computer Computations, (Proc. Sympos. IBM Thomas J. Watson Res. Center, Yorktown Heights, N.Y., 1972). New York: Plenum 1972, pp. 85-103.

11. M. Minoux, *Efficient greedy heuristics for Steiner tree problems using reoptimization and supermodularity*, INFOR 28 (1990), 221–233.

12. M. Thimm, *On the approximability of the Steiner tree problem*, manuscript, Humboldt-Universität zu Berlin, August 2000.

13. A. Zelikovsky, *Better approximation bounds for the network and Euclidean Steiner tree problems*, Technical report CS-96-06, University of Virginia, 1996.

$\log n$-Approximative NLC_k-Decomposition in $O(n^{2k+1})$ Time

(Extended Abstract)

Öjvind Johansson

Department of Numerical Analysis and Computer Science,
Royal Institute of Technology, SE-100 44 Stockholm, Sweden
ojvind@nada.kth.se, http://www.nada.kth.se/~ojvind

Abstract. NLC_k for $k = 1, \ldots$ is a family of algebras on vertex-labeled graphs introduced by Wanke. An *NLC-decomposition* of a graph is a derivation of this graph from single vertices using the operations in question. The *width* of such a decomposition is the number of labels used, and the *NLC-width* of a graph is the minimum width among its NLC-decompositions. Many difficult graph problems can be solved efficiently with dynamic programming if an NLC-decomposition of low width is given for the input graph. This paper shows that an NLC-decomposition of width at most $\log n$ times the optimal width k can be found in $O(n^{2k+1})$ time. Related concept: *clique-width*.

1 Introduction

A frequently proposed approach for dealing with graph problems is to decompose the input graph in such a way that dynamic programming can then be applied. In [15] Wanke introduced the algebra of k-NLC graphs for this purpose. Graph vertices are labeled with numbers in $[k] = \{1, \ldots, k\}$, and the operations are relabeling and union. The latter permits edges to be drawn between the two involved graphs. Importantly, this is done on the basis of vertex labels; once two vertices have acquired the same label, they can no longer be differentiated for edge-drawing. The class of k-NLC graphs, also called NLC_k, now consists of all graphs that can be derived from single vertices with these operations. Wanke went on to show that the maximum cut size of a graph can be computed in polynomial time if the graph is given as an NLC_k-decomposition (a derivation in NLC_k), and that under the same conditions, it can be determined whether a graph has a Hamiltonian circuit. More examples are given in [10].

A similar algebra was presented in [8] (alternatively, see [9]). For this it has been shown that all problems expressible in certain versions of monadic second-order logic can be solved in linear time for graphs given as decompositions with at most k labels [6,7]. (Maximum Clique and c-Colorability are two such problems.) The number of labels needed to produce a graph in this algebra is called its *clique-width*.

A. Brandstädt and V.B. Le (Eds.): WG 2001, LNCS 2204, pp. 229–240, 2001.
© Springer-Verlag Berlin Heidelberg 2001

Both of these algebras generalize the class of cographs, described in [2] for example. Moreover, they are both closely related to the kind of graph grammar introduced in [5]. (Essentially, they enable bottom-up construction corresponding to the top-down substitution mechanism of such a grammar.) The main difference between them is that in the latter, edges are not added when two graphs are united, but with a separate unary operator instead. Indeed, it turns out that a derivation according to one of these algebras easily can be turned into a derivation according to the other, with either no more labels (for the first algebra) or at most twice as many (for the second) [13]. And for both algebras it is unknown whether a potential decomposition with at most k labels can be found in polynomial time, except for very small values of k. For cographs, which correspond exactly to one and at most two labels, respectively, decomposition algorithms follow easily from [3]. It has thereafter been shown that an NLC_2-decomposition can be found in $O(n^4 \log n)$ time [14]. And a decomposition algorithm for the algebra in [8] with three labels has been outlined in [1].

Below is presented an algorithm that can find an NLC-decomposition of minimum width for any graph G. It works by building, according to the rules of the algebra, larger and larger *candidates* — graphs that are consistent with G in the sense that they can occur in derivations of it. Until G itself has been obtained, the number of labels used, k, is incremented whenever nothing more can be done with the current labels. Of course, the algorithm is not intended to be used for large values of k. A mechanism is provided for removing candidates that no longer contain any unique information. Yet, a problem with this exact algorithm is that it is hard to see how large the set of produced candidates may become. The algorithm has therefore been developed further, so that if G has NLC-width k, an NLC-decomposition of width at most $k \log n$ can be found in $O(n^{2k+1})$ time. Note that this decomposition can then be turned into a "clique-decomposition" of width at most $2 \log n$ times the clique-width of G.

The implications of this for dynamic programming will vary. Let us consider the c-Colorability problem, for instance. It is not hard to devise an algorithm that runs in $O(2^{2ck}n)$ time on an NLC_k-decomposition and determines if the corresponding graph has a c-coloring. (There are at most n union operations in which we combine and evaluate 2^{ck} possible color configurations for each operand. Relabeling operations require less work.) Thus, if k is $O(\log n)$, the running time will still be polynomial.

Some additional notes on the relationship with other graph classes may be of interest. To begin with, bounded tree-width implies bounded NLC-width and clique-width [4,9,13,15]. In contrast, bounded NLC-width/clique-width does not in general imply bounded tree-width. However, for graphs without large complete bipartite subgraphs, it does [12]. Finally, distance-hereditary graphs have clique-width at most 3, whereas unit interval graphs and permutation graphs have unbounded clique-width [11].

2   Basic Definitions

With a *labeled graph* G is meant a graph with a labeling function, lab_G, mapping each vertex v to a positive integer. We say v is labeled with (or has the label) $lab_G(v)$. The unlabeled graph one obtains by disregarding lab_G is denoted by unlab(G), and the vertex and edge sets of G are written V(G) and E(G).

Two labeled graphs G_1 and G_2 are disjoint if V(G_1) $\cap$ V(G_2) = $\emptyset$. Their *disjoint union* is then defined as the labeled graph G given by V(G) = V(G_1) $\cup$ V(G_2), E(G) = E(G_1) $\cup$ E(G_2), $lab_G(u) = lab_{G_1}(u)$ for all $u \in$ V(G_1), and $lab_G(u) = lab_{G_2}(u)$ for all $u \in$ V(G_2).

Definition 1 (Union). Let G_1 and G_2 be disjoint labeled graphs, and let $S \subseteq \mathbb{Z}^+ \times \mathbb{Z}^+$. Then $(G_1) \times_S (G_2)$ is the graph one obtains by forming the disjoint union of G_1 and G_2, and adding to that all edges $\{u, v\}$ satisfying $u \in$ V(G_1), $v \in$ V(G_2), and $(lab_{G_1}(u), lab_{G_2}(v)) \in S$. See Fig. 1.

Definition 2 (Relabeling). Let G be a labeled graph, and let R be a partial mapping from $\mathbb{Z}^+$ to $\mathbb{Z}^+$. Then $\circ_R(G)$ is the labeled graph G' defined by V(G') = V(G); E(G') = E(G); and for all $u \in$ V(G'), $lab_{G'}(u) = R(lab_G(u))$ if $lab_G(u)$ is in the domain of R, and $lab_{G'}(u) = lab_G(u)$ otherwise. See Fig. 1.

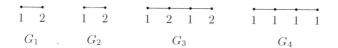

$$G_1 \qquad G_2 \qquad G_3 \qquad G_4$$

Fig. 1. Union and relabeling. $G_3 = (G_1) \times_{\{(2,1)\}} (G_2)$ and $G_4 = \circ_{\{(2,1)\}}(G_3)$.

Definition 3 (NLC$_k$, NLC$_k$-term). An NLC$_k$-term D expresses a valid use of the above operators $\times_S$, with $S \subseteq [k]^2$, and $\circ_R$, with R partial from $[k]$ to $[k]$, as well as the unary operator λ_i, with $i \in [k]$. The latter produces a graph consisting of a single vertex labeled with i, and for our purposes, the operand must be the name of the vertex. G(D) denotes the labeled graph produced by D, and NLC$_k$ is the class of (labeled) graphs produced by NLC$_k$-terms. An NLC-term is an NLC$_k$-term for some value of k.

Definition 4 (NLC$_k$-decomposition). An NLC$_k$-decomposition of a graph G is an NLC$_k$-term D such that $G = G(D)$ if G is labeled, or such that $G = $ unlab(G(D)) if G is unlabeled. An NLC-decomposition is an NLC$_k$-decomposition for some value of k.

NLC-terms are best though of as binary trees. The *width* of such a term is the number of labels used in it, and the *NLC-width* of a graph is the minimum width among all its NLC-decompositions.

For a graph G and a set of vertices V, $G \setminus V$ denotes the induced subgraph of G that remains when vertices in V have been removed. Similarly, for an NLC-term D, $D \setminus V$ denotes the subterm of D we obtain by deleting all subterms corresponding to single vertices in V, and removing also all operations that become superfluous. Evidently, $G(D \setminus V) = G(D) \setminus V$. In the following, we simply say that we *remove* vertices from a term or decomposition.

3   An Exact Algorithm

In this section we investigate how one can find an NLC-decomposition of minimum width for a given graph. For simplicity, all graphs are assumed to be labeled. (This is no restriction.) We first characterize those graphs that can occur in derivations of the input graph, and then take a brief look at the algorithmic idea.

Let u, v, and w be vertices in a graph G. We say that w *distinguishes* u from v in G if it has an edge in this graph to one of u and v, but not to the other.

Definition 5 (NLC$_k$-candidate). An NLC$_k$-candidate for a graph G is a labeled graph C satisfying (i) $V(C) \subseteq V(G)$; (ii) the edges of C are those induced by G; (iii) for any pair of vertices u and v in C with $lab_C(u) = lab_C(v)$, we have (a) $lab_G(u) = lab_G(v)$, and (b) no vertex in $V(G) \setminus V(C)$ distinguishes u from v in G; and (iv) $C \in$ NLC$_k$. An NLC-candidate (or candidate, simply) is an NLC$_k$-candidate for some value of k.

Specification 1 (Basic idea). To NLC-decompose a graph G, we are going to initialize a *candidate set* to the single vertices in G, each labeled with 1, and then add repeatedly to this, NLC-candidates for G that can be formed (using union and relabeling) from those already in the set (that is, from copies of these). More precisely, we will add NLC$_1$-candidates in a first *stage* until this leads no further, then NLC$_2$-candidates in stage two, and so on, and we shall record how each new candidate was formed. Thereby, we will find a decomposition of G of minimum width.

For this approach to be successful, we must be able to keep down the size of the candidate set. We shall first consider the need for differing candidates on the same vertex set. For $V \subseteq V(G)$, let $P_G(V)$ denote that partition of V in which two vertices belong to the same class if and only if (i) they have the same label in G, and (ii) they are not distinguished by any vertex in $V(G) \setminus V$. Also, let a *vertex class* of a labeled graph be a maximal, uniformly labeled set of its vertices. Then, if two vertex classes in a candidate C for G are subsets of the same class in $P_G(V(C))$, one of these vertex classes can in fact be merged into the other (by a relabeling) without affecting the role of C in a derivation of G. We shall make use of this. Let the *dimension* of C, $\dim C$, be the number of vertex classes it has. The candidate C is now said to be *label-compact* with respect to G if (i) the vertex classes of C are equal to the classes of $P_G(V(C))$, and (ii) the range of lab_C is $\{1, \ldots, \dim C\}$.

Specification 2 (Union step). When we try to form a new candidate through union of two old ones, we shall allow relabeling of these as part of a *union step*. Hence, for a particular vertex set, we need to save at most one candidate, and we will choose this to be label-compact. Thus, we let each union step end with a relabeling as well. The width of a union step is the number of labels involved. (Naturally, we use no more than necessary.)

We shall now inquire whether a candidate can replace smaller candidates that it "includes".

Definition 6 (Extension). Let C_1 and C_2 be NLC-candidates for a graph G. If C_1 is an induced subgraph of C_2 such that each vertex class of C_2 contains, as a subset, one vertex class of C_1, then C_2 is an *immediate extension* of C_1 with respect to G. More generally, an *extension* of C_1 with respect to G is any NLC-candidate for G that can be obtained by a relabeling of an immediate extension of C_1.

Lemma 1. *If C_1 and C_2 are NLC-candidates for a graph G, then C_2 is an extension of C_1 if and only if (i) each vertex class of C_1 is a subset of some vertex class of C_2, and (ii) each vertex class of C_2 contains, as a subset, at least one vertex class of C_1.*

Lemma 2. *Let C_1 be an NLC-candidate for a graph G. If C_2 is an extension of C_1 and C_3 is an extension of C_2 (in both cases with respect to G), then C_3 is also an extension of C_1.*

Lemma 3. *Let C_1 and C_2 be NLC-candidates for a graph G such that C_2 is an arbitrary extension of C_1, let D_1 and D_2 be NLC-decompositions of C_1 and C_2, and let D be an NLC-decomposition of G containing D_1 as a subterm. Then there exists a mapping R_2, such that we get a new NLC-decomposition D' of G by replacing D_1 by $\circ_{R_2}(D_2)$. (We first remove each vertex in $V(C_2) \setminus V(C_1)$ from D.)*

Lemma 3 suggests that when we add a candidate C to the candidate set $\mathcal{S}$, we may be able to remove from $\mathcal{S}$ any candidate which C is an extension of. This is not self-evident, though. Suppose that a decomposition hinges on a union step involving two particular graphs. What if $\mathcal{S}$ contains not both of these, but extensions of them which happen to overlap? Obviously, we must consider what we can do with nondisjoint candidates. If we extend the union step as described next, the suggestion above will work fine.

Specification 3 (Union step, extended application). Two nondisjoint NLC-candidates C_1 and C_2 for a graph G are *separable* if either $C_2 \setminus V(C_1)$ or $C_1 \setminus V(C_2)$ is also an NLC-candidate for G. When this is the case, we can get an NLC-candidate for G on the vertex set $V(C_1) \cup V(C_2)$ through a union step, either for C_1 and $C_2 \setminus V(C_1)$, or for C_2 and $C_1 \setminus V(C_2)$. We will call this a union step for C_1 and C_2. Naturally, if there is a choice, we minimize the width of this underlying operation.

Specification 4 (Replacement rule). Each time we have produced a new candidate C through a union step for two members of the candidate set S, we shall do as follows: If S already contains an extension of C, then we discard C immediately. Otherwise we remove from S each candidate which C is an extension of, and we then add C instead.

We need to prove that the algorithm is still correct. If we generalize Lemma 3 first, such that D there can be a decomposition of any NLC-candidate for G, we can then show:

Theorem 1. *For each* NLC_k*-candidate C for the input graph G, the candidate set S contains an extension of C at the end of stage k.*

Proof. Induction over candidate size. Sketch of the induction step: We have $C = \circ_R((C_1) \times_S (C_2))$ for some NLC_k-candidates C_1 and C_2. Assume that S contains an extension C_1' of C_1. Use the generalization of Lemma 3; in particular, note that $C_2' = C_2 \setminus \mathrm{V}(C_1')$ will be an NLC-candidate. It also belongs to NLC_k, as C_2 does. Thus, assume also that S contains an extension C_2'' of C_2'. Use the generalized lemma again; this time, note that $C_1'' = C_1' \setminus \mathrm{V}(C_2'')$ will be an NLC-candidate. Hence, C_1' and C_2'' are separable. Conclude, moreover, that a union step for them has width at most k and produces an extension of C. $\square$

The replacement rule does not by itself guarantee efficiency of the algorithm. Perhaps though, if we prioritize union steps that make the replacement rule come into effect, the algorithm will run in polynomial time up to any fixed stage. With the approximate algorithm in the next section, we will achieve this performance for sure. As a preparation for the presentation there, let us look now at some implementation issues which pertain to the exact algorithm also.

During the course of the algorithm, we need to find the vertex classes (including labels) of candidates. We call this *surveying*. At the end, we also need to reconstruct the decomposition of the input graph. In order to be space-efficient, we probably want to represent candidates mainly with their derivations, and to survey them by traversing these derivations to some reasonable extent. At a union step for two disjoint candidates, we then make full use of their existing representations; if C is defined as $\circ_R((\circ_{R_1}(C_1)) \times_S (\circ_{R_2}(C_2)))$, we create a node for C containing the *parameters* R, R_1, R_2, and S, as well as references to C_1 and C_2. We can then survey C by surveying C_1 and C_2. However, if C is obtained in a union step for two nondisjoint (but separable) candidates C_1 and C_2, it is not enough to refer to both of these, as such a practice could lead to undesirably long surveying times. Instead, if C_1, for instance, is the intact part in the underlying disjoint union step, we may keep, with the node for C, also a list of the vertex classes in $C_2 \setminus \mathrm{V}(C_1)$, so that we can survey C by surveying C_1 and this list. This will allow us to survey a candidate in $O(kn)$ time, where k is the stage of the algorithm and $n = |\mathrm{V}(G)|$, either by merging vertex lists in a bottom-up way, or by composing relabeling operators from the top and down.

One should note, of course, that when a candidate C is removed from the candidate set, it will not always be possible to delete its representation, as this may still be referred to by other candidates. However, these can easily be kept track of in a list belonging to C. If this list becomes empty, the node for C can be deleted.

Once the vertex classes are known for two candidates C_1 and C_2, the (minimum) width, as well as the parameters, of a potential union step for C_1 and C_2 can be determined in $O(kn)$ time, where k is the number of labels in either candidate, and $n = |V(G)|$. We say that we *match* C_1 and C_2.

4   A Polynomial-Time Approximative Algorithm

4.1   Abstract Derivations

To get a better grip of the candidate set, we shall now see what we can do when nondisjoint candidates are not separable.

Definition 7 (Abstract union step). Let C_1 and C_2 be label-compact NLC-candidates for a graph G, and let $V^* = V(C_1) \cup V(C_2)$. An *abstract union step* for C_1 and C_2 with respect to G produces a label-compact NLC-candidate C for G on the vertex set V^*. The *nominal width* of this abstract union step is the maximum of the dimensions of C_1, C_2, and C.

To distinguish between union steps as defined previously, and abstract union steps, the former will sometimes be called *concrete*, and their widths *actual*. Hence, the derivations we have been concerned with so far are *concrete union step derivations*, and the width of such a derivation is the highest actual width among its union steps. An *abstract union step derivation* will now be a derivation consisting of abstract union steps; its nominal width is the highest nominal width among these. Notice that a concrete union step always can be regarded as abstract.

An abstract union step producing a candidate C from candidates C_1 and C_2 may be represented by a node for C with the usual parameters R, R_1, R_2, and S, and links to C_1 and C_2. As in the nondisjoint but separable case, we also need a specific representation of either $C_1 \setminus V(C_2)$ or $C_2 \setminus V(C_1)$ in order to allow for efficient surveying (the same method and complexity as before). Note that in general, S can make sense only for edges between $C_1 \setminus V(C_2)$ and $C_2 \setminus V(C_1)$. We will not need S though; below, when we transform an abstract derivation into a concrete derivation, we will look up the edges in G instead. The value of an abstract union step derivation is now made clear by the next theorem.

Theorem 2. *Let G be a graph with n vertices, and let D be an abstract union step derivation of G with nominal width k. Then there exists a concrete union step derivation (and thus an NLC-decomposition) of G with width at most $k \log n$.*

Proof (abridged). Think of D as a tree; for a node x, let $V(x)$ denote the vertex set in G corresponding to all the leaves under x. Transform D as follows: for each node from the root and downwards, first exchange, if necessary, the left and right child l and r, so that $|V(l)| \geq |V(r)|$, and then remove from the right subtree each leaf whose corresponding vertex is already represented in the left subtree. (In practice, we have to avoid handling all the multiple leaves of D. By surveying as previously described, and passing down information on which vertices of G are valid for each child subtree, we can transform D in $O(kn^2)$ time.)

Informally, each node x in D now corresponds to an induced subgraph C' of the NLC-candidate C it formerly corresponded to. Importantly, if y is a leaf of D located to the right of the leaves under x, then y corresponds to a vertex which is not in $V(C)$, and which cannot distinguish between vertices within a vertex class of C'. Observe that on the path from the root of D down to any leaf y, there hangs at most $\log n$ subtrees to the left. Each of these corresponds to a graph with at most k vertex classes, and within each such class, vertices are uniformly labeled in G, as well as indistinguishable to any vertex whose leaf is y, or lies to the right of y in D. Hence, a label-compact NLC-candidate on the vertex set corresponding to the leaves to the left of y can have at most $k \log n$ vertex classes. As y was chosen arbitrarily, it immediately follows that we need at most $k \log n + 1$ labels for a concrete derivation of G that adds one vertex at a time in the order given by the leaves of D. With a more detailed argument one can show that $k \log n$ labels is in fact enough. One way (though not the fastest) to find the parameters of such a derivation is to perform an ordinary match for one vertex at a time, together with the NLC-candidate already obtained from the previous vertices. This takes a total of $O(kn^2 \log n)$ time.   □

Specification 5 (Approximation algorithm). We extend the exact algorithm by performing, at each stage k, also abstract union steps of nominal width at most k.

The approximation algorithm will give us an abstract union step derivation of minimal nominal width. We shall not prove this though, since, for our purposes, it suffices to note that Theorem 1 still holds; in the proof of that theorem, there is in fact no requirement that the extensions C_1' and C_2'' themselves belong to NLC_k. Thus, if the input graph G has NLC-width k, we will find an abstract derivation of G with nominal width at most k before the end of stage k. To reach this point quickly, we shall prioritize certain union steps that make the replacement rule come into effect. A new concept will help us then:

Definition 8 (Representative). A set of vertices R is a *representative* for an NLC-candidate C if R contains at least one vertex from each vertex class in C. Regarded as a representative, R is *introduced* when, for the first time, the candidate set is augmented with a candidate which R can represent.

It follows easily from Lemma 1 that a representative for a candidate C also is a representative for any extension of C. Hence, once a representative has been introduced, it will (that is, can) always represent some member of the candidate set. We shall now identify a particular situation in which the abstract union step will be of great importance. We then follow up with an observation which will prove useful. Together, the two lemmas reveal very clearly the nature of the algorithm below.

Lemma 4. *Let C_1 and C_2, of the dimensions d_1 and d_2 respectively, be label-compact NLC-candidates for a graph G, and suppose that they have a representative R in common. Then an abstract union step for C_1 and C_2 with respect to G has nominal width $\max\{d_1, d_2\}$, and the outcome C of such a step is an extension of both C_1 and C_2.*

Lemma 5. *Let C be an NLC-candidate for a graph G, and assume that the candidate set $\mathcal{S}$ does not contain any extension of C. Then, for any k satisfying $\dim C \leq k \leq |V(C)|$, either C, if now added to $\mathcal{S}$, introduces some representative of size k, or else C has common representatives of size k with at least two candidates in $\mathcal{S}$.*

4.2 The Abstract Derivation Algorithm

Linksearch. In order to keep down the number of candidates, we are going to "apply" Lemma 4 whenever possible, so that the members of the candidate set $\mathcal{S}$ do not have representatives in common. Before we go into the details of this, we need to look at a data structure that we will use to speed up the algorithm. We shall keep, at stage k, a table of all representatives of size at most k. With *registering* a representative R to a candidate C, we mean marking in the table that R is a representative for C. We register R at most once. Simply stated, when we remove a candidate C from $\mathcal{S}$, we keep the node for C with a *link* to the replacing candidate (which has more vertices). Once R has been registered, we will be able to find its current member of $\mathcal{S}$ by a *linksearch* in $O(n)$ time.

Inspection, Sweep. When a candidate C has been formed, the first thing we want to find out is whether $\mathcal{S}$ already contains a candidate which is an extension of C. It is the main purpose of the table of representatives to assist us in this. If the table contained all representatives introduced so far, things would be rather simple. However, it may be costly to make sure that the table is always "up to date". The *inspection* procedure for a new-formed candidate C is therefore a little complicated:

We begin to go through the representatives for C. As long as they are not registered, we register them to C. If we find one which is already registered, we follow the links to its current candidate C'. If this is an extension of C (case 1), we are not going to add C to $\mathcal{S}$. Yet, if we have registered representatives to C, we keep the node for C, with a link to C', so that we do not lose the new registrations. The inspection is then over. If C' is not an extension of C (case 2), no other member of $\mathcal{S}$ could be so either (such a candidate would have a representative in common with C'), so fundamentally, C should now be added to $\mathcal{S}$. However, as we do not want members of $\mathcal{S}$ to have common representatives, we are first going to apply Lemma 4 as long as possible, before the resulting extension of C should finally be considered "added" to $\mathcal{S}$. To find and carry out these Lemma 4 applications, we are mainly going to rely on the sweep procedure, defined below. First though, we might as well try to exploit the table of representatives once more. To begin with, we can remove C' from S and replace C with an extension of both. (The details are the same as in a sweep.) Now, as the label-compact C' is not an extension of C, C contains a vertex outside C'. Hence, C (and thus the common extension too) also has one or more representatives which are not representatives for C'. We can easily find at least one; if it is not registered, we register it; if it already is, we follow the links and apply Lemma 4 again. After this, it might no longer be useful to look in the table, so instead we send the most recent extension of C to the sweep procedure, which will find and take care of any remaining applications of the lemma. The inspection is then over. In summary, we have seen what to do if we find, via the table, a member of S sharing a representative with C. There remains the possibility that we find no previously registered representative for C (case 3). The table does then not tell us what to do with this candidate. This too we resolve with a call to the sweep procedure.

In general, a *sweep* for a candidate C goes as follows: We compare C with the members of S until we find one, C', sharing a representative (of any size) with C. If C' is an extension of C, we link C to C', and the sweep is over. If C is an extension of C', we link C' to C, remove C' from S, and continue to compare C with the remaining members of S. Otherwise, we let C'' be the outcome of a concrete or abstract union step for C and C'; we link both of these to C'', we remove C' from S, and we terminate the current sweep, starting instead a new sweep for C''. If, in this way, we reach the end of S (such that we would have continued with another member, if there was one), we add C to S.

The Main Procedure. Now that we have looked at the inspection procedure, we are ready for a description of the entire abstract derivation algorithm. To be able to match candidates in a systematic way, we will keep the candidate set S divided into two parts: a *core set*, whose members have all been matched against each other, and a *waiting list* containing remaining candidates.

We begin stage k by extending the table of representatives to encompass elements of size k also. For each candidate in S, we register all representatives of this size. As no two members of S have a common representative of any size, this registration is cheap, and it will probably pay off. We then empty the core set and put all the candidates of S in the waiting list. We are now ready to begin the main loop of stage k:

As long as there are candidates in the waiting list, we take one, C, and we match this against each member of the core set. If we find that a concrete or abstract union step can be performed with at most k labels, we stop for a moment, and we send the outcome of this to the inspection procedure. If the result of that is that a new candidate should be added to S, we place this in the waiting list. By now, C may have been removed from S. If so, we go on to the next candidate in the waiting list. Otherwise, we return to C and the next remaining member of the core set. If the entire core set has been matched with C already, C itself is added there, and we continue with another candidate in the waiting list. When this is empty, stage k is over.

Correctness and Analysis. To see that the abstract derivation algorithm is correct, just note that it implements Specifications 1 through 5. (The fact that a new candidate is not considered added to the candidate set until it has passed the inspection procedure does not change the process itself.)

We shall now analyze the abstract derivation algorithm. We will arrive at the following:

Theorem 3. *The abstract derivation algorithm runs in $O(n^{2k+1})$ time and $O(n^{k+1})$ space up to stage k.*

We can observe first that there are $A_k = \sum_{i=1}^{k} \binom{n}{i}$ representatives of size at most k. So obviously, the registration operations make up a diminishing part of the overall time consumption, and we shall therefore disregard them below.

In the sweep procedure, we compare two candidates directly in order to see if they have a common representative. If they have, we also determine if one of them is an extension of the other. We shall refer to either of these tasks as a *comparison*. Strictly speaking, a comparison can be conducted in $O(n)$ time. However, we naturally want to include the time to survey the candidates in question. We shall allow the latter to be $O(kn)$, as this will not impair the total result. Thus, both comparison and match will be $O(kn)$-time tasks.

Let us also observe that a match often is followed by a linksearch, which in its turn is generally followed by a comparison. We shall attribute the cost of all of this to the match. As each linksearch is caused by "its own" match, we need not mention linksearches and their associated comparisons again.

Finally, we note that a union step can be stored in $O(n)$ time and space. Theorem 3 now follows from the below lemma, which can be proved from an adaptation of Lemma 5.

Lemma 6. *The main procedure performs at most $A_k^2 + O(n^{2k-1})$ matches up to stage k. At most $2A_k$ of these add a candidate to $\mathcal{S}$. In connection with them, the inspection procedure may perform another $2A_k$ matches. Hence, at most $4A_k$ union steps may have to be stored. In addition, up to A_k^2 comparisons may be performed by the sweep procedure.*

4.3   Conclusion

For any given graph G of NLC-width at most k, the above algorithm gives us an abstract union step derivation of nominal width k or less in $O(n^{2k+1})$ time and $O(n^{k+1})$ space. Certainly within these resource bounds, we can convert this derivation to an NLC$_{k \log n}$-decomposition as in the proof of Theorem 2.

Acknowledgments

The author wishes to thank Stefan Arnborg, Johan Håstad, and Jens Lagergren for their advice, as well as the referees for their comments on the manuscript.

References

1. D. G. Corneil, M. Habib, J.-M. Lanlignel, B. Reed, and U. Rotics. Polynomial time recognition of clique-width ≤ 3 graphs. In *Proc. 4th Latin American Symposium on Theoretical Informatics*, volume 1776 of *Lecture Notes in Computer Science*, pages 126–134, Berlin, 2000. Springer.
2. D. G. Corneil, H. Lerchs, and L. S. Burlingham. Complement reducible graphs. *Discrete Appl. Math.*, 3:163–174, 1981.
3. D. G. Corneil, Y. Perl, and L. K. Stewart. A linear recognition algorithm for cographs. *SIAM J. Comput.*, 14:926–934, 1985.
4. D. G. Corneil and U. Rotics. On the relationship between clique-width and treewidth. These proceedings.
5. B. Courcelle, J. Engelfriet, and G. Rozenberg. Handle-rewriting hypergraph grammars. *J. Comput. System Sci.*, 46:218–270, 1993.
6. B. Courcelle, J. A. Makowsky, and U. Rotics. Linear time solvable optimization problems on graphs of bounded clique width. *Theory Comput. Systems*, 33:125–150, 2000.
7. B. Courcelle, J. A. Makowsky, and U. Rotics. On the fixed parameter complexity of graph enumeration problems definable in monadic second order logic. *Discrete Appl. Math.*, 108:23–52, 2001.
8. B. Courcelle and S. Olariu. Clique-width: A graph complexity measure — preliminary results and open problems. In *Proc. 5th Int. Workshop on Graph Grammars and Their Application to Computer Science*, pages 263–270, Williamsburg, VA, Nov. 1994.
9. B. Courcelle and S. Olariu. Upper bounds to the clique width of graphs. *Discrete Appl. Math.*, 101:77–114, 2000.
10. W. Espelage, F. Gurski, and E. Wanke. How to solve NP-hard graph problems on clique-width bounded graphs in polynomial time. These proceedings.
11. M. C. Golumbic and U. Rotics. On the clique-width of perfect graph classes. In *Proc. 25th Int. Workshop on Graph-Theoretic Concepts in Computer Science*, volume 1665 of *Lecture Notes in Computer Science*, pages 135–147, Berlin, 1999. Springer.
12. F. Gurski and E. Wanke. The tree-width of clique-width bounded graphs without $K_{n,n}$. In *Proc. 26th Int. Workshop on Graph-Theoretic Concepts in Computer Science*, volume 1928 of *Lecture Notes in Computer Science*, pages 196–205, Berlin, 2000. Springer.
13. Ö. Johansson. Clique-decomposition, NLC-decomposition, and modular decomposition — relationships and results for random graphs. *Congr. Numer.*, 132:39–60, 1998.
14. Ö. Johansson. NLC$_2$-decomposition in polynomial time. *Internat. J. Found. Comput. Sci.*, 11:373–395, 2000. Extended abstract in LNCS 1665, pages 110–121.
15. E. Wanke. k-NLC graphs and polynomial algorithms. *Discrete Appl. Math.*, 54:251–266, 1994.

On Subfamilies of AT-Free Graphs

(Extended Abstract)

Ekkehard Köhler[1], Derek G. Corneil[2], Stephan Olariu[3], and Lorna Stewart[4]

[1] Technische Universität Berlin, Germany
[2] University of Toronto, Canada
[3] Old Dominion University, USA
[4] University of Alberta, Canada

Abstract. We introduce two subfamilies of AT-free graphs, namely, path orderable graphs and strong asteroid free graphs. Path orderable graphs are defined by a linear ordering of the vertices that is a natural generalization of the ordering that characterizes cocomparability graphs. On the other hand, motivation for the definition of strong asteroid free graphs comes from the fundamental work of Gallai on comparability graphs. We show that cocomparability graphs $\subset$ path orderable graphs $\subset$ strong asteroid free graphs $\subset$ AT-free graphs. In addition, we settle the recognition question for the two new classes by proving that recognizing path orderable graphs is NP-complete whereas the recognition problem for strong asteroid free graphs can be solved in polynomial time.

1 Introduction

We say that a vertex in a graph $G = (V, E)$ *intercepts* a path in G if it is adjacent to at least one vertex of the path, and it *misses* the path otherwise. An *asteroidal triple* (*AT* for short) is an independent set of three vertices such that, between every pair, there is a path that is missed by the third. A graph is *AT-free* if it does not contain an AT.

One of the most compelling motivations for the study of AT-free graphs is the idea that these graphs exhibit a type of linear structure. Indeed, the linear structure exhibited by AT-free graphs is explained, in part, in [1], where it is shown that every connected AT-free graph contains a *dominating pair* (two vertices such that every path connecting them is a dominating set), and a type of linear "shelling sequence" called a *spine*.

The original motivation for the results of the present paper was the idea that AT-free graphs might be further characterized by the existence of a vertex ordering satisfying certain conditions.

Vertex orderings have proven to be useful algorithmic tools for several families of graphs. For example, chordal graphs (respectively, cocomparability graphs) are characterized by the existence of vertex orderings that do not contain the forbidden ordered configuration shown in Figure 1 (a) [2] (respectively, (b) [8]). A graph is an *interval graph* if and only if it has a vertex ordering that contains neither of the configurations of Figure 1 (see for example [10]). Such vertex

A. Brandstädt and V.B. Le (Eds.): WG 2001, LNCS 2204, pp. 241–253, 2001.
© Springer-Verlag Berlin Heidelberg 2001

orderings are referred to as chordal orderings, cocomparability orderings, and interval orderings, respectively.

(a)   (b)

Fig. 1. Forbidden ordered configurations.

In other words, in an interval ordering, for every path on two vertices (that is, for every edge), the left endpoint of the path is adjacent to all vertices between the two endpoints of the path. In a cocomparability ordering, each vertex between the two endpoints of a P_2 is adjacent to one or both endpoints of the P_2.

It is well-known that interval graphs are exactly those graphs that are both chordal and cocomparability [5] or, equivalently, both chordal and AT-free [9]. Furthermore, cocomparability graphs form a proper subclass of AT-free graphs [6].

An alternate characterization of the cocomparability ordering is given in Observation 1.

Observation 1. *A vertex ordering $v_1, \ldots, v_n$ of graph G is a cocomparability ordering if and only if $\forall\ v_i, v_j, v_k$ with $i < j < k$, vertex v_j intercepts each v_i, v_k-path of G.*

From this, one can easily see that a cocomparability graph must be AT-free since any independent triple occurs in some order, say $x < y < z$, in a cocomparability ordering, and thus, there cannot exist an x, z-path missed by y.

In an attempt to generalize the cocomparability ordering while retaining the AT-free property, we introduce the following definition.

Definition 1. *A graph $G = (V, E)$ is* path orderable *if there is an ordering of the vertices $v_1, \ldots, v_n$ such that for each triple of vertices v_i, v_j, v_k with $i < j < k$ and $v_i v_k \notin E$, vertex v_j intercepts each v_i, v_k-path of G; such an ordering $v_1, \ldots, v_n$ of V is called a* path ordering.

It is easily seen by Observation 1 and Definition 1 that cocomparability graphs are all path orderable. C_5, the chordless cycle on five vertices, is a path orderable graph which is not a cocomparability graph. It is clear that path orderable graphs must be AT-free; however, while it was hoped that Definition 1 might characterize AT-free graphs, we shall see later that path orderable graphs form a strict subset of AT-free graphs; as an example see the graph in Figure 2. It is obviously AT-free, however it has no path ordering, as will be shown later.

Nevertheless, since path orderable graphs are interesting in their own right, we attempted to provide a structural characterization of this graph class by identifying a type of forcing relation on the nonadjacent pairs of vertices and

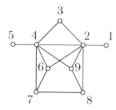

Fig. 2. AT-free graph G which is not path orderable.

the type of structure which would make the vertex ordering of Definition 1 impossible.

These investigations follow in Gallai's footsteps [3] in that they involve ideas similar to his forcing relation on the edges of a comparability graph (equivalently, the nonedges of a cocomparability graph) and his definitions of wreaths and asteroids. Specifically, we define strong asteroids and show that path orderable graphs are strong asteroid free. However, it turns out that the strong asteroid concept does not provide a characterization of path orderable graphs; we shall see that path orderable graphs form a proper subclass of strong asteroid free graphs which, in turn, form a proper subclass of AT-free graphs.

Thus, we will identify two distinct subclasses of AT-free graphs, both of which contain cocomparability graphs:

cocomparability $\subset$ path orderable $\subset$ strong asteroid free $\subset$ AT-free

The interest lies, in part, in the natural vertex ordering, in one case, and the relationship with Gallai's work, in the other case. Furthermore, the identification of these graph classes should allow us to narrow the gap between known polynomial and known NP-complete behaviour of problems in the domain of AT-free graphs.

We conclude the paper with a proof that the recognition of path orderable graphs is NP-complete, and a polynomial time recognition algorithm for strong asteroid free graphs. We note that the NP-completeness result settles an open problem stated in [12].

2   Background

In his paper on comparability graphs, Gallai studies the forcing between the edges imposed by a transitive orientation (to avoid misunderstandings, from now on we will refer to the transitive-forcing as t-forcing). Let G be a (not necessarily comparability) graph. Two edges which share a common endpoint and whose other endpoints are nonadjacent t-force each other directly. That is, in any transitive orientation, either both edges are directed away from the common endpoint, or both are directed towards it. The transitive closure of the direct t-forcing relation partitions the edges of a graph into its *t-forcing-classes*. Either there are exactly two different transitive orientations of the edges of a t-forcing-class, or there are none. The latter case occurs when some edge is t-forced in both directions, in which case the graph is not a comparability graph. Edges xy and xz are said to be *knotted* if y and z are connected in $\overline{N(x)}$, the complement of the subgraph of G induced by $N(x)$, where $N(x)$, the neighbourhood of x, is defined by $N(x) = \{u|ux \in E\}$.

To capture the t-forcing in a given graph G, Gallai uses the concept of a *knotting graph*: For a graph $G = (V, E)$ the corresponding *knotting graph* is given by $K[G] = (V_K, E_K)$ where V_K and E_K are defined as follows. For each vertex v of G there are copies $v_1, v_2, \ldots, v_{i_v}$ in V_K, where i_v is the number of connected components of $\overline{N(v)}$, the complement of the graph induced by $N(v)$. For each edge vw of E there is an edge $v_i w_j$ in E_K, where w is contained in the ith connected component of $\overline{N(v)}$ and v is contained in the jth connected component of $\overline{N(w)}$.

In this graph two edges are incident if they are knotted. The edges of the t-forcing classes of G are given by the connected components of $K[G]$. Using this structure, Gallai shows that a graph G is a comparability graph if and only if $K[G]$ is bipartite.

The following definitions describe structures which lead to t-forcing classes which cannot be transitively oriented, and knotting graphs which are not bipartite.

Definition 2. *An* odd wreath *of size k in a graph is a cycle of knotted edges, specifically, a sequence of vertices $v_0, v_1, v_2, \ldots, v_k$ where k is odd, $v_1, \ldots, v_k$ are distinct, $v_0 = v_k$, and $\forall 0 \leq i < k$, edges $v_i v_{i+1}$ and $v_{i+1} v_{i+2}$ exist in the graph and are knotted (addition modulo k).*

Definition 3. *An* odd asteroid *of size k in a graph is a sequence of vertices $v_0, v_1, v_2, \ldots, v_k$ where k is odd, $v_1, \ldots, v_k$ are distinct, $v_0 = v_k$, and $\forall 0 \leq i < k$, $\exists$ a $v_i v_{i+1}$-path in G which is missed by $v_{(i+\frac{k+1}{2}) \bmod k}$.*

Gallai points out that an asteroid is the complement of a wreath, and proves that a graph is a comparability graph if and only if it contains no odd wreath or, equivalently, a graph is a cocomparability graph if and only if it contains no odd asteroid.

3   Path Orderable Graphs and Strong Asteroid Free Graphs

t-forcing is a fundamental concept for comparability graphs, and thus for cocomparability graphs as well. Given the similarities of the linear ordering characterizations of path orderable graphs and cocomparability graphs, one might expect a similar forcing concept for path orderable graphs. In fact such is the case.

For a graph G and a vertex v of G let $C_1, \ldots, C_k$ be the connected components of $G \setminus (N(v) \cup \{v\})$ and let $B_i^1, \ldots, B_i^\ell$ be the connected components of the graph induced by the vertices of C_i in $\overline{G}$ $(1 \leq i \leq k)$; the B_i^j are called the *blobs* of v in G.

Lemma 1. *Let G be a path orderable graph and $v_1, \ldots, v_n$ a corresponding path ordering. For every vertex v of G and every blob B of v, the vertices of B either occur all before v in the path ordering or all after v in the path ordering.*

Proof. Suppose there is a vertex v and a blob B of v with $u, w \in B$ and $u < v < w$ in the path ordering of G. By the definition of blobs, u and w are in the same connected component C of $G \backslash N(v)$. Since u and w are also in the same connected component B of C in $\overline{G}$, there has to be a path of non-edges in B between u and w. Thus, there is a pair of vertices u', w' in B with $u'w' \notin E$ and $u' < v < w'$. But $u', w' \in C$; therefore there is a u', w'-path in $G \setminus N(v)$, contradicting the path ordering.   □

By Lemma 1, any path ordering has to fulfill the property that if one of the vertices u of a blob B of v precedes v in the ordering, then all of the vertices of B have to be in front of v as well. As an example consider the graph in Figure 2. Following the above definition of blobs, vertex 3 has the three blobs $\{6, 7, 8, 9\}$, $\{5\}$, $\{1\}$; vertex 7 has the blobs $\{3, 1, 9\}$, $\{2\}$, $\{5\}$; vertex 8 has the blobs $\{3, 5, 6\}$, $\{4\}$, $\{1\}$; vertex 5 has only the blob $\{1, 2, 3, 6, 7, 8, 9\}$ and vertex 1 has only the blob $\{3, 4, 5, 6, 7, 8, 9\}$. Suppose now there is a path ordering of G. By Lemma 1 we can, without loss of generality, assume that 1 is in front of all vertices of its blob and thus 5 appears after all vertices of its blob in the path ordering; in particular vertices 3, 6, 7, 8, 9 are between 1 and 5. Since 7 and 8 are in the same blob of 3, they either appear both before or both after 3 in the path ordering. However, if they both appear before 3, then, again by Lemma 1, we have a contradiction because 3 and 1 are in the same blob of 7, but on different sides in the path ordering. On the other hand, if both 7 and 8 appear after 3 in the path ordering we have again a contradiction, since 3 and 5 are in the same blob of 8 but on different sides in the path ordering. Hence there cannot be a path ordering for the graph in Figure 2.

When interpreting the constraints of Lemma 1 as orientations of the edges of $\overline{G}$, in the sense that edges from the same blob of a vertex v to v in $\overline{G}$ have to have the same orientation, one can define the following forcing on the edge set of $\overline{G}$.

Let G be a graph (not necessarily path orderable) and let $e_1 = uv$, $e_2 = vw$ be edges of $\overline{G}$ with a common end-vertex v. Then one can define a relation $\approx$ by $e_1 \approx e_2$ (e_1 and e_2 *force each other* or *are knotted* at v) if and only if u and w are in the same blob of v (possibly $u = w$) in G. The consecutive application of this relation defines a class partition of the edges of $\overline{G}$, where two edges e_a, e_b are in the same class (*forcing class*) of $\overline{G}$ if there is a sequence $e_1, e_2, \ldots, e_k$ of edges such that $e_a = e_1 \approx e_2 \approx \ldots \approx e_k = e_b$. Observe that the forcing classes are refinements of the t-forcing classes.

An orientation of the edges of $\overline{G}$ is said to *agree with the forcing* if for any vertex v and any blob B of v all edges between B and v are oriented in the same direction (either towards v or away from v). For a graph G a linear ordering $v_1, \ldots, v_n$ of the vertices of G is said to *agree with the forcing* if the corresponding implied orientation of the edges of $\overline{G}$ (uv is oriented from u to v if $u < v$ in the linear ordering) agrees with the forcing.

Note that when the orientation of one of the edges of a forcing class is fixed, then the orientation of all the edges of its forcing class is determined; hence, either there are exactly two different orientations of the edges of a forcing class that agree with the forcing, or none. In the latter case, some edge is forced to be oriented in both directions, meaning that there is no ordering consistent with the forcing.

Lemma 2. *A graph G is path orderable if and only if there is a linear ordering of the vertices of G agreeing with the forcing.*

Proof. If G is path orderable, then, by Lemma 1, the path ordering has to agree with the forcing relation.

Suppose there is a linear ordering of G, that agrees with the forcing relation and suppose there is a triple $u < v < w$ of vertices that violates the path ordering property, i.e. $uw \notin E$ and there is a u, w-path in $G \setminus N(v)$. Hence, u and w are in the same connected component C of $G \setminus N(v)$ and, since $uw \notin E$, u and w are also in the same blob B of v. But then this ordering does not agree with the forcing relation; contradiction.   □

Corollary 1. *A graph G is path orderable if and only if there is an acyclic orientation of $\overline{G}$, agreeing with the forcing relation.*

Proof. Determine a topological ordering, using the acyclic orientation of $\overline{G}$; then the corollary follows from Lemma 2.   □

One can define a graph, similar to Gallai's knotting graph, representing the forcing classes of $\overline{G}$. For a graph $G = (V, E)$ the *altered knotting graph* is given by $\mathrm{K}^*[G] = (V_K, E_K)$ where V_K and E_K are defined as follows. For each vertex v of G there are copies $v_1, \ldots, v_{i_v}$ in V_K, where i_v is the number of blobs of v in $\overline{G}$. For each edge vw of E there is an edge $v_i w_j$ in E_K, where w is contained in the ith blob of v in $\overline{G}$ and v is contained in the jth blob of w in $\overline{G}$.

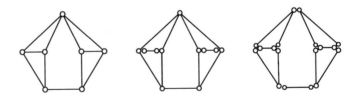

Fig. 3. A graph G together with $\mathrm{K}[G]$ and $\mathrm{K}^*[G]$.

As Gallai did for the knotting graph, we draw the altered knotting graph $\mathrm{K}^*[G]$ of a given graph G by putting different copies of the same vertex close together.

Our next task is to examine configurations which cannot occur in path orderable graphs. In a step toward this goal, we define restricted types of wreaths and asteroids.

Definition 4. *An* odd strong wreath *of size k in a graph G is a sequence of vertices $v_0, v_1, \ldots, v_k$ where k is odd, $v_1, \ldots, v_k$ are distinct, $v_0 = v_k$, and $\forall 0 \leq i < k$, edges $v_i v_{i+1}$ and $v_{i+1} v_{i+2}$ exist in the graph and are knotted in the altered sense, that is, v_i and v_{i+2} are in the same blob of v_{i+1} in $\overline{G}$ (addition modulo k).*

Definition 5. *An* odd strong asteroid *of size k in a graph G is a sequence of vertices $v_0, v_1, \ldots, v_k$ where k is odd, $v_1, \ldots, v_k$ are distinct, $v_0 = v_k$, and $\forall 0 \leq i < k$, v_i and v_{i+1} are in the same blob of $v_{(i + \frac{k+1}{2}) \bmod k}$ in G.*

The two notions are complementary, that is, a graph G has an odd strong wreath if and only if $\overline{G}$ contains an odd strong asteroid. Furthermore, strong asteroids and strong wreaths are restricted types of asteroids and wreaths. We also note that the asteroidal triples are the odd strong asteroids of size three.

Definition 6. *A graph G is* strong asteroid free *if it does not contain an odd strong asteroid.*

Similar to the t-forcing results, the following holds:

Lemma 3. *The forcing-classes of a graph G correspond exactly to the connected components of $K^*[G]$.*

The next two observations follow from the fact that an odd strong asteroid of size k in G corresponds to an odd cycle of size k in $K^*[\overline{G}]$.

Observation 2. *A graph G is strong asteroid free if and only if $K^*[\overline{G}]$ is bipartite.*

Observation 3. *A graph G is AT-free if and only if $K^*[\overline{G}]$ is triangle-free.*

Lemma 4. *If a graph G is path orderable then $K^*[\overline{G}]$ is bipartite.*

Proof. Let $v_1, \ldots, v_n$ be a path ordering of G. Now orient the edges of $K^*[\overline{G}]$ as follows: $v_i v_j$ is oriented from v_i to v_j if $i < j$. Now, by Lemma 1, each vertex of $K^*[\overline{G}]$ has either only incoming or only outgoing edges. Hence, it is bipartite. □

Not only does the graph in Figure 2 show that path orderable graphs are strictly contained in AT-free graphs, it also establishes that strong asteroid free graphs are strictly contained in AT-free graphs, as shown in the next lemma.

Lemma 5. *The class of strong asteroid free graphs is strictly contained in the class of AT-free graphs.*

Proof. Consider the graph of Figure 2. It is easy to check that 1 3 5 7 8 is an odd strong asteroid in G, and that G is AT-free.   □

In the case of comparability graphs, Gallai not only showed that the knotting graph K[G] of a comparability graph is bipartite, but also proved that a bipartite knotting graph K[G] is a sufficient condition for G being a comparability graph. The major tool that he used for proving this result is a lemma which shows the following. Given a bipartite knotting graph K[G] and consider a triangle of G with the property that at least two of the edges of the triangle are in the same t-forcing-class. Then in any orientation of G that agrees with the t-forcing the triangle is not oriented cyclically.

It turns out that a similar lemma holds for strong asteroid free graphs, too. Specifically, for a graph G with a bipartite altered knotting graph K*[$\overline{G}$], any orientation of $\overline{G}$ that agrees with the forcing relation does not contain a cyclically oriented triangle. However, contrary to the t-forcing relation, this lemma is not enough to imply that the orientation is acyclic and, indeed, we shall show that this is not necessarily the case.

Observation 4. *Given a vertex v in a graph H and vertices $u, w \in N(v)$, which are the endpoints of a P_4 in $N(v)$, then the edges uv and wv are forcing each other (see Figure 4).*

Fig. 4. Vertex v with P_4 in $N(v)$ together with corresponding altered knotting graph K*[H].

Remark 1. Using this observation one can create a *forcing path*, i.e. a path P, where each consecutive pair of edges of P is knotted at the common end-vertex by the help of an added P_4 as described in Observation 4—see Figure 5 (in the following, edges and vertices of the path P itself are called *original edges/vertices*, the added edges and vertices are denoted as *auxiliary edges/vertices*). By the forcing, the orientation of any original edge of P forces the orientation of all other original edges of P. Note that the knotting graph of a forcing path does not contain a triangle or any odd cycle. Furthermore, if P has even length then the end-edges of P are either both oriented towards the inner vertices of P or both oriented outwards from the inner vertices of P. Similarly, if P has odd length the end edges of P have opposite orientations with respect to the inner vertices of P.

Theorem 5. *The class of path orderable graphs is strictly contained in the class of strong asteroid free graphs.*

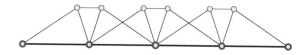

Fig. 5. A forcing path of length 4 (original edges and vertices are bold).

Proof. Consider the left graph in Figure 6. This graph is the complement of a strong asteroid free graph G. This is proved by constructing the altered knotting graph $K^*[\overline{G}]$ (see the right graph in Figure 6). By Observation 4, the thick edges force each-other, as shown in the altered knotting graph; and, without having a strong asteroid in G, there is a forced oriented cycle on the vertices $x_1, \ldots, x_k$ in $\overline{G}$. Consequently, by Corollary 1, G is not path orderable. This construction holds for any $k \geq 4$.   □

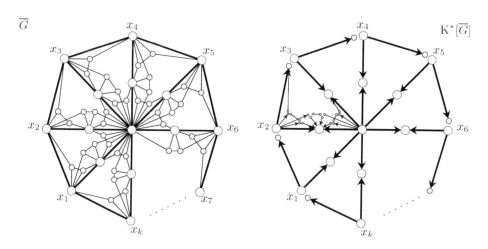

Fig. 6. Complement of a strong asteroid free graph, which is not path orderable (left), together with its altered knotting graph (right). To ease understanding of its structure, in the knotting graph only for one of the "arms" of the example the corresponding auxiliary P_4 vertices are drawn in the figure. One of the two possible forced orientations of the main forcing class is given in the right picture.

4   Recognition of Path Orderable and Strong Asteroid Free Graphs

In this section, we show that the recognition of path orderable graphs is NP-complete. This result answers a question which is posed by J. Spinrad in [12]. In contrast, we describe how to recognize strong asteroid free graphs in polynomial time.

If there is only one forcing class for the edge set of $\overline{G}$ one can check in polynomial time whether G is path orderable: Compute $K^*[\overline{G}]$, check whether it is bipartite, assign an orientation to $K^*[\overline{G}]$ by orienting all edges from one of the bipartition classes to the other and check whether this orientation is acyclic on $\overline{G}$.

Similarly one can check whether G is path orderable if the number of forcing classes of $\overline{G}$ is bounded by a constant.

For comparability graphs Gallai's results for the general case, i.e. where no assumption on the number of edge classes is made, leads to a polynomial time recognition algorithm. For this he introduced the (by now well-known) concept of modular decomposition and proved that, using this decomposition scheme, the problem of recognizing comparability graphs reduces to the problem of recognizing prime comparability graphs. But what about the recognition of path orderable graphs? Can one extend the decomposition scheme to this problem?

NOT-ALL-EQUAL 3SAT [4]
INSTANCE: Set U of variables, collection $\mathcal{C}$ of clauses over U such that each clause $c \in \mathcal{C}$ has $|c| = 3$.
QUESTION: Is there a truth assignment A for U such that each clause in C has at least one *true* literal and at least one *false* literal?

Remark 2. Without loss of generality one can assume that none of the clauses contains more than one literal of a variable.

Theorem 6. *Recognition of path orderable graphs is NP-complete.*

Due to space restrictions in this extended abstract we leave out the proof for this theorem but instead sketch the main ideas of the construction.

The proof is done using a transformation from NOT-ALL-EQUAL 3SAT. Given an instance I of NAE 3SAT, a graph G is constructed, which is the complement of a path orderable graph if and only if I is NAE 3 SAT-satisfiable. In particular it is shown that I is NAE 3 SAT-satisfiable if and only if there is an acyclic orientation of G that agrees with the forcing. By Corollary 1 this is equivalent with $\overline{G}$ being path orderable.

The basic construction of G is done as follows. For every variable x of U an edge e_x is created (in the following called a *variable edge*) and the two possible orientations of e_x are associated with the two possible values *true* and *false* of x.

For each clause $C = x \vee y \vee z$ with literals x, y, z a gadget is constructed, mainly consisting of two C_4s as shown in Figure 7. In each of the C_4s three of the edges (the *base edges*) correspond to the three literals x, y, z of C; a *true* literal of C corresponds to a clockwise orientation of the corresponding base edges in both of the C_4s, whereas a *false* literal corresponds to a counter-clockwise orientation of the corresponding base edges in both C_4s. Furthermore, in each orientation

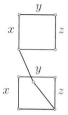

Fig. 7. Gadget for clauses.

that agrees with the forcing, the fourth edges of the two C_4s will be guaranteed to have opposite orientations in the two C_4s.

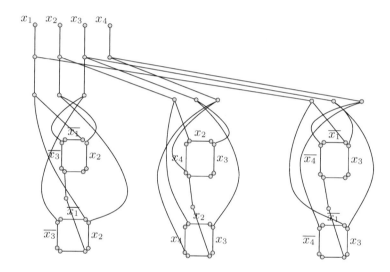

Fig. 8. General structure of $K^*[\overline{G}]$ for the instance $I = (\overline{x_1} \vee x_2 \vee \overline{x_3}) \wedge (x_2 \vee x_3 \vee x_4) \wedge (\overline{x_1} \vee x_3 \vee \overline{x_4})$ (auxiliary vertices and edges are omitted).

The general structure of the connection between variable edges and base edges by forcing paths is shown in Figure 8; for easier understanding, the auxiliary edges and vertices of the forcing paths are omitted in this picture. For a variable edge e_x (see the top of Figure 8) a downwards orientation corresponds to assigning *false* to variable x whereas an upwards orientation corresponds to assigning *true* to x. For each literal x or $\overline{x}$, there is a forcing path of length 4, having e_x and the corresponding base edge as its end-edges; depending on whether the literal is $\overline{x}$ or x, either the start- or the end-vertex of the base edge (with respect to a clockwise ordering in the C_4) is made the end-vertex of the forcing path.

Now, by Remark 1, assigning an upwards orientation to the variable edge e_x results in the desired clockwise orientation of the base edges of each instance of the literal x and a counter clock-wise orientation of the base edges of each instance of the literal $\overline{x}$ for any orientation agreeing with the forcing.

All we have to show now is that there is an acyclic orientation of G agreeing with the forcing relation if and only if C has a NOT-ALL-EQUAL 3SAT satisfying assignment.

By the forcing of the edges and the appropriate knotting of the forcing path from the variable representing edges to the edges representing the literals, each *true* literal in a clause C leads to a clockwise oriented edge and analogously, each *false* literal implies a counter-clockwise oriented edge in the corresponding C_4s. Since every clause has at least one *true* and one *false* literal, each of the C_4s has both an edge that is oriented clockwise and one that is oriented counter-clockwise. Hence, none of the C_4s is cyclically oriented and by the above observations that only the C_4s are possible candidates for oriented cycles, the orientation is acyclic.

Suppose now that there is an acyclic orientation of G that agrees with the forcing relation. We assign to a variable x of U the value *true*, if the edge representing variable x (edges on top of Figure 8) is oriented upwards and *false* otherwise. Since the orientation agrees with the forcing relation, all we have to show is that all of the clauses have at least one *true* and one *false* literal. Suppose there is a clause C, which has only *true* (*false*) literals. By the definition of G and the forcing relation, three edges in each of the C_4s in C's gadget are oriented counter-clockwise (clockwise). Since the "fourth-edges" have opposite orientations in the two C_4s of C, exactly one of the C_4s is oriented cyclically, contradicting the orientation of G being acyclic.

A polynomial time recognition algorithm for strong asteroid free graphs follows from Observation 2. Given graph G, the altered knotting graph of $\overline{G}$, $K^*[\overline{G}]$, can be computed in polynomial time: for each vertex v of G, the blobs of v in G can be computed in $O(n^2)$ time; each vertex has fewer than n blobs. Thus, $K^*[\overline{G}]$ has $O(n^2)$ vertices and $O(n^2)$ edges (since each edge of $\overline{G}$ corresponds to exactly one edge of $K^*[\overline{G}]$), and can be constructed in $O(n^3)$ time. To test whether $K^*[\overline{G}]$ is bipartite can be done in $O(n^2)$ time. Overall, the recognition algorithm requires $O(n^3)$ time.

5   Concluding Remarks

We have defined two graph classes and shown that cocomparability graphs $\subset$ path orderable graphs $\subset$ strong asteroid free graphs $\subset$ AT-free graphs. Furthermore, we have shown that the recognition problem for path orderable graphs is NP-complete, and the recognition of strong asteroid free graphs can be solved in polynomial time. We note that AT-free graph recognition is also in P [1,7].

Although it is somewhat disappointing that no two of these families are equivalent, these classes may give insight into some open problem complexities on AT-free graphs. By adding graph classes in the hierarchy between cocomparability graphs and AT-free graphs, we may be able to identify more precisely the boundary between polynomial and NP-complete behaviour of some of the problems which are known to be polynomially solvable on cocomparability graphs but either NP-complete or unresolved on AT-free graphs. Examples of such problems include graph colouring, clique cover, clique, and the Hamiltonian path and cycle problems. One step in this direction is the observation that the clique problem is NP-complete for path orderable graphs. This follows from the facts that the complements of triangle-free graphs are contained in path orderable graphs, and the independent set problem is known to be NP-complete on triangle-free graphs [11].

Acknowledgements

The authors wish to thank the Natural Science and Engineering Research Council of Canada for their financial support.

References

1. D. G. Corneil, S. Olariu, L. Stewart, Asteroidal triple-free graphs, SIAM J. Discrete Math. **10** (1997) 399–430.
2. D. R. Fulkerson, O. A. Gross, Incidence matrices and interval graphs, Pacific J. Math. **15** (1965) 835–855.
3. T. Gallai, Transitiv orientierbare Graphen, Acta Math. Acd. Sci. Hungar. **18** (1967) 25–66.
4. M. R. Garey, D. S. Johnson, Computers and Intractability: A Guide to the Theory of NP-Completeness, W. H. Freeman, New York 1979.
5. P. C. Gilmore, A. J. Hoffman, A characterization of comparability graphs and of interval graphs, Canad. J. Math. **16** (1964) 539–548.
6. M. C. Golumbic, C. L. Monma, W. T. Trotter, Tolerance graphs, Discrete Appl. Math. **9** (1984) 157–170.
7. Ekkehard Köhler, Graphs without asteroidal triples, PhD thesis, Technische Universität Berlin, 1999.
8. D. Kratsch, L. Stewart, Domination on Cocomparability Graphs, SIAM J. Discrete Math. **6** (1993) 400–417.
9. C. G. Lekkerkerker and J. Ch. Boland, Representation of a finite graph by a set of intervals on the real line, Fund. Math. **51** (1962) 45–64.
10. S. Olariu, An optimal greedy heuristic to color interval graphs, Inform. Process. Lett. **37** (1991) 65–80.
11. S. Poljak, A note on stable sets and colorings of graphs, Comment. Math. Univ. Carolinae **15** (1974) 307–309.
12. J. P. Spinrad, Representations of graphs, book manuscript, 2000.

Complexity of Coloring Graphs
without Forbidden Induced Subgraphs

Daniel Král'[1], Jan Kratochvíl[1,*], Zsolt Tuza[2,**], and Gerhard J. Woeginger[3]

[1] Department of Applied Mathematics and
Institute for Theoretical Computer Science[†],
Charles University,
Malostranské nám. 25, 118 00 Prague, Czech Republic
{kral,honza}@kam.ms.mff.cuni.cz
[2] Computer and Automation Institute,
Hungarian Academy of Sciences,
H–1111 Budapest, Kende u. 13–17, and
Department of Computer Science,
University of Veszprém, Hungary
tuza@sztaki.hu
[3] Technical university Graz, Austria
gwoegi@igi.tu-graz.ac.at

Abstract. We give a complete characterization of parameter graphs H for which the problem of coloring H–free graphs is polynomial and for which it is NP–complete. We further initiate a study of this problem for two forbidden subgraphs.

1   Preliminaries and Overview of Results

Graph coloring belongs to the most important and applied graph problems. It also belongs to the first identified NP–complete problems. Many classes of graphs were shown to allow polynomial-time solution (e.g., interval graphs, chordal graphs, etc.). In this paper we aim at classifying the computational complexity of this problem when restricted to graphs that do not contain certain forbidden induced subgraphs. Related results appear in [8], where 3-colorability is studied. We consider, on the other hand, the coloring problem with the number of colors being part of the input. For one forbidden subgraph we obtain a complete characterization of the complexity, which performs the polynomial-time/NP–complete dichotomy. First results in the direction of two forbidden subgraphs are gathered in the last section, but a complete characterization is not yet at hand.

We consider finite simple undirected graphs. We say that a graph G is H–*free*, where H is another graph, if G does not contain an induced subgraph

* This author acknowledges further partial support of Czech research grant GAUK 158/1999.
** Research supported in part by the Hungarian Scientific Research Fund, grants OTKA T–026575 and T–032969, and Czech Grant GAČR 201/99/0242 (DIMATIA).
† Project LN00A056 supported by The Ministry of Education of Czech Republic

A. Brandstädt and V.B. Le (Eds.): WG 2001, LNCS 2204, pp. 254–262, 2001.
© Springer-Verlag Berlin Heidelberg 2001

isomorphic to H. Typically H will be a fixed "small" graph while G will be part of the input of the problem. For a class of graphs $\mathcal{A}$, we say that G is $\mathcal{A}$–*free* if it is H–free for every $H \in \mathcal{A}$. The graphs of $\mathcal{A}$ are then also referred to as forbidden induced subgraphs in the sense that $\mathcal{A}$ is the set of forbidden induced subgraphs for the class of $\mathcal{A}$–free graphs.

As usual, K_n denotes the complete graph on n vertices, C_n denotes the cycle on n vertices and P_n denotes the path with n vertices (i.e., path of length $n-1$). By $G \oplus H$ we denote the disjoint union of G and H, and in this sense kG means the disjoint union of k copies of G. We use the notation $G \leq H$ when G is isomorphic to an induced subgraph of H. The graph C_k^+ is a cycle of length k with a pending edge, e.g. $C_3^+ = \overline{P_3 \oplus K_1}$. The complement of a graph G is denoted by $\overline{G}$. The chromatic number of G is denoted by $\chi(G)$. The clique covering number of G is denoted by $\sigma(G)$ $(\sigma(G) = \chi(\overline{G}))$.

We study the computational complexity of the following problem:

H-FREE COLORING
Input: An H–free graph G, and an integer k.
Question: Is $\chi(G) \leq k$?

Our main result is summarized in the next theorem.

Theorem 1. *The problem H-FREE COLORING is polynomial-time solvable if H is an induced subgraph of P_4 or of $P_3 \oplus K_1$, and NP–complete for any other H.*

We first discuss the polynomial part, then present a useful reduction which is easier to be formulated in the complementary version (i.e., as a result on clique covering), and complete the proof of the NP–completeness part in Section 4. Some ideas towards sets of more forbidden graphs conclude the last section. A more detailed account on partial results concerning 2 forbidden induced subgraphs will be given in the full version of the paper.

2 Polynomial Cases

All graphs H for which the H-FREE COLORING problem is polynomial-time solvable are depicted in Fig. 1. If H is an induced subgraph of H' then every H–free graph is also H'–free, and hence H-FREE COLORING $\propto H'$-FREE COLORING. Therefore it suffices to prove the polynomial part only for the maximal graphs from Fig. 1, i.e. for P_4 and $P_3 \oplus K_1$.

It is well known that P_4–free graphs are perfect (they are exactly the so called co-graphs), and hence their chromatic number can be determined in linear time [1].

For graphs which do not contain $P_3 \oplus K_1$ we use the following structural result of Olariu:

Proposition 1. ([7]) *If G is $(P_3 \oplus K_1)$–free then every connected component of its complement $\overline{G}$ is triangle-free or is a complete multipartite graph.*

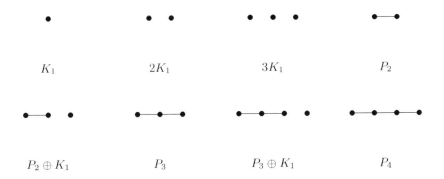

Fig. 1. The graphs H for which H-FREE COLORING is polynomially solvable

The chromatic number of a graph equals the clique covering number of its complement. In this way we determine the chromatic number of a $(P_3 \oplus K_1)$–free graph in polynomial time. The clique covering number of each triangle–free connected component of the complement equals the size of the component minus the maximum size of a matching, and thus can be determined in polynomial time (e.g., by Edmonds's algorithm). The clique covering number of complete multipartite graphs can also be computed in polynomial time (it equals the size of the largest part of the multipartition), and so it follows from Proposition 1 that the chromatic number of $(P_3 \oplus K_1)$–free graphs can be computed in polynomial time. This concludes the "easy" part of the theorem.

3 Clique Covering of Sparse Graphs

In this section we present an NP–completeness reduction for coloring graphs without forests which consist of short paths. It is more convenient to state the result for the complementary problem of clique covering. We denote by K_4^- the graph on 4 vertices with 5 edges.

Theorem 2. *The clique covering problem remains NP–complete even when restricted to planar $\{K_4, K_4^-, C_4, C_5\}$–free graphs.*

Proof. We use a reduction from the SATISFIABILITY problem restricted to planar formulas, namely a variant in which every clause of the input formula contains 2 or 3 literals and every variable occurs in exactly 3 clauses, once positive and twice negated. A formula is called planar if its bipartite incidence graph is planar. NP–completeness of planar formulas was first established by Lichtenstein [5], and NP–completeness of the above described restricted version can be found e.g. in [3].

Given a formula Φ with variable set X and clause set C, we construct a graph G as follows. Fix a planar noncrossing drawing of the incidence graph of Φ. For every variable $x \in X$, the variable gadget is the graph

$$G_x = (\{x, x^+, x_1^-, x_2^-\}, \{xx^+, xx_1^-, xx_2^-, x_1^- x_2^-\}).$$

For every clause c, the clause gadget G_c will be a cycle of length 7. Three nonconsecutive vertices of the cycle are used as ports for connecting with variable gadgets. If x, y, z are variables occurring in c such that in the fixed drawing, the edges cx, cy, cz leave the vertex c in this clockwise order, we denote $c(x), c(y), c(z)$ the ports on G_c also in this clockwise order. (If c contains only two literals, we use as ports two vertices of G_c which are at distance two on the cycle.) For every variable x, let c be the clause containing x positive and d_1, d_2 the clauses that contain $\neg x$. Then identify vertices $c(x)$ and x^+ into one vertex, vertices $d_1(x)$ and x_1^- into one vertex and vertices $d_2(x)$ and x_2^- into one vertex.

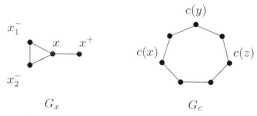

Fig. 2. The construction of variable and clause gadgets

It is clear that the graph G constructed in this way is planar. The only triangles occur in variable gadgets, and hence no two triangles of G share an edge. Thus G contains neither K_4 nor K_4^-. Similarly, G does not contain C_4 as an induced subgraph. In fact, since we may assume that no two clauses share two variables, the shortest cycles of G apart of the triangles in variable gadgets are the 7-cycles in the clause gadgets.

We claim that the clique covering number of G is always $\geq |X| + 3|C|$, with equality holding if and only if Φ is satisfiable.

For every variable x, we need a clique Q_x that would contain vertex $x \in V(G_x)$. This clique may further contain either x^+ (Q_x is the edge xx^+), or x_1^- and x_2^- (if Q_x is the triangle $xx_1^-x_2^-$). Obviously, $x^+ \in Q_x$ implies that Q_x contains neither x_1^- nor x_2^-. This establishes a 1-1 correspondence between truth assignments of X and the ways in which the vertex x is covered:

$$\phi(x) = \text{True iff } Q_x = \{xx^+\}.$$

Note that a variable x evaluates True in a clause c if and only if the port $c(x)$ in G_c is covered by Q_x.

For a clause c, we need at least 3 cliques to cover the vertices of G_c which are not ports. Hence $\sigma(G) \geq |X| + 3|C|$. Moreover, if none of the ports in G_c is covered by the appropriate Q_x, we need at least 4 cliques to cover this G_c. Therefore $\sigma(G) > |X| + 3|C|$ if Φ is not satisfiable.

To see the converse, note that C_7 without a vertex is the path P_6, which can be covered by 3 edges. Hence if at least one port in G_c is covered by the variable clique Q_x (i.e., x evaulates True in c), the remaining vertices of such G_c can be covered by 3 cliques. This shows that $\sigma(G) = |X| + 3|C|$ if Φ is satisfiable.

Corollary 1. *The* $\{2K_2, K_2 \oplus 2K_1, 4K_1, C_5\}$*-*Free Coloring *problem is NP–complete. In particular, the* H*-*Free Coloring *problem is NP–complete if* H *is a (not necessarily induced) spanning subgraph of* $2K_2$.

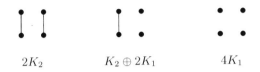

$$2K_2 \qquad\qquad K_2 \oplus 2K_1 \qquad\qquad 4K_1$$

Fig. 3. The minimal linear forests H for which H-Free Coloring is NP–complete

4 NP–Complete Cases

In this section we summarize the NP–completeness part of the proof of Theorem 1. We consider three cases.

Case 1 Suppose H contains a cycle, say C_k. Let G be an instance of vertex 3-colorability, which is well known to be NP–complete.

Let H' be a critical non-3-colorable graph of girth greater than k (i.e., H' does not contain cycles of length at most k; existence of graphs of large girth and chromatic number is a well known result in graph theory, cf. e.g. [2]). Take an edge ab of H', delete it and add a new extra vertex a' pending on b. Call this new graph $H'(a, a')$. Since the girth of H' was greater than k, $H'(a, a')$ does not contain C_k and also every path from a to a' has length greater than k. Since H' was critical non-3-colorable, $H'(a, a')$ is 3-colorable, but in every proper 3-coloring, a and b have the same color, i.e., a and a' have different colors.

Return to G and construct a graph G' by replacing every edge uv by a copy of $H'(u, v)$. Thus G' has girth greater than k and it is 3-colorable if and only if G is 3-colorable. Hence 3-coloring H–free graphs is NP–complete.

Case 2 Let H be a forest, and suppose H contains a vertex of degree at least 3. Then H contains $K_{1,3}$ (the claw). Since 3-coloring is NP–complete for line graphs by Holyer [4] and since line graphs are claw–free, $K_{1,3}$-Free Coloring is NP–complete, and so is H-Free Coloring in this case.

Case 3 Suppose finally that H is a linear forest, i.e., the disjoint union of paths.

If H has at least 4 connected components then H contains $4K_1$ and H-Free Coloring is NP–complete by Corollary 1.

If H has exactly 3 connected components then either $H = 3K_1$ (in which case $H \leq P_3 \oplus K_1$ and H-Free Coloring is polynomial by Proposition 1) or $K_2 \oplus 2K_1 \leq H$ (and H-Free Coloring is NP–complete by Corollary 1).

Suppose H has exactly 2 connected components. If none of them is an isolated vertex then $2K_2 \leq H$. If one of the components is a path of length at least 3 then $K_2 \oplus 2K_1 \leq H$. In these two cases H-FREE COLORING is NP–complete by Corollary 1. If, on the other hand, $H \leq P_3 \oplus K_1$ then H-FREE COLORING is polynomial by Proposition 1.

If H is connected then either $H \leq P_4$ and H-FREE COLORING is polynomial since every H–free graph is a co-graph, or $P_5 \leq H$ and H-FREE COLORING is NP–complete by Corollary 1 since $2K_2 \leq H$.

5   Towards Two Forbidden Subgraphs

A natural generalization of the question of coloring H–free graphs is as follows.

Meta-Problem   Given a finite set of graphs $\mathcal{A}$, what is the computational complexity of deciding the chromatic number of $\mathcal{A}$–free graphs?

We have completely classified the case of one-element sets $\mathcal{A}$, and the nearest goal may be the case of two-element sets, i.e., coloring graphs that do not contain two specified induced subgraphs. Here the situation seems more complex and we have only obtained partial results. Our proof of Theorem 1 was based on two different NP–completeness constructions (our construction in the proof of Theorem 2 and Holyer's reduction) which so far resist attempts to be combined.

We classify graphs H into four types (Type D is disjoint from Types A,B and C, but the latter three are not mutually disjoint):

Type A graphs containing cycles
Type B graphs containing a claw (induced copy of $K_{1,3}$)
Type C graphs containing an induced copy of a spanning subgraph of $2K_2$
    (these are depicted in Fig. 3)
Type D graphs which are induced subgraphs of P_4 or of $P_3 \oplus K_1$.

The following observation says that for determining the complexity of the $\{H_1, H_2\}$-FREE COLORING problem, we only need to care about cases when H_1 and H_2 are of different types among A, B and C. The three cases are then discussed in the subsequent subsections.

Proposition 2. *Let $\mathcal{A}$ be a finite set of graphs.*
   (i) If at least one of the graphs of $\mathcal{A}$ is of Type D then $\mathcal{A}$-FREE COLORING is polynomially solvable.
   (ii) If all graphs of $\mathcal{A}$ are of the same type A or B or C, then $\mathcal{A}$-FREE COLORING is NP–complete.

Proof. If one of the graphs $H \in \mathcal{A}$ determines a polynomially solvable instance, then so does the set $\mathcal{A}$, since every $\mathcal{A}$–free graph is also H–free. This implies (i) by Theorem 1.

If all graphs of $\mathcal{A}$ are of type B (or C), then every claw–free ($\{2K_2, K_2 \oplus 2K_1, 4K_1\}$–free) graph is $\mathcal{A}$–free and Holyer's reduction (or our Proposition 1) apply.

If all graphs of $\mathcal{A}$ are of type A, we perform the reduction described in Case 1 of Section 4, with k being the length of a longest cycle among all graphs of $\mathcal{A}$.

5.1 Types B versus C (Forests versus Forests)

Say H_1 is of Type B and H_2 is of Type C. Let us assume further that none of H_1, H_2 is of Type A (the open problems of these cases will be stated below). Every forest which contains a claw and at least one more vertex contains a spanning subgraph of $2K_2$, regardless of how and if the other vertex is connected to the claw. So if $H_1 \neq K_{1,3}$ then both H_1 and H_2 are of Type C and the $\{H_1, H_2\}$-FREE COLORING problem is NP–complete. Thus the only open question remains when $H_1 = K_{1,3}$:

Problem 1. Complexity of $\{H, K_{1,3}\}$-FREE COLORING in case H is a linear forest which contains a spanning subgraph of $2K_2$ as an induced subgraph.

5.2 Types A versus B (Cycles versus Claws)

Suppose $C_k \leq H_1$ and $K_{1,3} \leq H_2$. We first show that all cases with $k \geq 4$ determine difficult instances:

Proposition 3. *The problem $\{C_k, K_{1,3}\}$-FREE COLORING is NP–complete for every $k \geq 4$. Hence it is NP–complete for every pair H_1, H_2 such that $C_k \leq H_1$, $k \geq 4$, and $K_{1,3} \leq H_2$.*

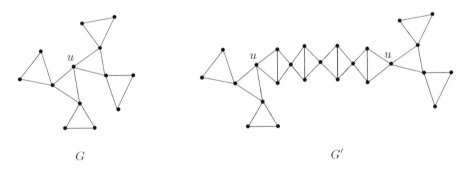

G G'

Fig. 4. Chain of K_4^-'s.

Proof. Holyer proved that 3-colorability of line graphs of cubic triangle–free graphs is NP–complete. Hence 3-colorability is NP–complete for 4-regular claw–free graphs such that any two triangles share at most one vertex. Take such a graph G, split each vertex into two (in such a way that both triangles containing the vertex survive) and connect these two vertices by a sufficiently long chain of copies of K_4^- as indicated in Fig. 4. The new graph G' does not contain short cycles (apart of triangles) nor it contains claws. The chain of K_4^-'s enforces that in any 3-coloring the split vertices receive the same color, and so the new graph is 3-colorable if and only if the original one was.

So it remains to consider the case when H_1 contains triangles but no cycle of greater length. Take again Holyer's reduction. Line graphs of cubic triangle–free graphs are claw–free and any two triangles share at most one vertex. Hence we can conclude:

Observation 3. The problem $\{K_4, K_4^-, K_{1,3}\}$-FREE COLORING is NP–complete.

This leaves the following open case:

Problem 2. Complexity of $\{H_1, H_2\}$-FREE COLORING for H_1 being a claw–free graph containing triangles but no longer cycles, such that every two cycles share at most one vertex, and H_2 being a forest of maximum degree at least 3.

Note that graphs H_1 from Problem 2 have a tree-like structure. If H_2 has more than 4 vertices it contains a spanning subgraph of $2K_2$, and hence the problem is NP–complete whenever H_1 is of Type B as well. So the interesting cases are if H_1 is a triangle, or a triangle plus an isolated vertex, or a triangle with one or two pendant edges. However, if $H_2 = K_{1,3}$ then the number of open cases is larger. Though many of them are shown NP–complete by the reduction in the proof of Proposition 3, many still remain open. A detailed description of the NP–complete and polynomial–time solvable cases known to us will appear in the full version of the paper. Note also that only cases with forests H_2 of maximum degree at most 4 remain open, because of Observation 5 below.

It may be surprising that excluding either C_3 or C_3^+ together with any other graph is essentially equivalent:

Observation 4. The problems $\{C_3, H\}$-FREE COLORING and $\{C_3^+, H\}$-FREE COLORING are polynomially equivalent.

Proof. Every C_3^+–free connected graph is a C_3–free graph or a complete multipartite graph due to Proposition 1. Both complete multipartite graphs and C_3–free graphs can be easily recognized in polynomial time. Thus if the problem $\{C_3, H\}$-FREE COLORING can be solved in polynomial time for a fixed k, we split a graph given as an instance of the problem $\{C_3^+, H\}$-FREE COLORING to its connected components, and we apply the algorithm of the problem $\{C_3, H\}$-FREE COLORING to the components which are C_3–free. The chromatic number of its other connected components can be determined easily in polynomial time, since each of them forms a complete multipartite graph. The chromatic number of the original graph is clearly the maximum chromatic number of its connected components. Thus the problem $\{C_3^+, H\}$-FREE COLORING can be solved in polynomial time in this case, too.

On the other hand, C_3 is an induced subgraph of C_3^+ and hence it holds that $\{C_3, H\}$-FREE COLORING $\propto \{C_3^+, H\}$-FREE COLORING, since any instance of the problem $\{C_3^+, H\}$-FREE COLORING is also an instance of the problem $\{C_3, H\}$-FREE COLORING.

We believe that at least the particular case of $H_1 = C_3$ in Problem 2 deserves separate interest. In other words, we investigate triangle–free graphs. The case of a triangle against a star is fully characterized:

Observation 5. The problems $\{C_3, K_{1,k}\}$-FREE COLORING and $\{C_3^+, K_{1,k}\}$-FREE COLORING are polynomially solvable for $k \leq 4$ and NP-complete otherwise.

Proof. It is enough to restrict to $\{C_3, K_{1,k}\}$-FREE COLORING due to Observation 4. Every triangle–free $K_{1,4}$–free graph has maximum degree 3 and by Brooks's theorem, it is 3-colorable, unless it has K_4 as a connected component, an easily distinguishable instance. The NP–completeness of $\{C_3, K_{1,5}\}$-FREE COLORING is proved in [6].

5.3   Types A versus C (Cycles versus Linear Forests)

In the last case we assume that $C_k \leq H_1$ and H_2 contains a spanning subgraph of $2K_2$ as induced subgraph. This case is solved if cycles of length greater than 4 appear in H_1:

Proposition 4. *The problem* $\{C_k, H_2\}$-FREE COLORING *is NP–complete if* $k \geq 5$ *and* H_2 *contains a spanning subgraph of* $2K_2$ *as an induced subgraph.*

Proof. If $k \geq 6$ then $2K_2 \leq C_k$ and the NP–completeness follows from the fact that both H_1 and H_2 are of the same Type C. For $k = 5$, NP–completeness follows from Corollary 1.

The last open case, however, includes a large number of subproblems and its complexity is not well understood. We note that many cases are polynomially solvable because of structural reasons (including Ramsey's Theorem), e.g., $\{K_4, 4K_1\}$-FREE COLORING allows only finite number of input graphs.

Problem 3. Complexity of $\{H_1, H_2\}$-FREE COLORING if H_1 contains only induced cycles of length 3 and/or 4, and H_2 contains a spanning subgraph of $2K_2$ as induced subgraph.

References

1. Chvátal, V.: Perfectly ordered graphs, in: Topics on Perfect Graphs (C.Berge, V.Chvátal, eds.), Annals of Discrete Mathematics 21 (1984) 63-65
2. Erdős, P.: Graph theory and probability, Canad. J. Math. 11 (1959) 34-38.
3. Fellows, M.R., Kratochvíl, J., Middendorf, M., Pfeiffer, F.: The complexity of induced minors and related problems, Algorithmica 13 (1995) 266-282
4. Holyer, I.: The NP–completeness of edge-coloring, SIAM J. Comput. 10 (1981) 718-720.
5. Lichtenstein, D.: Planar formulae and their uses, SIAM J. Comput. 11 (1982) 329-343.
6. Maffray, F., Preissmann, M.: On the NP–completeness of the k-colorability problem for triangle–free graphs, Discr. Math. 162 (1996), 313-317
7. Olariu, S.: Paw–free graphs, Inform. Process. Lett. 28 (1988), 53-54
8. Randerath, B., Schiermeyer, I.: 3-colorability in P for P_6–free graphs, Rutcor Research Report 39-2001, ext. abstract in Electronic Notes in Discrete Math., Vol. 8 (2001).

On Stable Cutsets in Line Graphs

Van Bang Le[1] and Bert Randerath[2]

[1] Fachbereich Informatik, Universität Rostock, D-18051 Rostock, Germany
le@informatik.uni-rostock.de
[2] Institut für Informatik, Universität zu Köln, D-50969 Köln, Germany
randerath@informatik.uni-koeln.de

Abstract. We answer a question of Brandstädt *et al.* by showing that deciding whether a line graph with maximum degree 5 has a stable cutset is NP-complete. Conversely, the existence of a stable cutset in a line graph with maximum degree at most 4 can be decided efficiently. The proof of our NP-completeness result is based on a refinement on a result due to Chvátal that recognizing decomposable graphs with maximum degree 4 is an NP-complete problem. Here, a graph is decomposable if its vertices can be colored red and blue in such a way that each color appears on at least one vertex but each vertex v has at most one neighbor having a different color from v. We also discuss some open problems on stable cutsets.

1 Introduction

In a graph, a *stable set* (a *clique*) is a set of pairwise non-adjacent (adjacent) vertices. A *cutset* (or *separator*) of a graph G is a set S of vertices such that $G - S$ is disconnected. A *stable cutset* (a *clique cutset*) is a cutset which is also a stable set (a clique).

Clique cutsets are a well-studied kind of separators in the literature, and have been used in divide-and-conquer algorithms for various graph problems, such as graph coloring and finding maximum stable sets; see [11]. Applications of clique cutsets in algorithm designing based on the fact that clique cutsets in arbitrary graphs can be found in polynomial time ([13]).

The importance of stable cutsets has been demonstrated first in [12,4] in connection to perfect graphs. Tucker [12] proved that if S is a stable cutset in G and if no induced cycle of odd length ≥ 5 in G has a vertex in S then the coloring problem on G can be reduced to the same problem on the smaller subgraphs induced by S and the components of $G - S$. Unfortunately, deciding whether a graph has a stable cutset is NP-complete (let STABLE CUTSET denote this problem). This is an easy consequence of a result due to Chvátal [3] on decomposable graphs; see also [1]. In fact, Chvátal's results from [3] implies the following

Theorem 1. STABLE CUTSET *is NP-complete, even if the input graph is restricted to line graphs with maximum degree* 6.

A. Brandstädt and V.B. Le (Eds.): WG 2001, LNCS 2204, pp. 263–271, 2001.
© Springer-Verlag Berlin Heidelberg 2001

In [1], Brandstädt *et al.* then asked whether the degree constraint in Theorem 1 is best possible. In this paper, we consider this question and prove that STABLE CUTSET is NP-complete, even for 5-regular line graphs of bipartite graphs, and is efficiently solvable for line graphs with maximum degree 4. Also, some open problems will be discussed.

2   Decomposable Graphs Revisited

A graph is *decomposable* if its vertices can be colored red and blue in such a way that each color appears on at least one vertex but each vertex v has at most one neighbor having a different color from v. In other words, a graph is decomposable if its vertices can be partitioned into two nonempty parts such that the edges connecting vertices of different parts form an induced matching.

Chvátal [3] proved that recognizing decomposable graphs is NP-complete, even for graphs with maximum degree 4. Based on Chvátal's original proof we extend this result. Thereby, we also extend a result of Moshi [9] that recognizing decomposable graphs is NP-complete, even if the input is restricted to bipartite graphs of minimum degree 2.

Theorem 2. *Recognizing decomposable graphs is NP-complete, even if the input is restricted to bipartite graphs with one color class consisting only of vertices of degree 3 and the other color class consisting only of vertices of degree 4.*

Proof. A hypergraph is called *bicolorable* if its vertices can be colored red and blue in such a way that no edge is monochromatic. Lovász [8] proved that recognizing bicolorable hypergraphs is an NP-complete problem, even if the input is restricted to hypergraphs with each edge having size 3. Given such a hypergraph H we shall construct a graph G such that G is decomposable if and only if H is bicolorable. Now observe that the complete bipartite graph $K_{2,3}$ is indecomposable. For each vertex v of H, belonging to d edges, take a chain A_v of $4d - 1$ different complete bipartite graphs $K_{2,3}$, say $D_1, ..., D_{4d-1}$ each with the partite set $C_j = \{a_j, b_j, c_j\}$ of degree two vertices for $j = 1, ..., 4d - 1$, such that we identify c_j with a_{j+1} for $j = 1, ..., 4d - 2$. (A_v is illustrated in Figure 1 for $d = 1$).

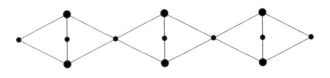

Fig. 1.

Label one of the vertices of degree 2 in this A_v by v^* and the remaining $4d$ vertices of degree two by the distinct labels $(v, e, i)_{(j)}$ with $i = 0, 1$, $j = 0, 1$

and e running through all the edges with $v \in e$. For each $i = 0, 1$ take a similar chain B_i of $2n + m - 2$ complete bipartite graphs $K_{2,3}$ with n and m standing for the number of vertices and edges in H, respectively. Label n of the vertices of degree two in B_i by $(v, i)_1$ and n of the vertices of degree two in B_i by $(v, i)_2$ with v running through all the vertices of H and label the remaining m vertices of degree two in B_i by (e, i) with e running through all the edges of H. For each $i = 0, 1$ and for each edge $e = \{x, y, z\}$ of H add the graph shown in Figure 2 in which the vertices of degree two are labeled.

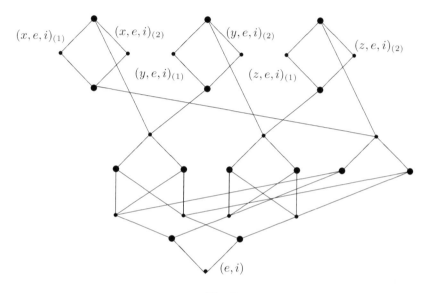

Fig. 2.

Furthermore, for each vertex v of H, add the graph shown in Figure 3 in which the vertices of degree two are labeled.

Note that in each graph described above, vertices of degree 2 are exactly the labeled vertices, and that each graph is bipartite with one color class of the bipartition consisting only of vertices of degree 3 and the other color class consisting only of vertices of degree 2 or 4.

Now, identify vertices with the same labels. The resulting graph is the desired G, a bipartite graph with minimum degree 3, maximum degree 4, all vertices of maximum degree 4 form a stable set and likewise all vertices of minimum degree 3 form another stable set.

Now consider a decomposition of G. As each complete bipartite graph $K_{2,3}$ is indecomposable, the n graphs A_v and the two graphs B_i are indecomposable: in each of them, all the vertices have the same color. Without loss of generality, say that all the vertices of B_0 are red. Now all the vertices in B_1 must be blue: otherwise all A_v would be red [since the (red colored) graph in Figure 3 would force red on v^*] and then it is not difficult to observe that all vertices of the graphs of Figure 2 type would receive the color red as well and therefore the entire graph would be red, a contradiction.

Next, consider an arbitrary edge $e = \{x, y, z\}$ of H. At least one of the graphs A_x, A_y or A_z must be red [otherwise the graph in Figure 2 would force blue on $(e, 0)$] and at least one of them must be blue [otherwise the graph in Figure 2 would force red on $(e, 1)$]. Hence, the colors on the n graphs A_v define a bicoloring of H. Conversely, each bicoloring of H may be used to color the n graphs A_v and then extended into a decomposition of G.   □

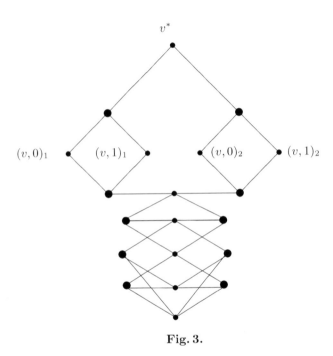

Fig. 3.

Remark 1. *It should be remarked that we could also use graphs of figure 1 of 'arbitrary' length. Hence, for every given $\delta > 0$ we can choose one color class containing all vertices of degree 3 and the other color class containing vertices of degree 2 and 4 s. t. the average degree is at most $(1 + \delta)3$.*

As Chvátal has shown, the NP-result is best possible with respect to degree constraints:

Theorem 3 (Chvátal [3]). *Decomposable graphs of maximum degree 3 can be recognized in polynomial time.*

We will make use of this theorem in proving that STABLE CUTSET can be solved in polynomial time for line graphs of maximum degree at most 4.

3 Stable Cutsets in Line Graphs

As suggested by Cunnigham (see [1]) and shown by Brandstädt *et al.* the STABLE CUTSET problem for line graphs can be derived by the result of Chvátal on decomposable graphs. Recall that the *line graph* $L(G)$ of a graph G has the edges of G as its vertices, and two distinct edges of G are adjacent in $L(G)$ if they are incident in G. The relationship between decomposability and a stable cutset is

Proposition 1 ([1]). *If $L(G)$ has a stable cutset, then G is decomposable. If G is decomposable and has minimum degree at least 2, then $L(G)$ has a stable cutset.*

Now observe that if G is a bipartite graph with one color class consisting only of vertices of degree 3 and the other color class consisting only of vertices of degree 4, then $L(G)$ is a 5-regular graph. With this last proposition, the extension of the complexity result of Chvátal as proven in the last section and the latter observation we derive the following strengthening of the NP-complete result given in Theorem 1.

Theorem 4. STABLE CUTSET *is NP-complete, even if the input is restricted to 5-regular line graphs of bipartite graphs.*

By Remark 1 at the end of the last section we also obtain the following related result.

Theorem 5. STABLE CUTSET *is NP-complete, even if the input is restricted to line graphs of order n and size $(2 + \epsilon)n$ of bipartite graphs with minimum degree at least 2, maximum degree 4, all vertices of degree 3 form an independent set and likewise all vertices of even degree form an independent set.*

These results are best possible, since we will demonstrate in the following that STABLE CUTSET is polynomial if the input is restricted to line graphs with maximum degree at most 4.

4 Observations

As a corollary of Theorem 3 and Observation 3 below, STABLE CUTSET can be solved in polynomial time on line graphs of maximum degree 3. We will see in the next sections that STABLE CUTSET can be solved in polynomial time for arbitrary graphs of maximum degree 3 and for line graphs of maximum degree 4. For this purpose, we need some observations.

Observation 1. *Let S be a stable cutset of G and let $v \in V(G)$. If $N(v)$ is not a stable set, then $G - v$ has a stable cutset.*

Proof. If $v \in S$, $S - v$ is a stable cutset in $G - v$. If v belongs to component A of $G - S$, then, as $N(v)$ is not a stable set, $|A| \geq 2$, hence S is a stable cutset in $G - v$. ☐

Observation 2. *Let $v \in V(G)$, and let S be a stable cutset of $G - v$. If $S \cap N(v) = \emptyset$ or $N(v) - S$ belongs to the same component in $(G - v) - S$, then G has a stable cutset.*

Proof. In the first case, $S \cup \{v\}$ is a stable cutset in G, in the second case, S is a stable cutset in G. ☐

A vertex v in G is *simplicial* if $N_G(v)$ is a clique. The observation below directly follows from the previous two.

Observation 3. *Assume that G has at least three vertices, and let v be a simplicial vertex of G. Then G has a stable cutset if and only if $G - v$ has a stable cutset.* ☐

The following observation admits an argument that will be used in further discussions.

Observation 4. *Let v be a vertex of a graph G such that $G[N(v)]$ consists of a clique C and a vertex a with $N(a) \cap C \subseteq \{c\}$ for a vertex $c \in C$. If $G - v$ has a stable cutset S, then G has a stable cutset, or we may assume that*

- $S \cap N(v) = \{c\}$,
- *there are two different components A, B of $(G - v) - S$ with $a \in A$ and $C - \{c\} \subseteq B$, and*
- $(G - v) - S = A \cup B$.

Proof. If $N(a) \cap C = \{c\}$ but $S \cap C \neq \{c\}$ then, by Observation 2, G has a stable cutset. If $N(a) \cap C = \emptyset$ but $S \cap C = \emptyset$ then, by Observation 2 again, G has a stable cutset. Thus, in any case, we may assume that $S \cap N(v) = \{c\}$.

If $N(v) - \{c\}$ belongs to the same component in $(G - v) - S$ then, by Observation 2, G has a stable cutset. Thus we may assume that a belongs to component A, $C - \{c\}$ belongs to another component B of $(G - v) - S$.

If $(G - v) - S$ has a third component then clearly S is a stable cutset in G. ☐

5   Stable Cutsets in Graphs with Maximum Degree 3

We show in this section that STABLE CUTSET can be solved in polynomial time on (not necessarily line) graphs with maximum degree at most 3. We discuss the non-trivial case where the input graph G has maximum degree 3 and at least 7 vertices. Furthermore, we may assume that every vertex of G has a non-stable neighborhood (otherwise we are done), and that G has no simplicial vertex (by Observation 3).

Consider a vertex v of degree 3, i. e. a vertex attaining the maximum degree 3. We will show that G has a stable cutset if and only if $G - v$ has a stable cutset. By Observation 1 we need only to show the if part. Thus, let S be an inclusion-minimal stable cutset in $G - v$. We will show that G has a stable cutset.

As $N(v)$ is not a stable set and not a clique, it consists of three vertices a, b, c such that b and c are adjacent and a is non-adjacent to b. By Observation 4, we may assume that $S \cap N(v) = \{c\}$, and $(G - v) - S$ has exactly two components. Let A, B be the components of $(G - v) - S$ with $a \in A$ and $b \in B$.

CASE 1: a is adjacent to c.

If $A \neq \{a\}$ then the neighbor of a in $A - \{a\}$ and the vertex b form a stable cutset in G disconnecting a and the neighbor of b different from v and c (such a neighbor of b exists because b is not simplicial). Similarly, if $B \neq \{b\}$, G has a stable cutset. Thus we may assume that $A = \{a\}$ and $B = \{b\}$. Then, as G has at least 5 vertices, $|S| \geq 2$, hence $\{a, b\}$ is a stable cutset in G disconnecting v and $S - \{c\}$. Case 1 is settled.

CASE 2: a is non-adjacent to c.

By the minimality of S, c has (exactly one) neighbor $c' \in A$. If a has no neighbor in S, $S \cup \{a\}$ is a stable cutset in G disconnecting c' and v. So, let us assume that a has (exactly one) neighbor $a' \in S$. By the minimality of S, a' has a neighbor $a'' \in B$.

If $a'' \neq b$, then b has exactly one neighbor $b' \in B - \{b\}$, and $\{a, c, b'\}$ is a stable cutset in G disconnecting b and c'. So, we may assume that $a'' = b$, hence $B = \{b\}$. Now, if a' and c' are non-adjacent, then $\{a', c', v\}$ is a stable cutset in G disconnecting a and b, and if a and c' are non-adjacent, then $\{a, b, c'\}$ is a stable cutset in G disconnecting v and a'. The case $a'c'$ and ac' are edges of G cannot occur because G has at least 7 vertices. Case 2 is settled.

Thus, for every vertex v of maximum degree 3, G has a stable cutset if and only if $G - v$ has a stable cutset. Repeat the same reduction on $G' = G - v$ if G' still has maximum degree 3 and at least 7 vertices. In this way, we can determine in linear time whether G has a stable cutset. Therefore we obtain

Theorem 6. STABLE CUTSET *can be solved in linear time on graphs with maximum degree at most* 3. □

6   Stable Cutsets in Line Graphs of Maximum Degree 4

Let $L = L(G)$ with maximum degree $\Delta = \Delta(L) = 4$ and at least 13 vertices. We shall reduce L to a line graph L^* such that $L^* = L(G^*)$, and

(i) L has a stable cutset if and only if L^* has a stable cut set, and
(ii) G^* has maximum degree at most 3 and minimum degree 2.

Thus, by Proposition 1 and Theorem 3, we obtain

Theorem 7. STABLE CUTSET *can be solved in polynomial time on line graphs of maximum degree at most* 4. □

THE REDUCTION

First, by Observation 3, we may assume that L has no simplicial vertex. Let v be a vertex of L of maximum degree $d(v) = 4$, and let xy be the corresponding edge in G. We call v *bad* if (exactly) one of x, y is of degree 4 in G (x or y cannot have degree 1 in G otherwise v would be a simplicial vertex in L). If no vertex of degree 4 in L is bad, then G has maximum degree 3 and we are done by Theorem 3.

CLAIM: If L has a bad vertex v of degree 4 then L has a stable cutset if and only if $L - v$ has a stable cutset.
PROOF: See the full version of the paper.

Note that $L' := L - v$ is again a line graph. Repeat the same reduction for L' instead of L if L' still has a bad vertex of degree 4 and at least 13 vertices. In this way we obtain the desired line graph L^*.

The root graph G^* with $L^* = L(G^*)$ can be determined in linear time (see [5,7,10]).

7   Concluding Remarks

We have shown that STABLE CUTSET is NP-complete for 5-regular line graphs of bipartite graphs, and that STABLE CUTSET can be solved in polynomial time for line graphs of maximum degree at most 4 and for arbitrary graphs of maximum degree at most 3. Thus, it would be interesting to fill the gap in the complexity status of STABLE CUTSET on arbitrary graphs of maximum degree 4.

Problem 1. *Is* STABLE CUTSET *in P or NP-complete on graphs with maximum degree 4?*

We are going to explain a connection between this problem and another interesting result on stable cutsets. The following nice theorem was conjectured by Caro and has been proved by Chen and Yu [2].

Theorem 8 ([2]). *Every graph with n vertices and at most $2n - 4$ edges has a stable cutset.* □

(This theorem implies that every graph with at least 8 vertices and maximum degree at most 3 has a stable cutset, therefore STABLE CUTSET is in P for graphs with maximum degree at most 3; cf. also our discussion in Section 5 without use of Theorem 8.)

Now, let G be a graph with n vertices and maximum degree 4. Then G has at most $2n$ edges. By Theorem 8, Problem 1 remains open only in four cases, namely for graphs with n vertices and m edges where $2n - 3 \leq m \leq 2n$.

Thus, the following problem is of interest.

Problem 2 (SCS(n,m)). *Given a graph G with n vertices and at most m edges. Does G have a stable cutset?*

Theorem 8 shows that SCS$(n, 2n - 4)$ is in P, while our Theorem 5 implies that, even on line graphs, SCS$(n, (2 + \epsilon)n)$ is NP-complete for every fixed $\epsilon > 0$. A question arises: What is the maximum number m (depending on n) for which SCS(n, m) can be solved efficiently? In particular, is SCS$(n, 2n - 3)$ in P?

References

1. A. Brandstädt, F. Dragan, V.B. Le and T. Szymczak, *On stable cutsets in graphs*, Discrete Appl. Math. 105 (2000), 39–50.
2. G. Chen and X. Yu, *A note on fragile graphs*, Preprint (1998).
3. V. Chvátal, *Recognizing decomposable Graphs*, J. Graph Theory 8, (1984), 51–53.
4. D.G. Corneil, J. Fonlupt, *Stable set bonding in perfect graphs and parity graphs*, J. Combin. Theory (B) 59 (1993), 1–14.
5. D. Degiorgi, *A new linear algorithm to detect a line graph and output its root graph*, Tech. Report 148, Inst. für Theoretische Informatik, ETH Zürich, Zürich, Switzerland (1990)
6. R.L. Hemminger, L.W. Beineke, *Line graphs and line digraphs*, In: *Selected Topics in Graph Theory I*, L.W. Beineke, R.T. Wilson, eds., Academic Press, London, (1978), 271–305.
7. P.G.H. Lehot, *An optimal algorithm to detect a line graph and output its root graph*, J. Assoc. Comput. Mach. 21 (1974), 569–575.
8. L. Lovász, *Covering and colorings of hypergraphs*, In *Proceedings, 4th Southeastern Conference on Combinatorics, Graph Theory and Computing*, Utilitas Mathematica, Winnipeg (1973), 3–12.
9. A.M. Moshi, *Matching cutsets in graphs*, J. Graph Theory 13, (1989), 527–536.
10. N.D. Roussopoulos, *A $\mathcal{O}(\max\{m,n\})$ algorithm for determining the graph H from its line graph G*, Inf. Process. Lett. 2 (1973), 108–112.
11. R.E. Tarjan, *Decomposition by clique separators*, Discrete Math. 55 (1985), 221–232.
12. A. Tucker, *Coloring graphs with stable cutsets*, J. Combin. Theory (B) 34 (1983), 258–267.
13. S.H. Whitesides, *An algorithm for finding clique cut-sets*, Inf. Process. Lett. 12 (1981), 31–32.

On Strong Menger-Connectivity of Star Graphs[*]

Eunseuk Oh and Jianer Chen

Department of Computer Science, Texas A&M University,
College Station, TX 77843-3112, USA
{eunseuko,chen}@cs.tamu.edu

Abstract. Motivated by parallel routing in networks with faults, we study the following graph theoretical problem. Let G be a d-regular graph. We say that G is *strongly Menger-connected* if for any copy G_f of G with at most $d - 2$ nodes removed, every pair of nodes u and v in G_f are connected by $\min\{deg_f(u), deg_f(v)\}$ node-disjoint paths in G_f, where $deg_f(u)$ and $deg_f(v)$ are the degrees of the nodes u and v in G_f, respectively. We show that the star graphs, which are a recently proposed attractive alternative to the widely used hypercubes as network models, are strongly Menger-connected. An algorithm of optimal running time is developed that constructs the maximum number of node-disjoint paths of nearly optimal length in star graphs with faults.

1 Introduction

Parallel routing (i.e., construction of node-disjoint paths) has been an important issue in the study of computer networks. Routing by node-disjoint paths between nodes can not only avoid communication bottlenecks, thus increase the efficiency of message transmission, but also provide alternative paths in case of node failures. By the famous Menger's Theorem [14], if a network G is a d-connected graph, then every pair of nodes in G are connected by d node-disjoint paths.

With the continuous increasing in network size, routing in networks with faults has become unavoidable. Suppose that the network G has a set S_f of faulty nodes. Let u and v be two non-faulty nodes in G. Based on local information, we know the numbers $deg_f(u)$ and $deg_f(v)$ of non-faulty neighbors of u and v, respectively, and are interested in constructing the maximum number of node-disjoint fault-free paths between u and v. Obviously, the number of node-disjoint fault-free paths between u and v cannot be larger than $\min\{deg_f(u), deg_f(v)\}$. We are interested in the precise bound on the size of the faulty node set S_f such that for any two non-faulty nodes u and v in G, there are $\min\{deg_f(u), deg_f(v)\}$ node-disjoint fault-free paths connecting u and v. Note that this problem has a strong similarity to that of Menger's Theorem.

Since most popular network topologies are regular graphs, we will concentrate on regular graphs. Suppose that the network G is a d-regular graph (i.e., all nodes of G have degree d), then in general the number of faulty nodes in the set S_f

[*] This work is supported in part by NSF under Grant CCR-0000206.

A. Brandstädt and V.B. Le (Eds.): WG 2001, LNCS 2204, pp. 272–283, 2001.
© Springer-Verlag Berlin Heidelberg 2001

should not exceed $d - 2$ to ensure $\min\{deg_f(u), deg_f(v)\}$ node-disjoint fault-free paths between any two nodes u and v. This can be seen as follows. Let u and v be two nodes in G whose distance is larger than 3. Pick any neighbor u' of u and let the other $d - 1$ neighbors of u' become faulty. Note that no neighbor of u' can be a neighbor of v since the distance from u to v is at least 4. Now each of the nodes u and v has d non-faulty neighbors. However, there are obviously no d node-disjoint fault-free paths in G from u to v since one of the d neighbors of u, the node u', leads to a "deadend". This motivates the following definition:

Definition 1. A d-regular graph G is *strongly Menger-connected* if for any copy G_f of G with at most $d - 2$ nodes removed, each pair u and v of nodes in G_f are connected by $\min\{deg_f(u), deg_f(v)\}$ node-disjoint paths in G_f, where $deg_f(u)$ and $deg_f(v)$ are the degrees of the nodes u and v in G_f, respectively.

Because of its motivation, in case the graph is strongly Menger-connected, we are also interested in constructing the node-disjoint paths so that the path lengths are minimized. To authors' knowledge, there has not been a systematic study on this issue on popular network structures. Some related research includes Galil and Yu's work on constructing short node-disjoint paths in a general graph [10] and Hsu's work on graph containers [13]. However, neither of them considered graphs with faulty nodes.

We will discuss the strong Menger-connectivity in detail on star graphs [2], which have received considerable attention recently as an attractive alternative to the widely used hypercubes as network models, because of their rich structure, smaller diameter, lower degree, and symmetry properties [2,19]. We provide another advantage for star graphs in this paper by showing that star graphs are strongly Menger-connected and developing an algorithm of optimal running time that constructs the maximum number of node-disjoint paths of nearly optimal length between any two nodes in the n-star graph with at most $n - 3$ nodes removed (note the n-star graph is $(n - 1)$-regular).

We briefly mention the related research on star graphs. Constructing a single path between two nodes in star graphs with faults has been studied extensively [1,2,3,4,11,18]. Constructing node-disjoint paths in star graphs without faulty nodes has also been studied [6,7,8,9,12]. In particular, an efficient algorithm has been developed [6] that constructs the maximum number of node-disjoint paths of optimal length for any two given nodes in star graphs. A randomized algorithm, based on the Information Dispersal Algorithm (IDA) [16], for constructing node-disjoint paths in star graphs with faults was proposed in [17].

2   Preliminaries

A permutation $u = \langle a_1 a_2 \cdots a_n \rangle$ of $\langle 12 \cdots n \rangle$ can be given by a product of disjoint cycles [5], which is called the *cycle structure* of u. A cycle is *trivial* if it contains only one symbol. The cycle containing the symbol 1 will be called the *primary cycle*. For example, the permutation $\langle 32541 \rangle$ has the cycle structure $(351)(2)(4)$, where the primary cycle (351) indicates that 3 is at 1's position, 5 is at 3's position, and 1 is at 5's position. The trivial cycles (2) and (4) indicate that 2 and 4 are in their "correct" positions.

We define a group of operations ρ_i on permutations for $1 \leq i \leq n$. For any permutation u, $\rho_i(u)$ is the permutation obtained from u by exchanging the first symbol and the ith symbol in u. Let us consider how these operations change the cycle structure of a permutation. Write u in its cycle structure

$$u = (a_{11} \cdots a_{1n_1} 1)(a_{21} \cdots a_{2n_2}) \cdots (a_{k1} \cdots a_{kn_k})$$

If the ith symbol a of u is not in the primary cycle, then $\rho_i(u)$ "merges" the cycle containing a into the primary cycle. More precisely, let $a = a_{21}$ (note each cycle can be cyclically permuted and the cycle order is irrelevant), then

$$\rho_i(u) = (a_{21} \cdots a_{2n_2} a_{11} \cdots a_{1n_1} 1)(a_{31} \cdots a_{3n_3}) \cdots (a_{k1} \cdots a_{kn_k})$$

If the ith symbol a is in the primary cycle, then $\rho_i(u)$ "splits" the primary cycle into two cycles. More precisely, suppose that $a = a_{1j}$, where $1 \leq j \leq n_1 + 1$ and we have let $a_{1n_1+1} = 1$, then

$$\rho_i(u) = (a_{11} \cdots a_{1j-1})(a_{1j} \cdots a_{1n_1} 1)(a_{21} \cdots a_{2n_2}) \cdots (a_{k1} \cdots a_{kn_k})$$

In particular, if $a = a_{12}$, then we say that the operation ρ_i "deletes" the first symbol a_{11} from the primary cycle.

The n-star graph S_n is an undirected graph consisting of $n!$ nodes labeled with the $n!$ permutations on $\langle 12 \cdots n \rangle$. A node u in S_n has $n - 1$ neighbors $\rho_i(u)$, $2 \leq i \leq n$. The n-star graph S_n is vertex-symmetric [2]. Thus, a set of node-disjoint paths from a node w to a node v can be mapped to a set of node-disjoint paths from a node u to ε in a straightforward way, where ε is the identity permutation $\langle 12 \cdots n \rangle$. Thus, we only need to concentrate on node-disjoint paths from u to ε in S_n. Denote by $dist(u)$ the distance from u to ε.

It has been known [6] that the following are necessary and sufficient rules for tracing a shortest path from a node u to the node ε in the n-star graph S_n:

Shortest Path Rules (briefly **sp-rules**)
1. If u has a trivial primary cycle, then in the next node on a shortest path from u to ε, a nontrivial cycle is merged into the primary cycle.
2. If u has a nontrivial primary cycle $(a_{11}a_{12} \cdots a_{1n_1} 1)$, then in the next node on a shortest path from u to ε, either another nontrivial cycle is merged into the primary cycle, or the first symbol a_{11} is deleted from the primary cycle.

Remark 1. If an edge $[u, v]$ in S_n does not follow the sp-rules, then $dist(v) = dist(u) + 1$ [6]. Thus, the length of a path from u to ε in which exactly k edges do not follow the sp-rules is equal to $dist(u) + 2k$.

For the n-star graph S_n, let $S_n[i]$ be the set of nodes in which the symbol 1 is at the ith position. It is well-known [1] that the set $S_n[1]$ is an independent set, and the subgraph induced by the set $S_n[i]$ for $i \neq 1$ is a $(n - 1)$-star graph. Note that a node is in the substar $S_n[i]$, $i \neq 1$, if and only if the primary cycle of the node is of form $(\cdots i1)$, and a node is in $S_n[1]$ if and only if the primary cycle of the node is trivial.

Two simple procedures [6] will be used in tracing a shortest path from a node u to ε. The first is the "Delete" procedure, written as $\rightarrow \overset{D}{\cdots} \rightarrow$, which repeatedly deletes the first symbol in the nontrivial primary cycle. The second is the "Merge-Delete" procedure, written as $\rightarrow \overset{M+D}{\cdots} \rightarrow$, which works in two stages: first repeatedly merges in an arbitrary order each of the nontrivial cycles into the primary cycle, then repeatedly deletes the first symbol in the primary cycle. It is easy to verify that both procedures follow the sp-rules strictly. A nice property of the Delete and Merge-Delete procedures is that if they start with a node u in the substar $S_n[i]$, $i \neq 1$, then all nodes, possibly except the last one, on the constructed shortest path are also in the substar $S_n[i]$.

3   Bridging Paths from a Node to a Substar

Our algorithm is based on the concept of *bridging paths* that connect a given node to a specific substar in the n-star graph. In this section, we give formal definitions for bridging paths, and study its properties.

Lemma 1. *Let $i > 1$ and $S_n^-[i]$ be the substar $S_n[i]$ with $k_i \leq n - 3$ nodes removed, $\rho_i(\varepsilon) \in S_n^-[i]$. For any node u in $S_n^-[i]$, a path P from u to $\rho_i(\varepsilon)$ in $S_n^-[i]$ can be constructed in time $O(k_i n + n)$ with at most two edges (in case the primary cycle of u is $(i1)$, one edge) not following the sp-rules.*

Definition 2. Let u be a node in the n-star graph S_n and u' be a neighbor of u in the substar $S_n[i]$, $i \neq 1$. For each neighbor v of u', $v \neq u$, a (u', j)-*bridging path* P from u to $S_n[j]$, $j \neq 1, i$, is defined as follows: if v is in $S_n[1]$ then $P = [u, u', v, \rho_j(v)]$, while if v is in $S_n[i]$ then $P = [u, u', v, \rho_i(v), \rho_j(\rho_i(v))]$.

Thus, from each neighbor u' in $S_n[i]$ of the node u, $i \neq 1$, there are $n - 2$ (u', j)-bridging paths of length bounded by 4 that connect the node u to the substar $S_n[j]$. See Figure 1 for an intuitive illustration for bridging paths.

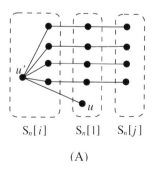
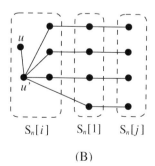

$S_n[i]$ $S_n[1]$ $S_n[j]$ $S_n[i]$ $S_n[1]$ $S_n[j]$

(A)   (B)

Fig. 1. Bridging paths from node u to substar $S_n[j]$: (A) u is in $S_n[1]$; (B) u is in $S_n[i]$

Since no two nodes in $S_n[i]$ share the same neighbor in $S_n[1]$ and no two nodes in $S_n[1]$ share the same neighbor in $S_n[j]$, for any neighbor u' of u, two (u', j)-bridging paths from u to $S_n[j]$ have only the nodes u and u' in common. Moreover, for any two neighbors u' and u'' of u in $S_n[i]$ (in this case, the node u must itself be also in $S_n[i]$), since u' and u'' have no other common neighbor except u (see, for example, [7,8]), a (u', j)-bridging path from u to $S_n[j]$ and a (u'', j)-bridging path from u to $S_n[j]$ share no nodes except u.

Definition 3. Let u in a node in S_n and u' be a neighbor of u in $S_n[i]$, $i \neq 1$. A (u', j)-bridging path P from u to $S_n[j]$ is *divergent* if in the subpath of P from u to $S_n[1]$, there are three edges not following the sp-rules.

Note that the subpath from u to $S_n[1]$ of a (u', j)-bridging path P contains at most three edges. In particular, if the subpath contains only two edges, then the path P is automatically non-divergent.

Lemma 2. A divergent (u', j)-bridging path P from a node u to a substar $S_n[j]$, where $u' \in S_n[i]$, $i \neq 1, j$, can be extended into a path Q from u to $\rho_j(\varepsilon)$ in time $O(n)$, such that at most 4 edges in Q do not follow the sp-rules and the extended part is entirely in $S_n[j]$. Moreover, for two divergent (u', j)-bridging paths, the two corresponding extended paths only share the nodes u, u', and $\rho_j(\varepsilon)$.

Proof. Let P be a divergent (u', j)-bridging path from the node u to the substar $S_n[j]$, where u' is a neighbor of u. Since P is divergent, it has length 4. Thus, the path can be written as $P = \{u, u', v, v', v''\}$, where u' is in $S_n[i]$, v is a neighbor of u' in $S_n[i]$, $v' = \rho_i(v)$ is in $S_n[1]$, and $v'' = \rho_j(v')$ is in $S_n[j]$.

Let $u' = (a_1 \cdots a_p i 1) * *$, where "$**$" stands for "other cycles". Since the edge $[u', v]$ does not follow the sp-rules and v is in $S_n[i]$, the node v must have the form either $v = (ba_1 \cdots a_p i 1) * *$, where (b) is a trivial cycle in u', or $v = (a_1 \cdots a_q)(a_{q+1} \cdots a_p i 1) * *$, where $2 \leq q \leq p$. Now since $[v, v']$ is an edge in S_n and v' is in $S_n[1]$, the node v' must be of the form either $v' = (ba_1 \cdots a_p i)(1) * *$, or $v' = (a_{q+1} \cdots a_p i)(1) * *$. Moreover, since the edge $[v, v']$ does not follow the sp-rules, when $v' = (a_{q+1} \cdots a_p i)(1) * *$, we must have $q + 1 \leq p$. In summary, if P is a divergent path, then the fourth node v' on P must be of form $(b_1 b_2 \cdots i)(1)$, where the cycle $(b_1 b_2 \cdots i)$ is nontrivial. Moreover, the (u', j)-bridging path P is distinguished from other divergent (u', j)-bridging paths by the symbol b_1 in the above format (i.e., two different divergent (u', j)-bridging paths will have two different symbols b_1 in the above format).

Now consider the fourth edge $[v', v'']$ on P, where v'' is in $S_n[j]$, $j \neq 1, i$. If j is in a trivial cycle in v', then the extended path Q is

$$u \to u' \to v \to v' = (b_1 b_2 \cdots i)(1) * * \to v'' = (b_1 b_2 \cdots i)(j1) * * \to$$

$$\to (b_2 \cdots i b_1 j 1) * * \xrightarrow{M+D} \to (b_1 j 1) \to (j1) = \rho_j(\varepsilon) \qquad (1)$$

The extended path Q has no common nodes in $S_n[j]$, except $\rho_j(\varepsilon)$, with the paths extended from the other (u', j)-bridging paths since the symbol b_1 distinguishes the path Q from other extended paths: the first part of Q has a unique cycle $(b_1 b_2 \cdots i)$ while the second part of Q has a cycle of the unique format $(\cdots b_1 j 1)$.

If j is in a nontrivial cycle in v', then there are two possible cases:
Case 1. j is not in the cycle $(b_1 b_2 \cdots i)$. Then the extended path Q is

$$u \to u' \to v \to v' = (b_1 b_2 \cdots i)(1) * * \to v'' = (b_1 b_2 \cdots i)(\cdots j1) * * \to \overset{D}{\cdots} \to$$
$$\to (b_1 b_2 \cdots i)(j1) * * \to (b_2 \cdots i b_1 j1) * * \to \overset{M+D}{\cdots} \to (b_1 j1) \to (j1) = \rho_j(\varepsilon) \quad (2)$$

Again, because of b_1, the extended path Q has no common nodes in $S_n[j]$, except $\rho_j(\varepsilon)$, with the paths extended from the other (u', j)-bridging paths.
Case 2. j is in the cycle $(b_1 b_2 \cdots i)$.
If $j = b_1$, then $(b_1 b_2 \cdots i) = (b_2 \cdots ij)$, and the path Q is

$$u \to u' \to v \to v' = (b_2 \cdots ij)(1) * * \to v'' = (b_2 \cdots ij1) * * \to$$
$$\to \overset{M+D}{\cdots} \to (ij1) \to (j1) = \rho_j(\varepsilon) \quad (3)$$

This path is node-disjoint from the paths extended from the other (u', j)-bridging paths because all nodes of it in $S_n[j]$ contain a cycle of form $(\cdots ij1)$.
If $j \neq b_1$, then $(b_1 b_2 \cdots i) = (b_1 \cdots j \cdots i)$, and the path Q is

$$u \to u' \to v \to v' = (b_1 \cdots j \cdots i)(1) * * \to v'' = (\cdots i b_1 \cdots j1) * * \to$$
$$\to (\cdots i b_1 \cdots)(j1) * * \to (\cdots i b_1 j1) * * \to \overset{M+D}{\cdots} \to (b_1 j1) \to (j1) = \rho_j(\varepsilon) \quad (4)$$

Again this path is node-disjoint from the paths extended from the other (u', j)-bridging paths because of the symbol b_1.

For all cases, we can easily verify that the path Q contains at most 4 edges not following the sp-rules, and that the part of Q extended from the (u', j)-bridging path P is entirely in the substar $S_n[j]$. Finally, from the sequences (1)-(4), it can be easily see that the construction of the path Q takes time $O(n)$ and is independent of the construction of the other paths. □

4   The Strong Menger-Connectivity of Star Graphs

We are ready to present our main algorithm. Let S_n^- be the n-star graph S_n with at most $n - 3$ nodes removed such that $\varepsilon \in S_n^-$. Denote by $deg_f(u)$ the degree of a node u in S_n^-. The algorithm is presented in Figure 2.

Theorem 1. *The algorithm* **M-connectivity** *runs in time* $O(n^2)$ *and constructs* $\min\{deg_f(u), deg_f(\varepsilon)\}$ *node-disjoint paths in* S_n^- *of length bounded by* $dist(u) + 8$ *from the node* u *to the node* ε.

Proof. Step 1 of the algorithm constructs paths between some neighbors of the nodes u and ε, Step 2 of the algorithm maximally pairs the rest neighbors of u and ε, and Step 3 constructs a path from u to ε for each of these pairs. Thus, the number of paths constructed by the algorithm **M-connectivity** is exactly $\min\{deg_f(u), deg_f(\varepsilon)\}$. What remain to discuss are the node-disjointness and the length of these paths, and the complexity of the algorithm.

First consider Step 1 of the algorithm. In case u is in $S_n[1]$, for each index $j \neq 1$ with both $\rho_j(u)$ and $\rho_j(\varepsilon)$ in S_n^-, by Lemma 1, we can construct in time $O(k_j n + n)$ a path Q_j from $\rho_j(u)$ to $\rho_j(\varepsilon)$ in S_n^- that is entirely in $S_n[j]$ such that at most two edges in Q_j do not follow the sp-rules, where $k_j \leq n-3$ is the number of nodes removed from $S_n[j]$. Now the path P_j from u through Q_j to ε has at most three edges do not follow the sp-rules (the edge $[\rho_j(\varepsilon), \varepsilon]$ always follows the sp-rules). By Remark 1, the length of P_j is bounded by $dist(u) + 6$. This path P_j is disjoint with other paths constructed in Step 1 because all internal nodes of P_j are in $S_n[j]$.

M-connectivity

Input: a node u in S_n^- (the n-star graph S_n with at most $n-3$ nodes removed)

Output: $\min\{deg_f(u), deg_f(\varepsilon)\}$ paths from u to ε in S_n^-.

1.   **if** the node u is in $S_n[1]$

1.1. **then** **for** each $j \neq 1$ with both $\rho_j(u)$ and $\rho_j(\varepsilon)$ in S_n^- **do**
         construct a path P_j of length $\leq dist(u) + 6$ from u to ε such
         that all internal nodes of the path are in $S_n[j]$;

1.2. **else** (* the node u is in a substar $S_n[i]$, $i \neq 1$ *)

1.2.1. **if** the node $\rho_i(\varepsilon)$ is in S_n^-
         then pick a neighbor v of u and construct a path P_v of length
         $\leq dist(u) + 4$ from u to ε via v such that all internal nodes of
         P_v are in $S_n[i]$ and P_v does not intersect a (u', j)-bridging
         path for any neighbor $u' \neq v$ of u and for any $j \neq 1, i$;

1.2.2. **if** the node $u_1 = \rho_i(u)$ is in S_n^-
         then find a j, $j \neq 1, i$, with both $\rho_j(u_1)$ and $\rho_j(\varepsilon)$ in S_n^-; extend the
         path $[u, u_1, \rho_j(u_1)]$ to a path P_1 of length $\leq dist(u) + 8$ from u
         to ε such that all nodes between $\rho_j(u_1)$ and $\rho_j(\varepsilon)$ are in $S_n[j]$;

2.   maximally pair the remaining neighbors of u and ε;

3.   **for** each pair $(\rho_h(u), \rho_j(\varepsilon))$ constructed in step 2 **do**

3.1. **if** there is a non-divergent $(\rho_h(u), j)$-bridging path with no used nodes
         then pick this non-divergent $(\rho_h(u), j)$-bridging path P
         else pick a divergent $(\rho_h(u), j)$-bridging path P with no used nodes;

3.2. extend the path P into a path P_j of length $\leq dist(u) + 8$ from u
     to ε such that the extended part of P_j is entirely in $S_n[j]$;

Fig. 2. Node-disjoint paths in S_n with $\leq n - 3$ nodes removed

---

In case u is in a substar $S_n[i]$, $i \neq 1$, Step 1 constructs at most two paths from u to ε. If the node $\rho_i(\varepsilon)$ is in S_n^-, by Lemma 1, we can construct in time $O(k_i n + n)$ a path Q_v in $S_n^- \cap S_n[i]$ from u to $\rho_i(\varepsilon)$ (where v is the second node on Q_v) such that at most two edges in Q_v do not follow the sp-rules, where $k_i \leq n - 3$ is the number of nodes removed from $S_n[i]$. The path Q_v plus the edge $[\rho_i(\varepsilon), \varepsilon]$ gives a path P_v of length bounded by $dist(u) + 4$ in which all internal nodes are in $S_n[i]$. We show that the path P_v can be constructed without intersecting a (u', j)-bridging path from u, for any neighbor $u' \neq v$ of

u and any $j \neq 1, i$. Suppose the contrast that P_v intersects such (u', j)-bridging paths. Let w be the last node on P_v that belongs to such a (u', j)-bridging path $Q_{u'}$ from u. Note that the neighbor u' of u is uniquely determined by the node w since for two different neighbors u' and u'' of u in $S_n[i]$, a (u', j')-bridging path and a (u'', j'')-bridging path have no common nodes except u. Now, if we let $P_{u'}$ be the subpath of $Q_{u'}$ from u to w plus the subpath of P_v from w to ε, then it is not hard to see that the length of the path $P_{u'}$ is not larger than the length of the path P_v, and that the path $P_{u'}$ does not intersect any (u'', j'')-bridging path from u for any neighbor $u'' \neq u'$ of u and for any $j'' \neq 1, i$.

If the neighbor $u_1 = \rho_i(u)$ of u is in S_n^-, consider the $n-2$ pairs $(\rho_j(u_1), \rho_j(\varepsilon))$ of neighbors of u_1 and ε, where $j \neq 1, i$. Since at most $n - 3$ nodes are removed from S_n, one of these pairs $(\rho_j(u_1), \rho_j(\varepsilon))$ has both nodes in S_n^-. By Lemma 1, a path Q_1 from $\rho_j(u_1)$ to $\rho_j(\varepsilon)$ can be constructed in the substar $S_n[j]$ in time $O(k_j n + n)$ such that at most two edges of Q_1 do not follow the sp-rules, where $k_j \leq n - 3$ is the number of nodes removed from $S_n[j]$. Now the concatenation of the path $[u, u_1, \rho_j(u_1)]$, the path Q_1, and the edge $[\rho_j(\varepsilon), \varepsilon]$ gives a path in S_n^- from u to ε of length bounded by $dist(u) + 8$ in S_n^-. Note that this path is obviously node-disjoint with the path constructed in Step 1.2.1.

Now we consider Step 3 of the algorithm.

Case 3.1. The node u is in $S_n[1]$.

For each pair $(\rho_h(u), \rho_j(\varepsilon))$ formed in Step 2, the nodes $\rho_h(\varepsilon)$ and $\rho_j(u)$ must be removed: otherwise the index h or j would have been picked in Step 1.1.

We construct a path Q_{hj} from u to $\rho_j(\varepsilon)$ by concatenating a $(\rho_h(u), j)$-bridging path from u to $S_n[j]$ with a path Q_j' entirely in the substar $S_n[j]$. Note that such a path Q_{hj} contains one node in $S_n[1]$, besides u, and all other nodes in $S_n[h]$ and $S_n[j]$. We say that a node $u_1 \neq u$ in $S_n[1]$ is *used* if u_1 has been used by a path constructed by the algorithm. Inductively, assume that for r pairs of neighbors of u and ε constructed in Step 2, the r node-disjoint paths have been constructed. We consider the $(r + 1)$st pair $(\rho_h(u), \rho_j(\varepsilon))$.

Each $(\rho_{h'}(u), \rho_{j'}(\varepsilon))$ of the previous r pairs implies two removed nodes, the nodes $\rho_{h'}(\varepsilon)$ and $\rho_{j'}(u)$, and one used node in $S_n[1]$. Also notice that the paths constructed in Step 1.1 do not use any nodes in $S_n[1]$. Thus, the number of removed nodes in the sets $S_n[1]$, $S_n[h]$, and $S_n[j]$ is at most $(n - 3) - 2r = n - 2r - 3$. Let k_j be the number of removed nodes in $S_n[j]$, $k_j \leq n - 2r - 3$.

case 3.1.A. there is a non-divergent $(\rho_h(u), j)$-bridging path P_{hj} for u with no used nodes. At least one of the first three edges on P_{hj} follows the sp-rules. Consider the last edge $[v', v'']$ on P_{hj}

If the edge $[v', v'']$ follows the sp-rules, then the path P_{hj} has at most two edges not following the sp-rules. By Lemma 1, we can construct a path Q_j' in the substar $S_n[j]$ from v'' to $\rho_j(\varepsilon)$ in time $O(k_j n + n)$ with at most two edges not following the sp-rules. Now the concatenation of the $(\rho_h(u), j)$-bridging path P_{hj}, the path Q_j', and the edge $[\rho_j(\varepsilon), \varepsilon]$ gives a path P_j from u to ε with at most 4 edges not following the sp-rules, whose length is bounded by $dist(u) + 8$.

If the edge $[v', v'']$ does not follow the sp-rules, then the path P_{hj} may have three edges not following the sp-rules. Since v' is in $S_n[1]$, its primary cycle is trivial. Now since v'' is in $S_n[j]$ and the edge $[v', v'']$ does not follow the sp-rules, the primary cycle of v'' must be $(j1)$. By Lemma 1, a path Q'_j from v'' to $\rho_j(\varepsilon)$ in $S_n[j]$ can be constructed in time $O(k_j n + n)$ with at most one edge not following the sp-rules. Now the concatenation of the $(\rho_h(u), j)$-bridging path P_{hj}, the path Q'_j, and the edge $[\rho_j(\varepsilon), \varepsilon]$ gives a path P_j from u to ε with at most 4 edges not following the sp-rules, whose length is bounded by $dist(u) + 8$.

Therefore, in case 3.1.A, for the pair $(\rho_h(u), \rho_j(\varepsilon))$, we can always construct, in time $O(k_j n + n)$, a path P_j with no used nodes and of length bounded by $dist(u) + 8$ from node u to node ε. This path is node-disjoint with all previously constructed paths since the part extended from the $(\rho_h(u), j)$-bridging path P_{hj} is entirely in the substar $S_n[j]$ that is not used by any other paths.

case 3.1.B. all non-divergent $(\rho_h(u), j)$-bridging paths contain used nodes.

There are totally $n - 2$ $(\rho_h(u), j)$-bridging paths from u in S_n. Suppose that q' of them contain either removed nodes or used nodes, and that $q = n - 2 - q'$ of them contain neither removed nodes nor used nodes. We show $q > 0$.

Assume $q = 0$, so $q' = n - 2$. Since any two $(\rho_h(u), j)$-bridging paths from u have only the nodes u and $\rho_h(u)$ in common and at most $n - 3$ nodes are removed from S_n, $q'_1 > 0$ of the $n - 2$ $(\rho_h(u), j)$-bridging paths contain only used nodes. Each of the rest $q'_2 = q' - q'_1 = n - q'_1 - 2$ $(\rho_h(u), j)$-bridging paths contains at least one removed node. Thus, at least q'_1 paths have been constructed by the algorithm for q'_1 pairs (note that each constructed path uses exactly one node in $S_n[1]$). Each $(\rho_{h'}(u), \rho_{j'}(\varepsilon))$ of these pairs implies two removed nodes $\rho_{j'}(u)$ and $\rho_{h'}(\varepsilon)$. Thus, the total number of removed nodes in the n-star graph S_n would have been at least $q'_2 + 2q'_1 = n + q'_1 - 2 > n - 3$, contradicting the assumption that at most $n - 3$ nodes are removed from S_n. This shows $q > 0$, i.e., there is at least one $(\rho_h(u), j)$-bridging path in S_n^- containing no used nodes.

By the assumption, the q $(\rho_h(u), j)$-bridging paths without used nodes in S_n^- are all divergent. By Lemma 2, these q $(\rho_h(u), j)$-bridging paths can be extended into q paths from u to $\rho_j(\varepsilon)$ with at most 4 edges not following the sp-rules. The constructed paths contain no used nodes since the extended part of each path is entirely in the substar $S_n[j]$. Moreover, no two of these q paths share a node that is not u, $\rho_h(u)$, and $\rho_j(\varepsilon)$.

We claim that at least one of these q extended paths contains no removed nodes. To the contrary, if each of these q extended paths contains at least one removed nodes, then the total number of removed nodes in the sets $S_n[1]$, $S_n[i]$, and $S_n[j]$ is at least $q + (q' - r) = n - r - 2 > n - 2r - 3$ (recall that r is the number of paths that have been constructed by the algorithm so far. Thus, among the q' $(\rho_h(u), j)$-bridging paths that contain either removed or used nodes, at least $q' - r$ of them must contain at least one removed node each). This contradicts the fact that there are at most $n - 2r - 3$ removed nodes in the sets $S_n[1]$, $S_n[i]$, and $S_n[j]$.

Thus, an extended path Q'_{hj} from u to $\rho_j(\varepsilon)$ in S_n^- without used nodes can be constructed. This path Q'_{hj} plus the edge $[\rho_j(\varepsilon), \varepsilon]$ gives a path P_j from u to ε with no used nodes and with at most 4 edges not following the sp-rules. Thus, the length of P_j is bounded by $dist(u) + 8$. Moreover, P_j can be constructed in time $O(k_j n + n)$ by tracing at most $k_j + 1$ of the extended paths from u to $\rho_j(\varepsilon)$. Finally, this path is node-disjoint with all previously constructed paths since its extended part is entirely in $S_n[j]$, which is not used by any other paths.

Case 3.2. The node u is in the substar $S_n[i]$, $i \neq 1$.

In this case, the node u has one neighbor in $S_n[1]$, and $n - 2$ neighbors in $S_n[i]$ (see Figure 1). Note that if the neighbor $\rho_i(u)$ of u in $S_n[1]$ is in S_n^-, then a path from u to ε via $\rho_i(u)$ has been constructed in Step 1.2.2. Thus, we only need to consider the neighbors of u that are in $S_n[i]$.

Again we assume that the algorithm has constructed r paths from u to ε by extending r bridging paths from u. Consider the $(r + 1)$st pair $(\rho_h(u), \rho_j(\varepsilon))$.

Since the n-star graph contains no cycle of length less than 6 [7,8], two neighbors of u share no common neighbors except u. Let u_1 and u_2 be two neighbors of u in $S_n[i]$. Since no two nodes in $S_n[i]$ have the same neighbor in $S_n[1]$ and no two nodes in $S_n[1]$ have the same neighbor in $S_n[j]$, a (u_1, j_1)-bridging path and a (u_2, j_2)-bridging path share no common nodes except u for any j_1 and j_2. Therefore, for the previous r paths from u to ε constructed by the algorithm by extending bridging paths from u, none of them would intersect a (u', j)-bridging path. Thus, no (u', j)-bridging path contains used nodes.

Thus, if there is a non-divergent $(\rho_h(u), j)$-bridging path P_{hj} without removed nodes, we can extend the path P_{hj}, in the way of Case 3.1.A, into a path P_j from u to ε such that the length of the path P_j is bounded by $dist(u) + 8$ and the extended part of P_j is entirely in $S_n[j]$. On the other hand, if all non-divergent $(\rho_h(u), j)$-bridging paths contain removed nodes, then, as in case 3.1.B, we can extend at least one divergent $(\rho_h(u), j)$-bridging path from u into a path P_j from u to ε such that the length of P_j is bounded by $dist(u) + 8$ and the extended part of P_j is entirely in $S_n[j]$.

Now we discuss the running time of the algorithm **M-connectivity**.

Each path is constructed by the algorithm by searching a proper path in a specific substar $S_n[j]$, which takes time $O(k_j n + n)$, where k_j is the number of removed nodes in the substar $S_n[j]$. No substar is used in extending more than one such a path. Therefore, the time complexity for constructing all these paths is bounded by $O(k_2 n + k_3 n + \cdots + k_n n + n(n - 1)) = O(n^2)$ since by our assumption $k_2 + k_3 + \cdots k_n \leq n - 3$. Thus, the time complexity of the algorithm **M-connectivity** is bounded by $O(n^2)$. $\qquad\square$

The following example shows that the bound on the path length in Theorem 1 is actually almost optimal. Consider the n-star graph S_n. Let the source node be $u = (21)$, here we have omitted the trivial cycles in the cycle structure. Then $dist(u) = 1$. Suppose that no neighbors of u and of ε are removed. By Theorem 1, there are $n - 1$ node-disjoint paths in S_n^- from u fro ε. Thus, for each i, $3 \leq i \leq n$, the edge $[u, u_i]$, where $u_i = (i21)$, leads to one P_i of these node-disjoint paths. Note that the edge $[u, u_i]$ does not follow the sp-rules. Now suppose that the $n - 3$ nodes $(i2)(1)$ are removed, for $i = 3, 4, \ldots, n - 1$. Then the third node on the path P_i must be $v_i = (ji21)$ for some $j \neq 1, 2, i$, and the edge $[u_i, v_i]$ does not follow the sp-rules. Since the only edge from v_i that follows the sp-rules is the edge $[v_i, u_i]$, the next edge $[v_i, w_i]$ on P_i again does not follow the sp-rules.

Now since all the first three edges on P_i do not follow the sp-rules, by Remark 1, $dist(w_i) = dist(u) + 3 = 4$, and the path P_i needs at least four more edges to reach ε. That is, the length of the path P_i is at least $7 = dist(u) + 6$. Thus, with $n - 3$ removed nodes, among the $n - 1$ node-disjoint paths from u to ε, at least $n - 3$ of them must have length larger than or equal to $dist(u) + 6$.

The situation above seems a little special since the distance $dist(u)$ from u to ε is very small. In fact, even for large distance nodes u, it is still possible to construct examples in which some of the node-disjoint paths connecting u and ε must have length at least $dist(u) + 6$. We leave this to the interested readers.

5   Conclusion

Strong Menger-connectivity is a natural extension of the study of d-connectivity on d-regular graphs, and has direct applications in network fault tolerance and network parallel routing. Since one of the motivations of network parallel routing is to provide alternative routing paths when failures occur, strong Menger-connectivity can also be regarded as the study of fault tolerance in networks with faults. In this paper, we have demonstrated that the recently proposed star graphs are strongly Menger-connected. We presented an algorithm of running time $O(n^2)$ that for two given non-faulty nodes u and v in the n-star graph S_n with at most $n - 3$ faulty nodes, constructs the maximum number (i.e., $\min\{deg_f(u), deg_f(v)\}$) of node-disjoint fault-free paths from u to v such that the length of the paths is bounded by $dist(u, v) + 8$. We showed that the length of the paths constructed by our algorithm is almost optimal. The time complexity of our algorithm is optimal since each path from u to v may have length as large as $\Theta(n)$, and there can be as many as $n - 1$ node-disjoint paths from u to v. In fact, since the n-star graph S_n has $n!$ nodes, our algorithm is actually "sub-logarithmic" in terms of the input graph size.

We should mention that Rescigno [17] recently developed a randomized parallel routing algorithm on star graphs with faults, based on the Information Disersal Algorithm (IDA) [16]. The algorithm in [17] is randomized thus it does not always guarantee the maximum number of node-disjoint paths. Moreover, in terms of the length of the constructed paths and running time of the algorithm, our algorithm seems also to have provided significant improvements.

The study of strong Menger-connectivity shows another advantage of the star graphs over other popular network structures. In particular, the *orthogonal partition* of the star graphs [7], which decomposes the n-star graph into $n-1$ ($n-1$)-substars plus an independent set, seems very convenient for construction of node-disjoint paths: we basically can construct a path in each separated substar, ensuring that the constructed path is node-disjoint with other paths. Other popular network topologies, such as the hypercubes, do not seem to have this nice decomposition structure. For example, in the construction of node-disjoint paths in a hypercube with faults, we seemed not able to construct each path using a separated subcube, which has made the construction more involved [15].

References

1. S. B. AKERS, D. HAREL, AND B. KRISHNAMURTHY, The star graph: an attractive alternative to the n-cube, *Proc. Intl. Conf. of Parallel Proc.*, (1987), pp. 393-400.
2. S. B. AKERS AND B. KRISHNAMURTHY, A group-theoretic model for symmetric interconnection networks, *IEEE Trans. on Computers 38*, (1989), pp. 555-565.
3. S. B. AKERS AND B. KRISHNAMURTHY, The fault tolerance of star graphs, *Proc. 2nd International Conference on Supercomputing*, (1987), pp. 270-276.
4. N. BAGHERZADEH, N. NASSIF, AND S. LATIFI, A routing and broadcasting scheme on faulty star graphs, *IEEE Trans. on Computers 42*, (1993), pp. 1398-1403.
5. G. BIRKHOFF AND S. MACLANE, *A Survey of Modern Algebra*, The Macmillan Company, New York, 1965.
6. C. C. CHEN AND J. CHEN, Optimal parallel routing in star networks, *IEEE Trans. on Computers 46*, (1997), pp. 1293-1303.
7. C. C. CHEN AND J. CHEN, Nearly optimal one-to-many parallel routing in star networks, *IEEE Trans. Parallel, Distrib. Syst. 8*, (1997), pp. 1196-1202.
8. K. DAY AND A. TRIPATHI, A comparative study of topological properties of hypercubes and star graphs, *IEEE Trans. Parallel, Distrib. Syst. 5*, (1994), pp. 31-38.
9. M. DIETZFELBINGER, S. MADHAVAPEDDY, AND I. H. SUDBOROUGH, Three disjoint path paradigms in star networks, *Proc. 3nd IEEE Symposium on Parallel and Distributed Processing*, (1991), pp. 400-406.
10. Z. GALIL AND X. YU, Short length versions of Menger's Theorem, *Proc. 27th Ann. ACM Symp. Theory of Computing* (STOC'95), (1995), pp. 499-508.
11. Q.-P. GU AND S. PENG, Fault tolerant routing in hypercubes and star graphs, *Parallel Processing Letters 6*, (1996), pp. 127-136.
12. Q.-P. GU AND S. PENG, An efficient algorithm for k-pairwise disjoint paths in star graphs, *Information Processing Letters 67*, (1998), pp. 283-287.
13. D. F. HSU, On container width and length in graphs, groups, and networks, *IEICE Trans. Fundamentals E77-A*, (1994), pp. 668-680.
14. K. MENGER, Zur allgemeinen kurventheorie, *Fund. Math. 10*, (1927), pp. 96-115.
15. E. OH AND J. CHEN, Parallel routing in hypercube networks with faulty nodes, *Tech. Report*, Dept. Computer Science, Texas A&M University, (2000).
16. M. O. RABIN, Efficient dispersal of information for security, load balancing, and fault tolerance, *Journal of ACM 36*, (1989), pp. 335-348.
17. A. A. RESCIGNO, Fault-tolerant parallel communication in the star network, *Parallel Processing Letters 7*, (1997), pp. 57-68.
18. A. A. RESCIGNO AND U. VACCARO, Highly fault-tolerant routing in the star and hypercube interconnection networks, *Parallel Processing Letters 8*, (1998), pp. 221-230.
19. S. SUR AND P. K. SRIMANI, Topological properties of star graphs, *Computers Math. Applic. 25*, (1993), pp. 87-98.

The Complexity of the Matching-Cut Problem[*]

Maurizio Patrignani and Maurizio Pizzonia

Dipartimento di Informatica e Automazione,
Università di Roma Tre,
Via della Vasca Navale 79, 00146 Roma, Italy
{patrigna, pizzonia}@dia.uniroma3.it

Abstract. Finding a cut or finding a matching in a graph are so simple problems that they are hardly considered problems at all. In this paper, by means of a reduction from the NAE3SAT problem, we prove that combining these two problems together, i.e., finding a cut whose split edges are a matching is an NP-complete problem. It remains intractable even if we impose the graph to be simple (no multiple edges allowed) or its maximum degree to be k, with $k \geq 4$. On the contrary, we give a linear time algorithm that computes a matching-cut of a series-parallel graph. It's open whether the problem is tractable or not for planar graphs.

1   Introduction

We consider two elementary graph-theoretic problems: finding a matching in a graph, i.e., a set of disjoint edges; and finding a cut in a graph, i.e., a set of edges whose removal disconnects the graph (see Figure 1(a,b)). These two problems can be separately solved in so a straightforward way that they are hardly considered problems at all: an arbitrary edge is a matching, and the set of edges incident to an arbitrary vertex is a cut. However, we show that combining these two problems together, that is, finding a cut that at the same time is a matching (see Figure 1(c)), yields an NP-complete problem.

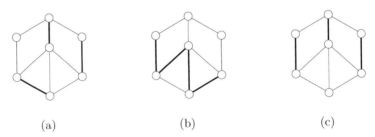

(a)   (b)   (c)

Fig. 1. (a) A matching. (b) A cut. (c) A matching cut.

The problem remains intractable even if we impose the graph to be simple (no multiple edges allowed) or its maximum degree to be k, with $k \geq 4$. On

---

[*] Research supported in part by the Murst Project: "Algorithms for Large Data Sets".

A. Brandstädt and V.B. Le (Eds.): WG 2001, LNCS 2204, pp. 284–295, 2001.
© Springer-Verlag Berlin Heidelberg 2001

the contrary, we give a linear time algorithm that computes a matching-cut of a series-parallel graph. It's open whether the problem is tractable or not for planar graphs.

This interesting combinatorial problem firstly arose in a practical application. In fact, in [6,3] is described a new approach to three-dimensional orthogonal graph drawing, based on generating the final drawing through a sequence of steps. The method starts from a "degenerate" drawing, in which all vertices and edges are overlapped on the same point. At each step the drawing "splits" into two pieces and finds a structure more similar to its final version. The observation that a cut involving non adjacent edges yields a more efficient (in terms of time of computation) and effective (in terms of volume occupation, number of bends introduced, and average edge length) performance, stimulated the investigation on the complexity of finding such cut.

The paper is organized as follows: in Section 2 basic definitions are given; in Section 3 the problem of finding a matching-cut is demonstrated to be NP-complete. Section 4 shows that the problem retains its complexity even if we impose the graph to be simple or its maximum degree to be k, with $k \geq 4$. In Section 5 we produce a linear time algorithm that finds a matching-cut in a series-parallel graph.

2   Preliminaries

A *graph* $G(V, E)$ consists of a set V of vertices and a set E of edges (u, v), such that u and v are vertices in V.

An edge (u, v) is said to be *incident* to vertices u and v. Two edges are *adjacent* if they are incident to the same vertex v. The *degree* of a vertex v is the cardinality of the set of its incident edges. In a *simple* graph every edge incides on two distinct vertices and two edges share at most a vertex. A *path* from u to v is a sequence of edges $e_1, \ldots, e_k$ such that, e_1 is incident to u, e_k is incident to v, and for every $i = 1, \ldots, k - 1$, e_i is adjacent to e_{i+1}. A graph is *connected* if a path exists between each pair of vertices. In the following we consider only connected graphs.

A *cut* of a graph is a partition of its vertex set V into two non-empty sets. Note that a cut of a graph with n vertices can be associated with an array of n elements with values in $\{0, 1\}$, where all the elements with the same value correspond to the vertices of the same block of the partition. Since $(0, 0, \ldots, 0)$ and $(1, 1, \ldots, 1)$ are forbidden configurations (they don't correspond to partitions), and since there are two configurations corresponding to the same partition, it follows that there are $2^{n-1} - 1$ distinct cuts.

If the graph is connected, a cut can be alternatively characterized by specifying the set of edges connecting vertices that do not belong to the same block of the partition. Observe that one description of the cut can be easily obtained from the other in linear time.

A *matching* in a graph G is a set $E' \subseteq E$ such that each pair of distinct edges e_i and e_j of E' are not adjacent.

The problem of finding a cut in a graph can be solved in linear time: it suffices a random labeling of its vertices with the two labels 0 and 1. Imposing the cut to be maximum the problem becomes NP-complete ([4]), but approximable within 1.1383 ([5]).

Finding a matching can be solved in $O(1)$ time: each single edge is a matching.

Although finding a cut and finding a matching are simple problems, we expect the problem of finding a cut that is a matching to be a more difficult task. In the following section we will demonstrate that this problem is NP-complete.

3   NP-Completeness of the Matching-Cut Problem

Formally, the Matching-Cut problem is defined as follows:

Problem: **Matching-Cut**
Instance: A connected graph $G(V, E)$.
Question: Does a set $E' \subseteq E$, such that E' is a cut and a matching, exist?

Given a cut, it's easy to check in linear time if it is a matching by marking the extremes of each edge belonging to the cut. If no vertex is marked twice, the cut is also a matching. A non-deterministic Turing machine may generate all possible cuts and check in linear time if the cuts are matching-cuts. It follows that the matching cut problem is in NP.

We will demonstrate that the Matching-Cut problem is NP-hard by means of a reduction from the NAE3SAT problem:

Problem: **Not-All-Equal-3-Sat (NAE3SAT)**
Instance: A set C of clauses, each containing 3 literals from a set of boolean variables.
Question: Can truth values be assigned to the variables so that each clause contains at least one true literal and at least one false literal?

The NAE3SAT problem was demonstrated to be NP-hard by Schaefer [7].

Given a formula ϕ in conjunctive normal form with v variables $x_1, \dots, x_v$ and m clauses $C_1, \dots, C_m$, with three literals each, we construct a graph $G(\phi)$ that admits a matching-cut if and only if ϕ is satisfiable within the NAE3SAT constraints.

We start by building two chains of $2m + v$ double linked vertices as shown in Figure 2.

In the following, the upper chain will be called *FALSE-chain*, and the lower one *TRUE-chain*. Since the vertices of each chain are linked with multiple edges, no matching-cut separates two vertices of the same chain, and each chain belongs to the same block of the partition that constitutes a matching-cut. We build the remaining part of the graph $G(\phi)$ adding subgraphs connected both to the upper and to the lower chains in such a way that, if a matching-cut exists, it necessarily separates the two chains cutting through all the inserted subgraphs.

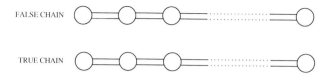

Fig. 2. First step of the construction of $G(\phi)$: we create two chains of $2m + v$ double linked vertices.

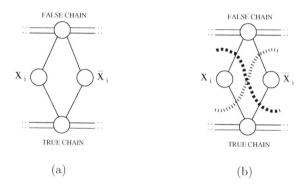

(a)   (b)

Fig. 3. The variable-gadget for variable x_i (a) and the only two possible matching-cuts (b).

We call *TRUE-set* the block of the partition which the TRUE-chain belongs to, and *FALSE-set* the other.

The subgraphs to be introduced are of two types: the variable-gadget and the clause-gadget. Figure 3-a shows the variable-gadget. We introduce a variable-gadget for each boolean variable x_i. The only way to cut the variable-gadget with a matching-cut is along the lines of Figure 3-b. Each cut leaves the x_i vertex and the $\bar{x}_i$ vertex on the two opposite sets. Any other cut is not allowed, since it is not a matching.

Figure 4 shows the clause-gadget. We introduce a clause-gadget for each clause of the ϕ formula. The three black vertices of Figure 4 correspond to the literals l_i, l_j, and l_k of the clause $(l_i \vee l_j \vee l_k)$. There are six ways to cut through the clause-gadget with a matching-cut, as Figure 5 shows. Each cut corresponds to the truth assignment for the literals illustrated in Table 1.

Note that there is no matching-cut for the clause-gadget that leaves all the three vertices corresponding to the literals on the same side, and that there is no other matching-cut that could separate a subgraph of the clause-gadget.

Finally, each vertex representing a literal l_i, l_j, or l_k of the clause-gadget is connected to the vertex representing the corresponding literal of the variable-gadget by means of two edges, in order to impose that they will belong to the same block of the partition determined by the matching-cut. Figure 6 shows the whole construction for a formula with three boolean variables and a single clause.

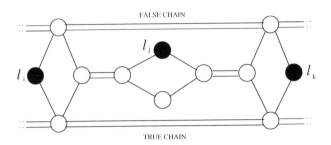

Fig. 4. The clause-gadget for clause $(l_i \vee l_j \vee l_k)$.

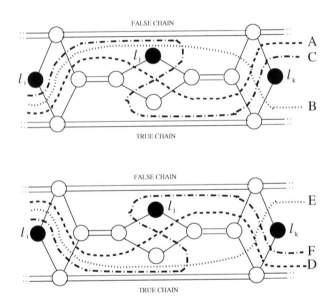

Fig. 5. The six matching-cuts of the clause-gadget of Fig. 4.

Table 1. Truth assignment corresponding to the six cuts showed in Figure 5.

l_i	l_j	l_k	cut
false	false	false	no cut
false	false	true	A
false	true	false	B
false	true	true	C
true	false	false	D
true	false	true	E
true	true	false	F
true	true	true	no cut

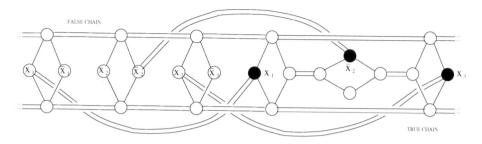

Fig. 6. Construction for the first three variables and a clause $(X_1 \vee \overline{X}_2 \vee X_3)$.

We claim that a matching-cut can be found in the graph $G(\phi)$ if and only if the corresponding instance ϕ of the NAE3SAT problem has a solution.

Suppose that such cut exists. Since the chains can not be cut and since no lesser part of the gadgets can be separated from the rest of the graph, the matching-cut has to separate the two chains, cutting through all the gadgets that tie them together. All the variable-gadgets are attached to the FALSE-chain and to the TRUE-chain, so the matching-cut must cut through their edges according to one of the two ways shown in Figure 3(b), determining a truth assignment of the boolean variables. Leaving a variable-vertex on a set implies that the corresponding literals in the clauses belong to the same set, and the negate literals belong to the other set. The clause-gadgets are also cut into two parts by the matching-cut, and necessarily according to one of the six ways illustrated in Figure 5, implying that at least one of the literal-vertices belongs to the TRUE-set and the other to the FALSE-set. So each clause, and the ϕ formula, is satisfied as requested by the NAE3SAT problem.

Conversely, if a truth assignment exists such that each clause has at least a true literal and a false one then a matching-cut can be constructed so that

- it cuts each variable-gadget leaving the variable-vertex in the set corresponding to its boolean value and the negate literal is in the other set,
- it cuts each clause-gadget using the cut corresponding to the given truth assignment of the literals as shown in Table 1.

It's easy to verify that the cut so obtained is a matching-cut. The following theorem is therefore demonstrated:

Theorem 1. *The Matching-Cut problem for a connected graph G is NP-complete.*

4   Simple Graphs and Graphs with Maximum Degree Four

The construction described in Section 3 can be easily modified in order to demonstrate that the problem retains its complexity even for simple graphs. Namely, we note that between a pair of vertices of the graph $G(\phi)$ there can be a single

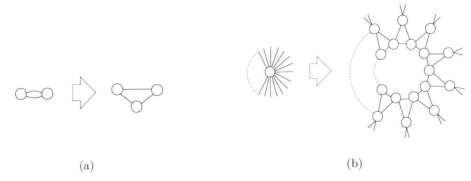

Fig. 7. (a) Replacing an edge with a path of two edges to eliminate double edges. (b) Replacing a vertex with a star-shaped graph to eliminate vertices with degree greater that 4.

edge or two, and in the latter case we can replace one of the two edges with a two edges long path, as shown in Figure 7(a).

It's easy to see that the three vertices of the triangle are forced to be in the same set of the bipartition induced by a matching-cut. So, the following corollary is demonstrated:

Corollary 1. *The Matching-Cut problem for a simple connected graph G is NP-complete.*

We modify the same construction to demonstrate that the problem remains NP-complete even imposing a maximum degree k, with $k \geq 4$ for the vertices of the graph. Observe that the vertices of the proposed construction have even degree. Also, edges are introduced two-by-two in order to impose that the double linked vertices will belong to the same set of the bipartition determined by a matching-cut. So, each vertex with degree greater than 4 can be replaced with the star construction of Figure 7(b), in which every vertex has degree four. It's easy to see that a matching-cut can not separate the vertices of the star-shaped subgraph of Figure 7(b).

Corollary 2. *The Matching-Cut problem for a connected graph G (simple or not) of degree k, with $k \geq 4$ is NP-complete.*

5   Matching-Cuts of Series-Parallel Graphs

In this section we describe a linear time algorithm that decides whether an SP-graph admits a matching-cut. Further, we modify the algorithm in order to return a matching-cut if exists.

A graph G with no self-loops is *series-parallel* (or *SP-graph*) if two of its vertices s and t are selected to be respectively the *source* and the *sink* of G (denoted $G(s,t)$), and G can be built by recursively applying the following rules:

basic step graph $G(s, t)$ consists of a single edge between s and t;

serial composition graph $G(s, t)$ is the union of two SP-graphs $G_u(s_u, t_u)$ and $G_d(s_d, t_d)$ such that $s = s_u$, $t_u = s_d$, and $t = t_d$;

parallel composition graph $G(s, t)$ is the union of two SP-graphs $G_l(s_l, t_l)$ and $G_r(s_r, t_r)$ such that $s = s_l = s_r$ and $t = t_l = t_r$.

Each SP-graph can be represented by a binary tree [8] in which nodes are of three types: Q, S, and P. Nodes types have the following meanings:

- A Q-node represents a single edge
- An S-node represents the serial composition of the graphs corresponding to its children, with the convention that the left child represents G_u and the right child represents G_d
- A P-node represents the parallel compositions of the graphs corresponding to its children, with the convention that the left child represents G_l and the right child represents G_r

Such tree is called *parse tree*. In order to guarantee the uniqueness of the parse tree for a given SP-graph, it is further required [2] that, if a node has the same type of its parent, then is its right child. It is possible to recognize an SP-graph (and build its parse tree) in linear time [8].

We associate each node of the parse tree with an SP-graph. For S-nodes and P-nodes the associated SP-graph is obtained by applying the rule represented by the node to the SP-graphs associated with the children. For brevity, in the following we do not distinguish between the node of an parse tree and the graph associated with it.

Our algorithm takes as input an SP-graph $G(s, t)$ and decides if it admits a matching-cut. It is based on a post-order traversal of the parse tree. In the following we say that for a given SP-graph a matching-cut is *st-separating* if it leaves s and t in distinct sets. Also, we say that vertex s (t) is *engaged* by a st-separating matching-cut if one of the edges of the matching-cut is incident to s (t).

We associate two labels $l_1(G')$ and $l_2(G')$ with each node $G'(s, t)$ of the parse tree of G. Label $l_2(G')$ signals if G' admits at least one non st-separating matching-cut ($l_2(G') = 1$) or not ($l_2(G') = 0$). Note that, if $l_2(G') = 1$ then for each ancestor G'' of G' it must hold $l_2(G'') = 1$.

Label $l_1(G')$ signals whether G' has at least one st-separating matching-cut and whether vertices s and t are necessarily engaged. The six possible values of label $l_1(G')$ are listed below along with their meanings.

0 There does not exist an st-separating matching-cut for such node.

$s \cdot t$ There exists at least one st-separating matching-cut and in all matching-cuts both s and t are engaged.

s There exists at least one st-separating matching-cut. In all matching-cuts s is engaged. Further, in some of them, t is not engaged.

t There exists at least one st-separating matching-cut. In all matching-cuts t is engaged. Further, in some of them, s is not engaged.

$s \oplus t$ There exists at least one st-separating matching-cut that engages s and not t and there exists at least one st-separating matching-cut that engages t and not s. Further, no matching-cut that does not engage either s or t exist.

1 There exists at least one st-separating matching-cut in which neither s nor t are engaged.

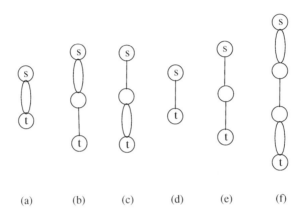

Fig. 8. Examples of SP-graphs with their l_1 labeling. (a) No matching-cut exists ($l_1 = 0$). (b) There exists at least one matching-cut engaging t only ($l_1 = t$). (c) There exists at least one matching-cut engaging s only ($l_1 = s$). (d) All matching-cuts engage both s and t ($l_1 = s \cdot t$). (e) There exists at least a matching-cut that engages s and not t and at least a matching-cut that engages t and not s ($l_1 = s \oplus t$). (f) There exists at least a matching-cut that does not engage neither s nor t ($l_1 = 1$).

The labels l_1 and l_2 may be computed recursively for each node G, applying one of the following rules, depending on the type of the node:

Q-node $l_1(G) = s \cdot t$ and $l_2(G) = 0$.
S-node Let $G_u(s_u, t_u)$ and $G_d(s_d, t_d)$ be the children of G. The value of $l_1(G)$ is given in the table shown in Figure 9(b) which summarizes the following

G_r \ G_l	0	$s \cdot t$	s	t	$s \oplus t$	1
0	0	0	0	0	0	0
$s \cdot t$	0	0	0	0	0	$s \cdot t$
s	0	0	0	$s \cdot t$	$s \cdot t$	s
t	0	0	$s \cdot t$	0	$s \cdot t$	t
$s \oplus t$	0	0	$s \cdot t$	$s \cdot t$	$s \cdot t$	$s \oplus t$
1	0	$s \cdot t$	s	t	$s \oplus t$	1

G_d \ G_u	0	$s \cdot t$	s	t	$s \oplus t$	1
0	0	s	s	1	1	1
$s \cdot t$	t	$s \oplus t$	$s \oplus t$	1	1	1
t	t	$s \oplus t$	$s \oplus t$	1	1	1
s	1	1	1	1	1	1
$s \oplus t$	1	1	1	1	1	1
1	1	1	1	1	1	1

(a) Parallel composition   (b) Serial composition

Fig. 9. Tables for computing $l_1(G)$ from the labels of the children of G.

observations. If $l_1(G_u) = l_1(G_d) = 0$ then G does not have an st-separating matching-cut. In all other cases the conditions on $t_u = s_d$ are not propagated to $l_1(G)$, while the conditions on s_u and t_d are. Note that, since an st-separating matching-cut either is st-separating for G_u or is st-separating for G_d, it can never be $l_1(G) = s \cdot t$.

The label $l_2(G) = 1$ if and only if $l_2(G_u) = 1$, or $l_2(G_d) = 1$, or it is possible to find two st-separating matching-cuts, one in G_u and one in G_d, such that $t_u = s_d$ is not engaged by both of them. Such observations are summarized by the table in Figure 10:

G_d \ G_u	0	$s \cdot t$	s	t	$s \oplus t$	1
0	0	0	0	0	0	0
$s \cdot t$	0	0	1	0	1	1
t	0	1	1	1	1	1
s	0	0	1	0	1	1
$s \oplus t$	0	1	1	1	1	1
1	0	1	1	1	1	1

Fig. 10. The table for computing $l_2(G)$ from the labels of the children of G.

P-node Let G_l and G_r be the children of G. The value of $l_1(G)$ is given in the table shown in Figure 9(a) which summarizes the following observations. If one between G_l and G_r does not admit an st-separating matching-cut, so does G. If G_l (G_r) admits an unconditioned st-separating matching-cut then $l_1(G) = l_1(G_r)$ ($l_1(G) = l_1(G_l)$). In all other cases G may have an st-separating matching-cut if and only if it is possible to find two matching-cuts, one in G_l and one in G_r, that do not engage the same vertex (s or t). Hence, in such cases, $l_1(G) = s \cdot t$.
The label $l_2(G)$ is 1 if and only if $l_2(G_l) = 1$ or $l_2(G_r) = 1$.

Theorem 2. *There exists a linear time algorithm to verify if an SP-graph admits a matching-cut.*

Proof. (sketch) The correctness of the algorithm follows from the correctness of the recursive computation of the labels l_1 and l_2 described above. For the time complexity, it suffices observing that the number of the nodes in the parse tree associated with an SP-graph of m edges is bounded by $2m$, and therefore a post-order traversal can be performed in linear time. For each node G of the parse tree the values of $l_1(G)$ and $l_2(G)$ can be computed in constant time, since they depend only on the values of its two children.

The algorithm can be easily modified as follows in order to return a matching-cut if it exists. Given the input SP-graph, we associate with each node $G(s,t)$ of its parse tree three sets S_1, S_1', and S_2. Such sets are subsets of the edges of the input SP-graph and their are used as follows. If $l_1(G) = s \cdot t$ then subset S_1

is a matching-cut for G. If $l_1(G) = s$ ($l_1(G) = t$) then subset S_1 is a matching-cut that engages s (t) only. If $l_1(G) = s \oplus t$ then S_1 is a matching-cut that engages s only, and subset S_1' is an alternative matching-cut that engages t only. If $l_1(G) = 1$ then subset S_1 is a matching-cut that engages neither s nor t. If $l_2(G) = 1$ then S_2 is a non st-separating matching-cut.

For each node G of the parse tree the three sets S_1, S_1', and S_2 can be updated in the same post-order traversal that calculates the values of the labels l_1 and l_2 by suitably combining the sets associated with the children of G. Since at each steps the sets of the children are not needed anymore, the involved set operations (union and copy) may be performed in constant time. It follows that the problem of finding a matching-cut in an SP-graph can be solved in polynomial (in fact, linear) time.

6   Conclusions and Open Problems

We studied the problem of finding a matching-cut in a graph, and showed that in the general case the problem is NP-complete, and that it retains its complexity even if the graph is simple or its maximum degree is bounded by a constant $k \geq 4$.

As for other classes of graphs, we described an algorithm to find a matching cut (if any) in a series-parallel graph in linear time, so demonstrating that the problem is in P for this particular class of graphs.

It's open whether the problem is tractable or not for planar graphs.

Acknowledgements

We thank Giuseppe Di Battista and Walter Didimo for their encouragement and support.

We are expecially grateful to Van Bang Le who pointed out the fact that the matching-cut problem was already known in the literature with the name of *decomposable graphs recognition*, and was proven to be NP-complete by Chvátal by reducing the problem of the bicolorability of a 3-uniform hypergraph ([1]). Also, in the same paper, it is shown that the problem is easy for graphs of maximum degree three. We think that the alternative proof presented in this paper can be useful to improve the understanding of the problem.

References

1. V. Chvátal. Recogniziong decomposable graphs. *J. Graph Theory*, 8:51–53, 1984.
2. G. Di Battista, P. Eades, R. Tamassia, and I. G. Tollis. *Graph Drawing*. Prentice Hall, Upper Saddle River, NJ, 1999.
3. G. Di Battista, M. Patrignani, and F. Vargiu. A Split&Push approach to 3D orthogonal drawing. *Journ. Graph Alg. Appl.*, 4:105–133, 2000.
4. M. R. Garey and D. S. Johnson. *Computers and Intractability: A Guide to the Theory of NP-Completeness*. W. H. Freeman, New York, NY, 1979.

5. M. X. Goemans and D. P. Williamson. Improved approximation algorithms for maximum cut and satisfiability problems using semidefinite programming. *J. ACM*, 42:1115–1145, 1995.
6. M. Patrignani and F. Vargiu. 3DCube: A tool for three dimensional graph drawing. In G. Di Battista, editor, *Graph Drawing (Proc. GD '97)*, volume 1353 of *Lecture Notes Comput. Sci.*, pages 284–290. Springer-Verlag, 1997.
7. T. J. Schaefer. The complexity of satisfiability problems. In *Proc. 10th Annu. ACM Sympos. Theory Comput.*, pages 216–226, 1978.
8. J. Valdes, R. E. Tarjan, and E. L. Lawler. The recognition of series-parallel digraphs. *SIAM J. Comput.*, 11(2):298–313, 1982.

De Bruijn Graphs and DNA Graphs
(Extended Abstract)

Rudi Pendavingh[1], Petra Schuurman[1], and Gerhard J. Woeginger[2]

[1] Department of Mathematics and Computing Science
Eindhoven University of Technology, The Netherlands
[2] Department of Mathematics, Universiteit Twente, The Netherlands

Abstract. In this paper we prove the NP-hardness of various recognition problems for subgraphs of De Bruijn graphs. In particular, the recognition of DNA graphs is shown to be NP-hard; DNA graphs are the vertex induced subgraphs of De Bruijn graphs over a four letter alphabet. As a consequence, two open questions from a recent paper by Błażewicz, Hertz, Kobler & de Werra [Discrete Applied Mathematics 98, 1999] are answered in the negative.

Keywords: graph theory, recognition algorithm, computational complexity, NP-hardness, De Bruijn graph, DNA graphs, DNA computing.

1 Introduction

Błażewicz, Hertz, Kobler & de Werra [3] introduced the concept of DNA graphs to model the computing and reconstruction phase of DNA chain sequencing by hybridization. For more information on the biological background, we refer the reader to Błażewicz, Hertz, Kobler & de Werra [3] or to Bains & Smith [1]. The graph theoretical background is fairly easy to describe: Let $G = (V, A)$ be a directed and simple graph that may contain loops. An (α, k)-*labeling* of G assigns a label $L(v) = [\ell_1(v), \ell_2(v), \ldots, \ell_k(v)]$ to every vertex $v \in V$ such that

(L1) All entries $\ell_i(v)$ in the labels of all vertices $v \in V$ are from an alphabet of size α (e.g., from the set $\{1, \ldots, \alpha\}$).

(L2) For $u, v \in V$ with $u \neq v$, $L(u) \neq L(v)$. Thus, different vertices get different labels.

(L3) For $u, v \in V$, $[\ell_2(u), \ldots, \ell_k(u)] = [\ell_1(v), \ldots, \ell_{k-1}(v)]$ holds if and only if there is an arc $(u, v) \in A$. In other words, an arc is encoded by the fact that the last $k - 1$ entries of the label of the tail-vertex are equal to the first $k - 1$ entries of the label of the head-vertex.

For integers $\alpha \geq 2$ and $k \geq 1$, we denote by $\mathcal{L}_k^\alpha$ the class of all directed simple graphs that possess an (α, k)-labeling. The set $\bigcup_{k=1}^\infty \mathcal{L}_k^\alpha$ is denoted by $\mathcal{L}_\infty^\alpha$, and the set $\bigcup_{\alpha=2}^\infty \mathcal{L}_k^\alpha$ is denoted by $\mathcal{L}_k^\infty$. Błażewicz, Hertz, Kobler & de Werra [3] call a graph G a *DNA graph* if and only if $G \in \mathcal{L}_\infty^4$. The four letters in the underlying alphabet correspond to the four nucleotide bases adenine (A), guanine (G), cytosine (C), and thymine (T).

A. Brandstädt and V.B. Le (Eds.): WG 2001, LNCS 2204, pp. 296–305, 2001.
© Springer-Verlag Berlin Heidelberg 2001

Fifty years ago and working in a somewhat different line of research, De Bruijn [4] studied the subwords of certain circular sequences. To this end he investigated the combinatorial structure of directed graphs whose vertex set consists of the α^k possible words of length k over an alphabet of size α. Two vertices are connected by an arc if and only if the last $k-1$ letters of tail-word are equal to the first $k-1$ letters of the head-word. Such a graph is nowadays called a *De Bruijn graph*, and it is denoted by $B(\alpha, k)$. De Bruijn graphs are used in communication networks and in VLSI design; cf. e.g. Samathan & Pradhan [8]. It is straightforward to see that a graph is a member of the above defined class $\mathcal{L}_k^\alpha$ if and only if it is the vertex induced subgraph of the De Bruijn graph $B(\alpha, k)$ with word length k and alphabet size α. Moreover, a graph is a DNA graph if and only if it is the vertex induced subgraph of $B(4, k)$ for some k.

In this paper we will study the computational complexity of the membership problems for the classes $\mathcal{L}_k^\alpha$, $\mathcal{L}_\infty^\alpha$, and $\mathcal{L}_k^\infty$ for various values of $\alpha \geq 2$ and $k \geq 1$. Let us start our discussion with the classes $\mathcal{L}_k^\alpha$.

- For $k = 1$, the membership problem for $\mathcal{L}_1^\alpha$ is trivial: A directed simple graph $G = (V, A)$ is in $\mathcal{L}_1^\alpha$ if and only if $A = V \times V$ and $|V| \leq \alpha$.
- For $k = 2$, Błażewicz, Hertz, Kobler & de Werra [3] design a polynomial time algorithm that decides for any input graph G and any input parameter α, whether G is a member of $\mathcal{L}_2^\alpha$.
- For any fixed number $k \geq 3$, we will prove in this paper (cf. Theorem 5) that it is NP-hard to decide for an input graph G and an input parameter α whether $G \in \mathcal{L}_k^\alpha$.
- For $\alpha = 2$, the complexity of the membership problem for $\mathcal{L}_k^2$ is unknown. We conjecture that it is polynomially solvable.
- For any fixed number $\alpha \geq 3$, we will prove in this paper (cf. Theorem 8) that it is NP-hard to decide for an input graph G and an input parameter k whether $G \in \mathcal{L}_k^\alpha$.

Note that if α and k both are not part of the input, then the membership problem for $\mathcal{L}_k^\alpha$ is easy. In this case the size of class $\mathcal{L}_k^\alpha$ is a fixed constant that does not depend on the input, and we simply may search through all of it. Next, we turn to the membership problems for the classes $\mathcal{L}_\infty^\alpha$ and $\mathcal{L}_k^\infty$. Some of these problems are fairly close to the membership problems for the classes $\mathcal{L}_k^\alpha$.

- Błażewicz, Hertz, Kobler & de Werra [3] give a polynomial time algorithm that takes a graph G and a parameter k as input, and correctly decides whether G is in $\mathcal{L}_k^\infty$.
- For $\alpha = 2$, the complexity of the membership problem for $\mathcal{L}_\infty^2$ is unknown.
- For any fixed number $\alpha \geq 3$, we will prove in this paper (cf. Theorem 9) that it is NP-hard to decide for an input graph G whether $G \in \mathcal{L}_\infty^\alpha$.

At the end of [3], the authors pose five open questions on the computational complexity of recognizing graphs in various classes $\mathcal{L}_k^\alpha$. **Question 1** considers a graph G with a given (α, k)-labeling, and it asks to find the largest possible label length q such that G is in $\mathcal{L}_q^\beta$ for some appropriate value of β. This question has been answered by Błażewicz, Formanowicz, Kasprzak & Kobler [2] who design a polynomial time algorithm for it. **Question 2** concerns the complexity of the membership problem in $\mathcal{L}_k^2$. This question remains open. **Question 3** asks for a polynomial time algorithm for the following problem: Given an integer k and a graph G, find the smallest integer α such that G is in $\mathcal{L}_k^\alpha$. Our Theorem 8 answers this question in the negative; the problem is NP-hard. **Question 4** asks for a polynomial time algorithm for the membership problem for $\mathcal{L}_k^\alpha$. Theorems 5 and 8 answer this question in the negative, and they show that the problem is NP-hard. **Question 5** asks for a polynomial time algorithm for the following problem: Given a graph G with a given (α, k)-labeling, determine all integers q such that G is in $\mathcal{L}_q^\alpha$. This question remains open.

Organization of the paper. In Section 2 we state some notation, we recall the definition of the graph coloring problem, and we derive some facts on De Bruijn graphs. Section 3 deals with the problem variants for fixed label lengths, and Section 4 deals with the problem variants for fixed sizes of the alphabet.

2   Preliminaries

Throughout the paper, labels will sometimes be considered as words over an appropriate alphabet, and then the entries of the labels will be considered as the letters of these words. For words w_1 and w_2, we define in the usual way their *concatenation* $w_1 \cdot w_2$ that results from appending word w_2 at the right end of word w_1.

All our NP-hardness proofs will be done by reductions from the graph coloring problem (cf. Garey & Johnson [5]). This coloring problem remains NP-hard even if the color bound $\gamma \geq 3$ is not part of the input.

GRAPH COLORING
Input: An undirected graph $H = (X, E)$ with $|X| = n$ vertices, and a color bound $3 \leq \gamma \leq n$.
Question: Does H have a feasible γ-coloring, i.e., does there exist a function $f : X \to \{1, \dots, \gamma\}$ such that $f(x) \neq f(y)$ for all edges $(x, y) \in E$?

In the rest of this section, we investigate when a De Bruijn graph $B(\alpha, k)$ is a subgraph of another De Bruijn graph $B(\beta, h)$. This will lead to a gadget for our NP-hardness proof in Theorem 9.

Lemma 1 *Let $\alpha, \beta \geq 2$ and $k, h \geq 2$ be integers such that $2\alpha > \beta$. If the De Bruijn graph $B(\alpha, k)$ is a subgraph of $B(\beta, h)$, then $B(\alpha, k-1)$ is a subgraph of $B(\beta, h-1)$.*

Proof. To simplify presentation, we will sometimes identify vertices with their labels. Assume that $B(\alpha, k)$ is a subgraph of $B(\beta, h)$, and let $f : \{1, \dots, \alpha\}^k \to \{1, \dots, \beta\}^h$ denote the corresponding injection between the two label sets. Consider an arbitrary label v of length $k-1$ over $\{1, \dots, \alpha\}$. There are α distinct vertices in $B(\alpha, k)$ whose labels end with the $k - 1$ entries in v; we denote this vertex set by S_v. Moreover, there are α distinct vertices in $B(\alpha, k)$ whose labels start with the $k - 1$ entries in v; we denote this vertex set by T_v. Note that the sets S_v and T_v are not necessarily disjoint. However, for $u \neq v$ we have $S_v \cap S_u = \emptyset$ and $T_v \cap T_u = \emptyset$.

In the graph $B(\alpha, k)$ there is an arc from every vertex in S_v to every vertex in T_v. Now consider the vertices in $f(S_v) = \{f(s)|s \in S_v\}$ and in $f(T_v) = \{f(t)|t \in T_v\}$. Since $B(\alpha, k)$ is a subgraph of $B(\beta, h)$, all labels of vertices in $f(S_v)$ must end with the same $h - 1$ entries $[\ell_2, \ldots, \ell_h] \doteq w$, and all labels of vertices in $f(T_v)$ must start with the same $h - 1$ entries w. Summarizing, these observations define for every label v in $\{1, \ldots, \alpha\}^{k-1}$ a unique corresponding label $w \doteq g(v)$ in $\{1, \ldots, \beta\}^{h-1}$. We claim

(i) that the function $g : \{1, \ldots, \alpha\}^{k-1} \to \{1, \ldots, \beta\}^{h-1}$ is an injection, and
(ii) that g certifies that $B(\alpha, k - 1)$ is a subgraph of $B(\beta, h - 1)$.

To see (i), suppose that for $u, v \in \{1, \ldots, \alpha\}^{k-1}$ with $u \neq v$ and for $w \in \{1, \ldots, \beta\}^{h-1}$, we have $g(u) = g(v) = w$. As f is an injection, $f(S_u \cup S_v)$ is a set of 2α labels, all with last $h - 1$ entries equal to w, thus with 2α distinct first entries. This contradicts the assumption that $2\alpha > \beta$.

To see (ii), consider an arc from vertex u to vertex v in the graph $B(\alpha, k-1)$. Then the label of u starts with an entry x followed by a sequence y of $k - 2$ entries, and the label of v starts with the sequence y followed by an entry z. Consider the vertex w in $B(\alpha, k)$ with label $x \cdot y \cdot z$, and let the label of $f(w)$ be $x' \cdot y' \cdot z'$ where x' and z' are single entries and where y' is a sequence of $h - 2$ entries. With this notation we have $f(w) = f(x \cdot y \cdot z) = x' \cdot y' \cdot z'$, and hence $g(u) = g(x \cdot y) = x' \cdot y'$ and $g(v) = g(y \cdot z) = y' \cdot z'$. Consequently, there is also an arc in the graph $B(\beta, h - 1)$ going from vertex $g(u)$ to vertex $g(v)$.   □

Theorem 2 *Let $\alpha, \beta \geq 2$ and $k, h \geq 1$ be integers such that $2\alpha > \beta$. If the De Bruijn graph $B(\alpha, k)$ is a subgraph of the De Bruijn graph $B(\beta, h)$, then $h = k$.*

Proof. Omitted in this extended abstract.   □

We note that since $B(2, 2)$ is a subgraph of the complete directed graph on 4 vertices $B(4, 1)$, the requirement that $2\alpha > \beta$ in the statement of Theorem 2 is necessary.

3   The Problem Variant with a Fixed Label Length

Let $H = (X, E)$ and γ be an arbitrary instance of GRAPH COLORING. Let $x_1, \ldots, x_n$ be an enumeration of all vertices in X and let $e_1, \ldots, e_m$ be an enumeration of all edges in E. Without loss of generality we assume that H does not contain any isolated vertices. We will now construct in polynomial time a directed graph $G = (V, A)$ and an alphabet size α, such that G is in $\mathcal{L}_3^\alpha$ if and only if H is γ-colorable.

- For every vertex $x_i \in X$ with $1 \leq i \leq n$, the graph G contains two corresponding vertices $v(x_i)$ and $v'(x_i)$. There is a loop at every vertex $v(x_i)$. From every vertex $v(x_i)$ with $1 \leq i \leq n$ there is an outgoing arc to $v'(x_i)$.

- For all $1 \leq i \neq j \leq n$, there is a directed path with three arcs from $v(x_i)$ to $v(x_j)$ through the two intermediate vertices $v_1(x_i, x_j)$ and $v_2(x_i, x_j)$. From every vertex $v_2(x_i, x_j)$ there is an outgoing arc to $v'(x_j)$, and an outgoing arc to every vertex $v_1(x_j, x_k)$ with $1 \leq k \leq n$ and $k \neq j$.
- For every edge $e_s \in E$ with $1 \leq s \leq m$, the graph G contains a corresponding vertex $v(e_s)$. There is a loop at every $v(e_s)$.
- For all $1 \leq s \neq t \leq m$, there is a directed path with three arcs from $v(e_s)$ to $v(e_t)$ through the two intermediate vertices $v_1(e_s, e_t)$ and $v_2(e_s, e_t)$. From every vertex $v_2(e_s, e_t)$ there is an outgoing arc to every vertex $v_1(e_t, e_u)$ with $1 \leq u \leq m$ and $u \neq t$.
- If vertex x_i is incident to edge e_s in H, then G contains a directed path from $v'(x_i)$ to $v(e_s)$ via vertices $v_1(x_i, e_s)$ and $v_2(x_i, e_s)$. From every vertex $v_2(x_i, e_s)$ there is an outgoing arc to every $v_1(e_s, e_t)$ with $1 \leq t \leq m$ and $t \neq s$.

This completes the construction of graph $G = (V, A)$. Finally, we define the alphabet size $\alpha = n + m + \gamma$. To simplify the presentation of the following arguments, we assume that the α letters in the alphabet are the vertices $x_1, \ldots, x_n$ and the edges $e_1, \ldots, e_m$ in H, together with γ colors $c_1, \ldots, c_\gamma$.

Lemma 3 *If $G = (V, A)$ has an $(\alpha, 3)$-labeling, then the graph H is γ-colorable.*

Proof. Consider an $(\alpha, 3)$-labeling of G. It is easy to see that the label of a vertex with a loop must consist of three identical letters. Without loss of generality we assume that every vertex $v(x_i)$ is labeled by $[x_i, x_i, x_i]$ and that every vertex $v(e_s)$ is labeled by $[e_s, e_s, e_s]$. From this we derive that vertex $v_1(x_i, x_j)$ has label $[x_i, x_i, x_j]$, and that vertex $v_2(e_t, e_s)$ has label $[e_t, e_s, e_s]$.

Now consider vertex $v'(x_i)$. Since it is a successor of $v(x_i)$, its label must be of the form $[x_i, x_i, \sigma]$ where σ is some letter from the alphabet. The letter σ can not be equal to x_i, since then $v'(x_i)$ and $v(x_i)$ would have the same label in contradiction to property (L2). The letter σ can also not be equal to x_j with $i \neq j$, since then vertices $v'(x_i)$ and $v_1(x_i, x_j)$ would have the same label. Since x_i is not an isolated vertex in H, there is a path from $v'(x_i)$ to some vertex $v(e_s)$ through the intermediate vertices $v_1(x_i, e_s)$ and $v_2(x_i, e_s)$. Then the label of vertex $v_1(x_i, e_s)$ equals $[x_i, \sigma, e_s]$, and the label of vertex $v_2(x_i, e_s)$ equals $[\sigma, e_s, e_s]$. We conclude that σ can neither be equal to e_s (since then $v_2(x_i, e_s)$ and $v(e_s)$ had the same label) nor can it be equal to some e_t with $t \neq s$ (since then $v_2(x_i, e_s)$ and $v_2(e_t, e_s)$ had the same label). The only remaining possibility for σ is that it is one of the colors $c_1, \ldots, c_\gamma$.

We now define $f(x_i) = \sigma_i$ for $1 \leq i \leq n$, where $[x_i, x_i, \sigma_i]$ is the label of vertex $v'(x_i)$. By the above discussion, every σ_i is one of the colors $c_1, \ldots, c_\gamma$. We claim that this yields a feasible coloring. Indeed, suppose that $f(x_i) = f(x_j) = \sigma$ where x_i and x_j are connected to each other by an edge e_s in H. Then the label of $v_2(x_i, e_s)$ equals $[\sigma, e_s, e_s]$, and the label $v_2(x_j, e_s)$ also equals $[\sigma, e_s, e_s]$, a contradiction to property (L2). □

Lemma 4 *If the graph H is γ-colorable, then $G = (V, A)$ has an $(\alpha, 3)$-labeling.*

Proof. Consider a feasible γ-coloring of H that assigns to every vertex $x_i \in X$ a color σ_i from the colors $c_1, \ldots, c_\gamma$. We define the following labeling:

- For $1 \leq i \leq n$, the label of $v(x_i)$ is $[x_i, x_i, x_i]$ and the label of $v'(x_i)$ is $[x_i, x_i, \sigma_i]$.
- For all $1 \leq i \neq j \leq n$, the label of $v_1(x_i, x_j)$ is $[x_i, x_i, x_j]$ and the label of $v_2(x_i, x_j)$ is $[x_i, x_j, x_j]$.
- For $1 \leq s \leq m$, the label of $v(e_s)$ is $[e_s, e_s, e_s]$.
- For all $1 \leq s \neq t \leq m$, the label of $v_1(e_s, e_t)$ is $[e_s, e_s, e_t]$ and the label of $v_2(e_s, e_t)$ is $[e_s, e_t, e_t]$.
- If vertex x_i is incident to edge e_s, then the label of $v_1(x_i, e_s)$ is $[x_i, \sigma_i, e_s]$, and the label of $v_2(x_i, e_s)$ is $[\sigma_i, e_s, e_s]$.

Clearly, this labeling assigns distinct labels to distinct vertices. To verify that the labeling also fulfills property (L3), we divide the vertices of G into five classes: The first class V_1 contains all vertices $v(x_i)$, $v_1(x_i, x_j)$, and $v_2(x_i, x_j)$; the entries in the labels of all these vertices are from $x_1, \ldots, x_n$. The second class V_2 contains all vertices $v'(x_i)$; the first two entries in the labels of these vertices are from $x_1, \ldots, x_n$, and the last entry is from $c_1, \ldots, c_\gamma$. The third class V_3 contains all vertices $v_1(x_i, e_s)$; their labels consist of some vertex x_i, followed by a color, followed by an edge. The fourth class V_4 contains all vertices $v_2(x_i, e_s)$; their labels consist of some color, followed by two edges. The fifth class V_5 contains all vertices $v(e_s)$, $v_1(e_s, e_t)$, and $v_2(e_s, e_t)$; their labels are triples of edges.

By the definition of the labeling, there can only be arcs within V_1, arcs from V_1 to V_2, arcs from V_2 to V_3, arcs from V_3 to V_4, arcs from V_4 to V_5, and arcs within V_5. It is now straightforward but somewhat tedious to verify that all arcs within V_1 and within V_5, from V_1 to V_2, and from V_4 to V_5 are correctly encoded. Moreover, every vertex x_i has a unique color σ_i, and therefore every label $[x_i, \sigma_i, e_s]$ of a vertex in V_3 only allows a unique predecessor and a unique successor. $\square$

Combining the statements of Lemma 3 and 4 now yields the statement of Theorem 5 below for $k = 3$. The construction is easily extended to the cases with a fixed $k \geq 4$. The main idea is to replace in our construction all the connecting paths with three arcs by new connecting paths with k arcs. The details are left to the reader.

Theorem 5 *For any fixed $k \geq 3$ the following problem is NP-hard: Decide for an input graph G and an input parameter α whether $G \in \mathcal{L}_k^\alpha$.*

4   The Problem Variant with a Fixed Alphabet Size

Let $H = (X, E)$ and γ be an arbitrary instance of GRAPH COLORING. By adding at most $|E| - 1$ independent edges to H, we can make $|E|$ a perfect power of two without increasing the chromatic number of H. Then by adding some isolated vertices (again without increasing the chromatic number of H), we can make $|X| = |E|$. Summarizing this yields that we may assume without loss of generality that $n = |X| = |E| = 2^z$ holds for some integer $z \geq 4$.

Let $x_1, \ldots, x_n$ be an enumeration of all vertices in X and let $e_1, \ldots, e_n$ be an enumeration of all edges in E. We will now construct in polynomial time a directed graph $G = (V, A)$ and a label length k, such that G is in $\mathcal{L}_k^\gamma$ if and only if H is γ-colorable.

- For every vertex $x_i \in X$ with $1 \leq i \leq n$, the graph G contains a corresponding *complete binary out-tree* $T(x_i)$ of height z. The root of this tree is the vertex $v(x_i)$. Every interior vertex in $T(x_i)$ has a left and a right out-going arc that connect it to its two children. All 2^z leaves in $T(x_i)$ are at the same distance z from the root. Enumerating the leaves in $T(x_i)$ from left to right, they are called $v(x_i, e_1), v(x_i, e_2), \ldots, v(x_i, e_n)$.
- For every edge $e_s \in E$ with $1 \leq s \leq n$, the graph G contains a corresponding vertex $v(e_s)$.
- If vertex x_i is incident to edge e_s in H, then G contains a directed path $P(x_i, e_s)$ from $v(x_i, e_s)$ to $v(e_s)$. The path $P(x_i, e_s)$ consists of $z + 2$ arcs and of $z+1$ new vertices. The interior vertices on this path are not connected to any other parts of graph G.

This completes the construction of graph $G = (V, A)$. We define the label length $k = 2z + 3$.

To simplify the presentation of the following arguments, we assume that the alphabet Σ consists of the letters $0, 1, \ldots, \gamma - 1$. Some of the labels will be concatenated from the *binary* representations of certain integers. For an integer i in the range $1 \leq i \leq 2^z$, we define $\text{BIN}(i)$ to be the z-digit binary representation of $i - 1$. This representation always contains an appropriate number of leading zeroes so that its length exactly equals z. For non-negative integers i and for letters $\sigma \in \Sigma$, we will use σ^i to represent the word consisting of exactly i letters σ.

Lemma 6 *If $G = (V, A)$ has a (γ, k)-labeling, then the graph H is γ-colorable.*

Proof. Consider a (γ, k)-labeling of G. For any vertex $x_i \in X$, we define its color $f(x_i)$ to be the $(2z + 2)$th digit of the label of the root $v(x_i)$. Since the alphabet size is γ, this coloring uses only γ colors. We claim that this yields a feasible coloring.

Indeed, suppose that $f(x_i) = f(x_j) = \sigma$ where x_i and x_j are connected to each other by an edge e_s in H. We consider the predecessors of $v(e_s)$ on the paths $P(x_i, e_s)$ and $P(x_j, e_s)$, and call these vertices $w(x_i, e_s)$ and $w(x_j, e_s)$, respectively. There is a directed path of $2z + 1$ arcs from $v(x_i)$ to $w(x_i, e_s)$. As a consequence of property (L3), the $(2z + 2)$th digit of the label of $v(x_i)$ must be equal to the first digit of the label of $w(x_i, e_s)$. Hence, this first digit equals σ. By analogous reasoning we get that the first digit of the label of $w(x_j, e_s)$ also equals σ. Since $w(x_i, e_s)$ and $w(x_j, e_s)$ both are predecessors of $v(e_s)$, the last $k - 1$ digits of their labels must agree with the first $k - 1$ digits of the label of $v(e_s)$. But now the labels of $w(x_i, e_s)$ and $w(x_j, e_s)$ agree in the first digit and also in the last $k-1$ digits, and hence they are equal to each other. This contradicts property (L2). □

Lemma 7 *If the graph H is γ-colorable, then $G = (V, A)$ has a (γ, k)-labeling.*

Proof. Consider a feasible γ-coloring of H that assigns to every vertex $x_i \in X$ a color σ_i from Σ. We define the following partial labeling:

- For $1 \le i \le n$, the label of $v(x_i)$ is $2^{z-1} \cdot 0 \cdot \text{BIN}(i) \cdot 2 \cdot \sigma_i \cdot 1$.
- For $1 \le s \le n$, the label of $v(e_s)$ is $1 \cdot \text{BIN}(s) \cdot 2^{z+2}$.

All other vertices are on some path with $2z + 2$ arcs from some root $v(x_i)$ to some $v(e_s)$. If a vertex v is j arcs away from $v(x_i)$ and $2z + 2 - j$ arcs away from $v(e_s)$, then its first $2z + 3 - j$ letters are the last $2z + 3 - j$ letters of the label of $v(x_i)$, and its last $j + 1$ letters are the first $j + 1$ letters of the label of $v(e_s)$. This completely determines the label of v. Summarizing, the label of every vertex on path $P(x_i, e_s)$ is an appropriate subword of length $2z + 3$ of the word

$$2^{z-1} \cdot 0 \cdot \text{BIN}(i) \cdot 2 \cdot \sigma_i \cdot 1 \cdot \text{BIN}(s) \cdot 2^{z+2}.$$

For an illustration, the reader may want to verify that the label of the leaf vertex $v(x_i, e_s)$ in $T(x_i)$ equals $\text{BIN}(i) \cdot 2 \cdot \sigma_i \cdot 1 \cdot \text{BIN}(s)$. Now we state some simple but important observations on these labels.

First: By considering a label $L(v)$, one can easily determine whether the corresponding vertex v is in one of the trees $T(x_i)$, or lies on one of the paths $P(x_i, e_s)$, or is one of the vertices $v(e_s)$. Indeed, if a label $L(v)$ has a 0 or a 1 as last digit, then the vertex v is contained in some tree $T(x_i)$. And if the label $L(v)$ ends with a 2, then the vertex v lies on one of the paths $P(x_i, e_s)$, or it is one of the vertices $v(e_s)$; and v is one of the vertices $v(e_s)$ if and only if its label $L(v)$ ends with a block of 2's of length $z + 2$.

Second: If a vertex v is contained in some tree $T(x_i)$, then its label $L(v)$ starts with a block of 2's that is followed by a 0 that in turn is followed by z binary digits that uniquely identify the index i. For a non-leaf vertex v, the only possible labels of successors of v result by removing the first digit from $L(v)$ and by appending a 0 or a 1 at its right end; this correctly encodes the tree structure of $T(x_i)$. For the leaves $v(x_i, e_s)$, there is at most one possible successor label that results by removing the first digit (0 or 1) and by appending a 2. Such a successor vertex exists if and only if x_i is incident to e_s. Hence, also the initial arcs of the paths $P(x_i, e_s)$ are correctly encoded by the labeling.

Finally: Let us consider a vertex v that lies on one of the paths $P(x_i, e_s)$. Its label $L(v)$ ends with a non-empty block of 2's that is preceded by z binary digits that uniquely identify the index s. These z binary digits are preceded by a digit 1, which in turn is preceded by the digit σ_i. Since the edge e_s is incident to two vertices x_i and x_j with different colors in the γ-coloring, we have $\sigma_i \ne \sigma_j$. Hence, the index s together with the letter σ_i uniquely identifies the path on which the vertex v lies. The only possible label for a successor results by removing the first digit, and by appending a 2. Hence, all the arcs on all the paths $P(x_i, e_s)$ are correctly encoded. □

Combining the statements of Lemma 6 and 7 now yields that there exists a γ-coloring for H if and only if there exists a (γ, k)-labeling for G. Since γ-coloring remains NP-hard for every fixed $\gamma \geq 3$, we get the following theorem.

Theorem 8 *For any fixed $\alpha \geq 3$ the following problem is NP-hard: Decide for an input graph G and an input parameter k whether $G \in \mathcal{L}_k^\alpha$.*

In fact our reduction also proves that it is hard to find labelings whose alphabet size is close to optimal, since all the inapproximability results from graph coloring (cf. eg. Håstad [6] and Khanna, Linial & Safra [7]) immediately carry over to the labeling problem. E.g. since it is NP-hard to find a 4-coloring for a 3-colorable graph [7], it is NP-hard to find a $(4, k)$-labeling for a graph G in $\mathcal{L}_k^3$. Since (unless P=NP) there is no constant factor approximation algorithm for graph coloring [6], there also is no constant factor approximation algorithm for minimizing the alphabet size α under a given label length k.

Theorem 9 *For any fixed $\alpha \geq 3$ the following problem is NP-hard: Decide for an input graph G whether $G \in \mathcal{L}_\infty^\alpha$.*

Proof. Let $H = (X, E)$ be an arbitrary instance of GRAPH γ-COLORING; throughout the proof we assume that $\gamma \geq 3$ is a fixed constant. We repeat the polynomial time construction of above: We make $n = |X| = |E| = 2^z$ for some integer z, and we construct a directed graph $G = (V, A)$, such that G is in $\mathcal{L}_{2z+3}^\gamma$ if and only if the undirected graph H is γ-colorable. Finally, we define the graph $G' = (V', A')$ to be the disjoint union of this graph G and of the De Bruijn graph $B(\gamma - 1, 2z + 3)$. The De Bruijn graph $B(\gamma - 1, 2z + 3)$ has

$$(\gamma - 1)^{2z+3} \leq \gamma^{2 \log n + 3} = \gamma^3 n^{2 \log \gamma}$$

vertices. Hence, the size of G' is polynomial in the size of $H = (X, E)$, and the whole construction can be done in polynomial time.

We claim that G' is in $\mathcal{L}_\infty^\gamma$ if and only if G is in $\mathcal{L}_{2z+3}^\gamma$. First, assume that G' is in $\mathcal{L}_\infty^\gamma$. Then G' is a vertex induced subgraph of $B(\gamma, k)$ for some k. Since on the other hand $B(\gamma - 1, 2z + 3)$ is a subgraph of G', Theorem 2 now yields that $k = 2z + 3$ must hold. Therefore, the vertex induced subgraph G of G' indeed is in $\mathcal{L}_{2z+3}^\gamma$. Next, assume that G is in $\mathcal{L}_{2z+3}^\gamma$. Then we can reuse the $(\gamma, 2z + 3)$-labeling that we defined in Lemma 7 for G. Moreover, the De Bruijn graph $B(\gamma - 1, 2z + 3)$ can be $(\gamma, 2z + 3)$-labeled in the natural way by all words of length $2z + 3$ over the alphabet $\{0, 1, \ldots, \gamma - 1\} \setminus \{2\}$. In the labels for vertices in G, every substring of length $z + 3$ contains at least once the digit 2. In the labels for vertices in $B(\gamma - 1, 2z + 3)$, the digit 2 does not show up at all. Consequently, these labelings do not generate any cross-edges between the graphs G and $B(\gamma - 1, 2z + 3)$, and they together yield a $(\gamma, 2z + 3)$-labeling for G'.

If we combine these statements with the statements in Lemma 6 and 7, we get that G' is in $\mathcal{L}_\infty^\gamma$ if and only if the graph H is γ-colorable. □

Corollary 10 *It is NP-hard to decide whether an input graph G is a DNA graph.*   □

References

1. W. BAINS AND G.C. SMITH [1988]. A novel method for nucleic acid sequence determination. *Journal of Theoretical Biology 135*, 303–307.
2. J. BŁAŻEWICZ, P. FORMANOWICZ, M. KASPRZAK, AND D. KOBLER [1999]. On the recognition of De Bruijn graphs and their induced subgraphs. Manuscript, September 1999.
3. J. BŁAŻEWICZ, A. HERTZ, D. KOBLER, AND D. DE WERRA [1999]. On some properties of DNA graphs. *Discrete Applied Mathematics 98*, 1–19.
4. N.G. DE BRUIJN [1946]. A combinatorial problem. *Koninklijke Nederlandse Akademie van Wetenschappen te Amsterdam. Proceedings 49*, 758–764.
5. M.R. GAREY AND D.S. JOHNSON [1979]. *Computers and Intractability: A Guide to the Theory of NP-Completeness*. Freeman, San Francisco.
6. J. HÅSTAD [1997]. Some optimal inapproximability results. *Proceedings of the 29th Annual ACM Symposium on the Theory of Computing (STOC'97)*, 1–10.
7. S. KHANNA, N. LINIAL, AND S. SAFRA [1993]. On the hardness of approximating the chromatic number. *Proceedings of the 2nd Israeli Symposium on Theory and Computing Systems (ISTCS'93)*, 250–260.
8. M.R. SAMATHAN AND D.K. PRADHAN [1989]. The De Bruijn multiprocessor network: A versatile parallel processing and sorting network for VLSI. *IEEE Transactions on Computers 38*, 567–581.

A Generic Greedy Algorithm, Partially-Ordered Graphs and NP-Completeness

Antonio Puricella and Iain A. Stewart

Department of Mathematics and Computer Science,
University of Leicester, Leicester LE1 7RH, U.K.

Abstract. Let π be any fixed polynomial-time testable, non-trivial, hereditary property of graphs. Suppose that the vertices of a graph G are not necessarily linearly ordered but partially ordered, where we think of this partial order as a collection of (possibly exponentially many) linear orders in the natural way. We prove that the problem of deciding whether a lexicographically first maximal subgraph of G satisfying π, with respect to one of these linear orders, contains a specified vertex is **NP**-complete.

1 Introduction

Miyano [6] proved that the problem of computing the lexicographically first maximal subgraph of a given graph, where this subgraph should satisfy some fixed polynomial-time testable, non-trivial, hereditary property π, is **P**-hard (even when the given graph is restricted to be either bipartite or planar and π is non-trivial on the class of bipartite or planar graphs, respectively). Because of the stipulations on π, the lexicographically first maximal subgraph satisfying the property π can be computed by a generic greedy algorithm. Note that Miyano's result is widely applicable; to *any* polynomial-time testable, non-trivial, hereditary property π, such as whether a graph is planar, bipartite, acyclic, of bounded degree, an interval graph, chordal, and so on. Miyano states that his work was inspired by that of Asano and Hirata [1], Lewis and Yannakakis [5], Watanabe, Ae and Nakamura [7] and Yannakakis [8] on node- and edge-deletion problems in **NP**. Typical in this work is the result of Lewis and Yannakakis [5] that the problem of finding the minimum number of nodes needing to be deleted from a graph so that the graph satisfies a fixed polynomial-time testable, non-trivial, hereditary property π is **NP**-hard.

Of course, a tacit assumption in Miyano's work is that the vertices of any graph are linearly ordered. In this paper, inspired by Miyano's results, we return to the setting of **NP** in that we consider computing lexicographically first maximal subgraphs of given graphs, where these subgraphs should satisfy some given polynomial-time testable, non-trivial, hereditary property π, except that now the graphs come equipped with not just one linear ordering of their vertices but several. Hence, for a given graph we will be involved with a collection of lexicographically first maximal subgraphs and not just one. Note that if we

A. Brandstädt and V.B. Le (Eds.): WG 2001, LNCS 2204, pp. 306–316, 2001.
© Springer-Verlag Berlin Heidelberg 2001

gave our linear orderings explicitly then a graph on n vertices could only come with a polynomial (in n) number of such linear orderings (as otherwise it would be unreasonable to define that the whole input has size n) and we would still be working within **P**. In order to work with an exponential number of linear orderings, we present our collection of linear orderings in the form of a partial order, *i.e.*, an acyclic digraph, with a source vertex providing the (common) least element of any of the linear orderings. As in Miyano's deterministic scenario, a non-deterministic polynomial-time greedy algorithm computes all lexicographically first maximal subgraphs.

Our main result is that the problem of deciding whether a lexicographically first maximal subgraph of a given partially ordered graph, where this subgraph should satisfy some fixed polynomial-time testable, non-trivial, hereditary property π, contains some specified vertex is **NP**-complete (even when the given graph is restricted to be planar bipartite and π is non-trivial on this class of graphs). We use similar techniques to Miyano although our proofs are comparatively simpler and the combinatorics is very different.

It is not at all obvious as to how we might use Miyano's result for **P**-completeness directly to prove an analogous result for **NP**-completeness but with partial orderings replacing linear orderings (indeed, we have failed with this approach and have had to revert back to 'first principles'). Furthermore, the **NP**-completeness results from [1,5,7,8] can not be formulated in our framework. For example, the result of Lewis and Yannakakis [5], mentioned above, is concerned with sizes of maximal subgraphs satisfying a specific property and as such is unrelated to our problems.

2   Basic Definitions

For standard graph-theoretic definitions, the reader is referred to [2]. A property π on graphs is *hereditary* if whenever we have a graph with the property π, the deletion of any vertex and its incident edges does not produce a graph violating π, *i.e.*, π is preserved by induced subgraphs. A property π is called *non-trivial* on a class of graphs if there are infinitely many graphs from this class satisfying π but π is not satisfied by all graphs of the class.

Let π be some property of graphs. Let G be a graph, let H be a partial order of the vertices of G, and let s and t be vertices of G. We assume that the partial order H is given in the form of an acyclic digraph detailing the immediate predecessors, *i.e.*, the *parents*, and the immediate successors, *i.e.*, the *children*, of each vertex. We think of a partial order H as encoding a collection of linear orders of the form $s = s_0, s_1, s_2, \ldots, s_k$, where s_{j+1} is a child of s_j and s_k has no children. Note that a partial order can encode an exponential number of linear orders.

The algorithm GREEDY(π) is as follows:

```
input(G, H, s)
   S := ∅
   current-vertex := s
```

```
(*)     if π(S∪{current-vertex},G) then
            S := S∪{current-vertex}
        fi
        while current-vertex has at least one child in H do
            current-vertex := a child of current-vertex in H
(**)        if π(S∪{current-vertex},G) then
                S := S∪{current-vertex}
            fi
        od
    output(S)
```

where $\pi(S \cup \{current\text{-}vertex\}, G)$ is a predicate evaluating to 'true' if, and only if, the subgraph of G induced by the vertices of $S \cup \{current\text{-}vertex\}$ satisfies π. Essentially, GREEDY(π) non-deterministically chooses a linear order from those encoded in the partial order and performs a greedy execution on G with respect to the property π and the chosen linear order. We say that a vertex v is the *current-vertex* if we have 'frozen' an execution of the algorithm GREEDY(π) immediately prior to executing either line (*) or line (**) and the value of the variable *current-vertex* at this point is v. Note that in general the algorithm GREEDY(π) is non-deterministic and produces a collection of sets of vertices as outputs. If the property π is hereditary and can be checked in polynomial-time then the algorithm GREEDY(π) non-deterministically computes, in polynomial-time, the lexicographically first maximal subgraphs of the graph G satisfying π with respect to the linear orders encoded within the partial order H.

Let $\mathcal{C}$ be a class of graphs and let π be some property of graphs. The problem GREEDY(partial order, $\mathcal{C}$, π) has: as its instances tuples (G, H, s, t), where G is a graph from $\mathcal{C}$, H is a partial order of the vertices of G and s and t are vertices of G; and as its yes-instances those instances for which there exists an execution of the algorithm GREEDY(π) on input (G, H, s) resulting in the vertex t being output. The problem GREEDY(linear order, $\mathcal{C}$, π) is defined similarly. Miyano's result from [6] can be stated as follows.

Theorem 1. *Let π be a polynomial-time testable, non-trivial, hereditary property on the class of graphs $\mathcal{C}$, where $\mathcal{C}$ is the class of all graphs, the class of planar graphs or the class of bipartite graphs. Then the problem GREEDY(linear order, $\mathcal{C}$, π) is **P**-complete.* □

3   Our Results

In order to prove our main result, we need to first establish a completeness result for the specific problem GREEDY(partial order, planar bipartite, independent set) (we only sketch the proof due to space limitations).

Theorem 2. *The problem GREEDY(partial order, planar bipartite, independent set) is **NP**-complete.*

Proof. (*Sketch*) We reduce from the known **NP**-complete problem Directed Hamiltonian Path (DHP): whose instances are triples (G, s, t), where G is a digraph and s and t are vertices of G; and whose yes-instances are instances for which there is a Hamiltonian path in G from s to t (see [3]).

Let $(G = (V, E), s, t)$ be an instance of DHP of size n. W.l.o.g. we assume that $|V| > 2$, that the vertex set of G is $\{1, 2, \dots, n\}$ and that $s = 1$ and $t = n$. Corresponding to this instance, we build an instance (G', H', s', t') of GREEDY(partial order, planar bipartite, independent set). The vertex set V' of G' and H' is

$$\{u_{i,j}, v_{i,j}, w_{i,j}, z_j : i, j = 2, 3, \dots, n - 1\} \cup \{x, s', t'\}.$$

The edges of G' are

$$\begin{aligned}
&\{(u_{i,j}, v_{i,j}), (u_{i,j}, w_{i,j}) : i, j = 2, 3, \dots, n - 1\} \\
&\cup \{(w_{i,j}, u_{i+1,j}) : i = 2, 3, \dots, n - 2; j = 2, 3, \dots, n - 1\} \\
&\cup \{(w_{n-1,j}, z_j) : j = 2, 3, \dots, n - 1\} \\
&\cup \{(z_j, w_{n-1,j+1}) : j = 2, 3, \dots, n - 2\} \\
&\cup \{(z_{n-1}, t')\}
\end{aligned}$$

and the edges of H' are

$$\begin{aligned}
&\{(v_{i,j}, v_{i+1,j'}) : i = 2, 3, \dots, n - 2; j, j' = 2, 3, \dots, n - 1; (j, j') \in E\} \\
&\cup \{(s', v_{2,j}) : j = 2, 3, \dots, n - 1; (1, j) \in E\} \\
&\cup \{(v_{n-1,j}, x) : j = 2, 3, \dots, n - 1; (j, n) \in E\} \\
&\cup \{(x, u_{2,2})\} \\
&\cup \{(u_{i,j}, w_{i,j}) : i = 2, 3, \dots, n - 2; j = 2, 3, \dots, n - 1\} \\
&\cup \{(w_{i,j}, u_{i+1,j}) : i = 2, 3, \dots, n - 2; j = 2, 3, \dots, n - 1\} \\
&\cup \{(u_{n-1,j}, u_{2,j+1} : j = 2, 3, \dots, n - 2\} \\
&\cup \{(u_{n-1,n-1}, w_{n-1,2})\} \\
&\cup \{(w_{n-1,j}, z_j) : j = 2, 3, \dots, n - 1\} \\
&\cup \{(z_j, w_{n-1,j+1}) : j = 2, 3, \dots, n - 2\} \\
&\cup \{(z_{n-1}, t')\}
\end{aligned}$$

The construction of the instance (G', H', s', t') is illustrated in Figs. ?? and 3 which depict: a digraph G; the resulting graph G'; and the resulting partial order H', respectively. We omit the proof that (G, s, t) is a yes-instance of DHP if, and only if, (G', H', s', t') is a yes-instance of GREEDY(partial order, planar bipartite, independent set). □

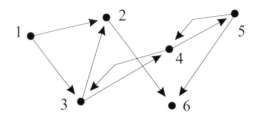

Fig. 1. A digraph G.

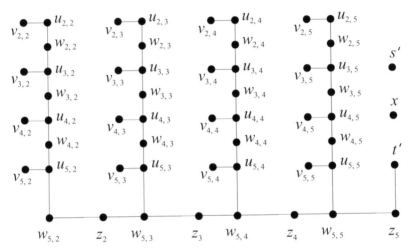

Fig. 2. The graph G' corresponding to G.

Now we consider the problem GREEDY(partial order, planar bipartite, π) where π is *any* polynomial-time testable, non-trivial, hereditary property. We begin with some specific graph-theoretic definitions.

A *cut-point* of a connected graph G is a vertex c such that its removal (along with its incident edges) from G results in a graph with at least 2 connected components. A *component relative to a cut-point* c is a subgraph consisting of c, one of the derived connected components and all those edges of G joining c and a vertex of the component. If a connected graph does not have any cut-points then it is *biconnected*.

Let $\mathbf{a} = (a_1, a_2, \ldots, a_s)$ and $\mathbf{b} = (b_1, b_2, \ldots, b_t)$ be two tuples of positive integers. We order these tuples lexicographically as follows. We say that $\mathbf{a} >_L \mathbf{b}$ if either:

- there exists some $i \in \{1, 2, \ldots, \min\{s, t\}\}$ such that $a_j = b_j$, for all $j \in \{1, 2, \ldots, i-1\}$, and $a_i > b_i$; or
- $s > t$ and $a_j = b_j$, for all $j \in \{1, 2, \ldots, t\}$.

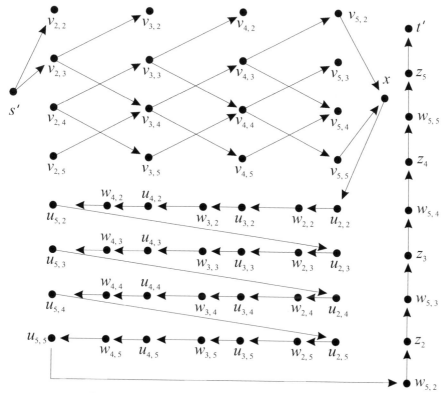

Fig. 3. The partial order H' corresponding to G.

The *α-sequence* α_G of a connected graph G is defined as follows. Suppose that G is not biconnected. If c is a cut-point of G whose removal results in a graph with k connected components then define $\alpha_{c,G} = (n_1, n_2, \ldots, n_k)$, where $n_1 \geq n_2 \geq \ldots \geq n_k$ are the numbers of vertices in the components relative to c. We define α_G to be the lexicographically-minimal tuple of the (non-empty) set $\{\alpha_{c,G} : c$ is a cut-point of $G\}$, and we define c_G to be any cut-point for which $\alpha_G = \alpha_{c_G,G}$. If G is biconnected then removing any vertex will result in one connected component relative to the removed vertex, and so we define $\alpha_G = (|G|)$ and c_G as any vertex.

Given a graph G with connected components $G_1, G_2, \ldots, G_k$, the *β-sequence* β_G of G is defined as $(\alpha_{G_1}, \alpha_{G_2}, \ldots, \alpha_{G_k})$, where $\alpha_{G_1} \geq_L \alpha_{G_2} \geq_L \ldots \geq_L \alpha_{G_t}$. A β-sequence is therefore a tuple of tuples of integers.

Theorem 3. *Let π be a property satisfying the following conditions:*

- *π is non-trivial on planar bipartite graphs;*
- *π is hereditary on induced subgraphs;*
- *π is satisfied by all sets of independent edges; and*
- *π is polynomial-time testable.*

The problem GREEDY(partial order, planar bipartite, π) is **NP**-*complete.*

Proof. For brevity, we refer to the problem GREEDY(partial order, planar bipartite, π) as $\mathcal{G}$. The property π is, by assumption, non-trivial on planar bipartite graphs. It follows that amongst all planar bipartite graphs violating π, there must be (at least) one with smallest β-sequence, where β-sequences are ordered lexicographically and where the comparison of components, i.e., α-sequences, is according to $\geq_L$. Let us call such a graph J; that is,

$$\beta_J = \min\{\beta_G : G \text{ is a planar bipartite graph violating } \pi\}.$$

Let $J_1, J_2, \ldots, J_k$ be the connected components of J ordered according to $\alpha_{J_1} \geq_L \alpha_{J_2} \geq_L \cdots \geq_L \alpha_{J_k}$. It follows that J has β-sequence $\beta_J = (\alpha_{J_1}, \alpha_{J_2}, \ldots, \alpha_{J_k})$. Let $c = c_{J_1}$ and let the connected components of J_1 relative to c be $I_0 \cup \{c\}, I_1 \cup \{c\}, \ldots, I_m \cup \{c\}$, where $|I_0| \geq |I_1| \geq \ldots \geq |I_m|$. Denote by I_* the subgraph of J_1 induced by the vertices of $\{c\} \cup I_1 \cup \ldots \cup I_m$. As π is non-trivial on planar bipartite graphs and it is hereditary on induced subgraphs, it follows that π is satisfied by any independent set of vertices, and so I_0 must contain at least one edge (otherwise J would be a set of independent vertices).

To prove the **NP**-completeness of the problem $\mathcal{G}$, we reduce from the problem GREEDY(partial order, planar bipartite, independent set), which, for brevity, we call $\mathcal{H}$ (it was proven **NP**-complete in Theorem 2). That is, from an instance (G, H, s, t) of $\mathcal{H}$, we create an instance (G', H', s', t') of $\mathcal{G}$.

We will divide the construction of G' from G into three phases. For any subset of vertices U of J, we denote by $\langle U \rangle$ the subgraph of J induced by the vertices of U. Note that as $\langle I_0 \cup \{c\} \rangle$ contains at least one edge and is connected, there exists a vertex d of $I_0 \cup \{c\}$ such that (c, d) is an edge of $\langle I_0 \cup \{c\} \rangle$.

Phase 1 For each vertex u of G, we attach a copy of $\langle I_* \cup \{c\} \rangle$ by identifying u with c (all such copies are disjoint). Call the resulting graph $\tilde{G}$. Note that the vertex set of $\tilde{G}$ consists of the vertices of G, which we call the G-*vertices*, together with disjoint copies of the vertices of I_*. As both $\langle I_* \rangle$ and G are planar and bipartite, $\tilde{G}$ maintains these properties.

Phase 2 We replace each edge (u, v) of $\tilde{G}$, where u and v are G-vertices, by a copy of $\langle I_0 \cup \{c\} \rangle$ by identifying u with c and v with d (all such copies are disjoint). Note that our choice of d results in the graph so formed being planar and bipartite.

Phase 3 We add disjoint copies of $J_2, J_3, \ldots, J_k$ to obtain G', which is clearly planar and bipartite.

The partial order H' consists of a linear order onto which is concatenated the partial order H (of the G-vertices). The linear order consists of: all vertices of G' that are vertices of some copy of $\langle I_0 \rangle$; followed by all vertices in the copies of $\langle I_* \rangle$; followed by all vertices of $J_2, J_3, \ldots, J_k$. It does not matter how we order the vertices of some copy of $\langle I_0 \rangle$, for example, in the linear order. We concatenate this linear order prior to H by including an edge from the last vertex of the linear order to the vertex s of H. Denote the vertex s' to be the first vertex of the above linear order, and denote the vertex t' to be the G-vertex of G' formerly known as t. Our construction can be visualised in Fig. 4.

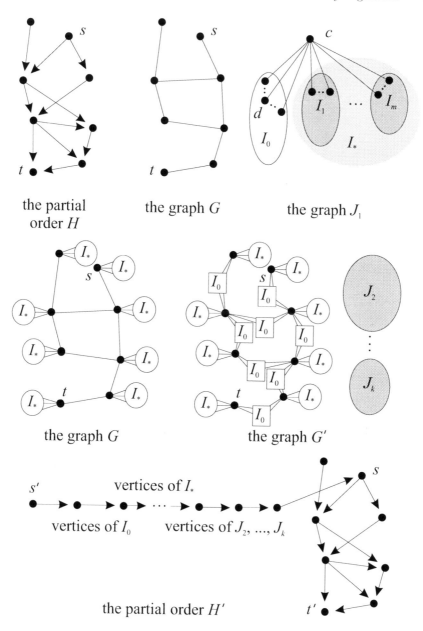

Fig. 4. Our basic construction.

We will now state three lemmas to be used in the remainder of the proof.

Lemma 1. *Any graph K consisting of any number of disjoint copies of $\langle I_0 \setminus \{d\}\rangle$ plus any number of disjoint copies of $\langle I_* \rangle$ plus a disjoint copy of each of $J_2, J_3, \ldots, J_k$ satisfies π.*

Proof. The connected components of K consist of $J_2, J_3, \ldots, J_t$ together with the connected components of the copies of $\langle I_0 \setminus \{d\} \rangle$ and $\langle I_* \rangle$. Consider the α-sequence α of a connected component of either $\langle I_0 \setminus \{d\} \rangle$ or $\langle I_* \rangle$. All components of α are strictly less than $|I_0| + 1$; and so α is strictly less than α_{J_1}. Hence, β_K has one less component equal to α_{J_1} than β_J, with all other components strictly less than α_{J_1}; and so K satisfies π by minimality of β_J.   □

Lemma 2. *Take a single copy of $\langle I_* \cup \{c\} \rangle$ and any number of disjoint copies of $\langle (I_0 \setminus \{d\}) \cup \{c\} \rangle$, and identify the vertices named c in all of these graphs. Then the resulting graph M satisfies π.*

Proof. Let M' be the connected component of M containing c. We begin by remarking that any other connected component of M has an α-sequence strictly less than the α-sequence $(|I_0| + 1)$; and so strictly less than α_{J_1}.

Suppose that c is a cut-point of M'. Then $\alpha_{c,M'}$ has components $|I_1|+1, |I_2|+1, \ldots, |I_m|+1$ as well as possibly some other components which are all strictly less than $|I_0| + 1$. Hence, by arguing similarly to as in the proof of Lemma 1, $\alpha_{M'}$ is strictly less than α_{J_1}. By the remark above, β_M is strictly less than β_J and so M satisfies π by the minimality of β_J.

Suppose that c is not a cut-point of M'. Then $I_* = I_1$, i.e., $m = 1$, and $M' = \langle I_* \rangle$; hence, $\alpha_{M'}$ is at most $(|I_1| + 1)$. Any connected component of M different from M' has size at most $|I_0| - 2$, and so $\alpha_{J_1} = (|I_0|+1, |I_1|+1)$ is strictly greater than the α-sequence of any connected component of M. Consequently, β_M is strictly less than β_J; and M satisfies π by the minimality of β_J.   □

Lemma 3. *Any graph N consisting of disjoint copies of $J_2, J_3, \ldots, J_k$ plus any number of disjoint copies of the graph M from Lemma 2 satisfies π.*

Proof. By the proof of Lemma 2, the graph M is such that the maximal component of β_M is strictly less than α_{J_1}. By reasoning as we did in the proof of Lemma 1, it follows that β_N is strictly less than β_J and so N satisfies π by the minimality of β_J.   □

Throughout, we refer to a G-vertex in G' and the corresponding vertex in G by the same name (and also to a vertex of H and the corresponding vertex in the portion of the partial order H' corresponding to H by the same name).

Consider the algorithm GREEDY(π) on input (G', H', s', t'). The partial order H' consists of a linear order, whose vertices are S_0, say, concatenated with the partial order H. The subgraph of G' induced by the vertices of S_0 is as is the graph K of Lemma 1 and consequently every vertex of S_0 is always placed in every output from GREEDY(π). Note that the algorithm GREEDY(π) on input (G', H', s', t') with current-vertex s is working with exactly the same partial order, namely H, as is the algorithm GREEDY(independent set) on input (G, H, s, t) with current-vertex s.

Suppose, as our induction hypothesis, that:

- the algorithm GREEDY(independent set) on input (G, H, s, t) has current-vertex u, for some ancestor u of s in H, and has so far output the set of vertices S;
- the algorithm GREEDY(π) on input (G', H', s', t') has current vertex u in H' and has so far output the set of vertices $S_0 \cup S$; and
- the subgraph of G' induced by the vertices of $S_0 \cup S'$ is in the form of a subgraph of the graph N in Lemma 3.

Note that the induction hypothesis clearly holds, in the base case, when the vertex u is actually s.

Suppose that the algorithm GREEDY(π) outputs the vertex u. If u is such that adding u to $S_0 \cup S'$ completes a copy of I_0 then we would have a copy of J within the subgraph of G' induced by the vertices of $S_0 \cup S \cup \{u\}$. This would yield a contradiction because this subgraph satisfies π (by definition), π is hereditary on induced subgraphs, and J would then have to satisfy π. Hence, the vertex u is not joined to any vertex of S in G and so u is output by the algorithm GREEDY(independent set).

Conversely, if the algorithm GREEDY(independent set) outputs u then this is because $S \cup \{u\}$ is an independent set in G; and consequently $S_0 \cup S \cup \{u\}$ induces in G' a subgraph of the form of a subgraph of the graph N in Lemma 3. Hence, by Lemma 3, u is output by the algorithm GREEDY(π).

By induction, we obtain that if S is a set of vertices output by the algorithm GREEDY(independent set) on input (G, H, s, t) then $S_0 \cup S$ is output by the algorithm GREEDY(π) on input (G', H', s', t'), and conversely. Hence, we have a log-space reduction from $\mathcal{H}$ to $\mathcal{G}$. □

Just as Miyano did in [6], we can actually remove the necessity in Theorem 3 for π to be satisfied by all sets of independent edges. Ramsey theory can be applied to show that any graph property that is non-trivial and hereditary on a class of graphs is either satisfied by all independent sets or by all cliques. We can then use this fact to eliminate the need for π to be satisfied by all independent edges (we omit the details here but point out that we proceed exactly as Miyano did). Hence, we obtain the following.

Corollary 1. *Let π be a polynomial-time testable, hereditary graph property that is non-trivial on planar bipartite graphs. The problem GREEDY(partial order, planar bipartite, π) is complete for* **NP**. □

We conclude by remarking that in the full version of this paper, we extend Corollary 1 so that it applies to directed graphs and we examine degree bounds on graphs so as to delineate when a problem GREEDY(partial order, graphs, π) becomes solvable in polynomial-time (for specific properties π).

References

1. Asano, T., Hirata, T.: Edge-deletion and edge-contraction problems. Proceedings of 14th Annual ACM Symposium on the Theory of Computing. (1982) 245–254

316   Antonio Puricella and Iain A. Stewart

2. Berge, C.: Graphs and Hypergraphs. North-Holland (1973)
3. Garey, M.R., Johnson, D.S.: Computers and Intractability: A Guide to the Theory of NP-Completeness. Freeman (1979)
4. Gibbons, A.M., Rytter, W.: Efficient Parallel Algorithms. Cambridge University Press (1988).
5. Lewis, J.M., Yannakakis, M.: The node deletion problem for hereditary properties is NP-complete. Journal of Computer and System Sciences. **20** (1980) 219–230
6. Miyano, S.: The lexicographically first maximal subgraph problems: P-completeness and NC algorithms. Mathematical Systems Theory. **22** (1989) 47–73
7. Watanabe, T., Ae, T., Nakamura, A.: On the removal of forbidden graphs by edge-deletion or by edge-contraction. Discrete Applied Mathematics. **3** (1981) 151–153
8. Yannakakis, M.: Node-deletion problems on bipartite graphs. SIAM Journal of Computing. **10** (1981) 310–327

Critical and Anticritical Edges in Perfect Graphs

Annegret Wagler

Konrad-Zuse-Zentrum für Informationstechnik Berlin
Takustraße 7, 14195 Berlin, Germany
wagler@zib.de

Abstract. We call an edge e of a perfect graph G critical if $G - e$ is imperfect and say further that e is anticritical with respect to the complementary graph $\overline{G}$. We ask in which perfect graphs critical and anticritical edges occur and how to find critical and anticritical edges in perfect graphs. Finally, we study whether we can order the edges of certain perfect graphs such that deleting all the edges yields a sequence of perfect graphs ending up with a stable set.

1   Introduction

The present paper is devoted to the investigation of critical and anticritical edges with respect to perfectness, a rich and well-studied graph property. Berge proposed to call a graph $G = (V, E)$ *perfect* if, for each of its induced subgraphs $G' \subseteq G$, the chromatic number $\chi(G')$ equals the clique number $\omega(G')$. I.e., for every induced subgraph G' of a perfect graph G we need as many stable sets to cover all nodes of G' as a maximum clique of G' has nodes (a set $V' \subseteq V$ is a clique (stable set) of G if the nodes in V' are pairwise (non-)adjacent).

Berge [1] conjectured two characterizations of perfect graphs. His first conjecture was that a graph G is perfect iff the clique covering number $\overline{\chi}(G')$ equals the stability number $\alpha(G')$ $\forall G' \subseteq G$ (i.e., that we need as many cliques to cover all nodes of G' as a maximum stable set of G' has nodes). Since complementation transforms stable sets into cliques and colorings into clique coverings, we have $\alpha(G) = \omega(\overline{G})$ and $\chi(G) = \overline{\chi}(\overline{G})$ where $\overline{G}$ denotes the complement of G. Hence, Berge conjectured that a graph G is perfect if and only if its complement $\overline{G}$ is. This was proven by Lovász [11] and is nowadays known as *Perfect Graph Theorem*.

The second Berge conjecture concerns a characterization of perfect graphs via forbidden subgraphs. It is a simple observation that chordless odd cycles C_{2k+1} with $k \geq 2$, termed *odd holes*, and their complements $\overline{C}_{2k+1}$, called *odd antiholes*, are imperfect. Clearly, each graph containing an odd hole or an odd antihole as induced subgraph is imperfect as well. Berge conjectured in [1] that a graph is perfect if and only if it contains neither odd holes nor odd antiholes as induced subgraphs; such graphs are nowadays called *Berge graphs*. This still open *Strong Perfect Graph Conjecture* has already been verified for several classes of F-free Berge graphs, e.g., if F is a claw (Parthasarathy and Ravindra [16], see Fig. 1(a)), a diamond (Tucker [19], see Fig. 1(b)), a clique K_4 of size 4 (Tucker [18]), or a

A. Brandstädt and V.B. Le (Eds.): WG 2001, LNCS 2204, pp. 317–327, 2001.
© Springer-Verlag Berlin Heidelberg 2001

bull (Chvátal and Sbihi [4], see Fig. 1(c)). Padberg [15] introduced the notion of *minimally imperfect graphs*: imperfect graphs with the property that removing any of its nodes yields a perfect graph. Using this term, the Strong Perfect Graph Conjecture reads that odd holes and odd antiholes are the only minimally imperfect graphs. Therefore, minimally imperfect Berge graphs are termed *monsters*. In order to verify or falsify the Strong Perfect Graph Conjecture, many structural properties of minimally imperfect graphs have been discovered. To mention only a few of them, the following node pairs must not occur in minimally imperfect graphs G. It is well-known that G has no *comparable pair* (two nodes x and y with $N(x) - y \subseteq N(y)$, i.e., all neighbors of x except eventually y belong to the neighborhood of y). G does not admit *twins* (*antitwins*), i.e., two nodes x and y such that all remaining nodes of G are adjacent to both or to none of x and y (to either x or to y) due to the *Replacement Lemma* [11] (*Antitwin Lemma* [14]). Note that the property of being a comparable pair, twins, or antitwins does not depend on whether or not x and y are adjacent. Furthermore, no minimally imperfect graph G contains an *even pair* (two nodes x and y such that all chordless paths connecting x and y have even length) by the *Even Pair Lemma* [13] and a *star-cutset* (a cutset S containing one node that is adjacent to all remaining nodes of S) by the *Star-Cutset Lemma* [3].

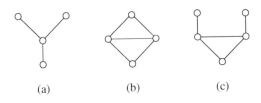

(a)   (b)   (c)

Fig. 1. Some special graphs: (a) claw, (b) diamond, (c) bull.

Although several, in general $\mathcal{NP}$-hard, combinatorial optimization problems can be solved in polynomial time for perfect graphs [5], the structure of perfect graphs is not well understood. The recognition problem for perfect graphs is unsolved and, in particular, the Strong Perfect Graph Conjecture still seems to be out of reach. On the other hand, the investigation of minimally imperfect graphs has revealed that these graphs have quite strong properties. This motivated us to consider new extremal graphs with respect to perfectness and lead to the concept of critical and anticritical edges in perfect graphs. The starting point of our investigation was the following graph evolution process. Imagine you have an arbitrary perfect graph and you consecutively delete one edge until you get a stable set, or you consecutively add one edge until a clique is reached. So you create a sequence of graphs starting and ending up with a perfect graph. But, if you choose the edges to be deleted or added randomly, most graphs of your sequence will be imperfect. The aim is to avoid the occurrence of imperfect graphs in our sequence and, thereby, to give some information which perfect graphs may be constructed by consecutively deleting or adding edges.

Problem 1. Given a certain perfect graph, is there a rule how to choose an edge to be deleted or added in order to keep perfectness?

We call an edge e of a perfect graph G *critical* if $G - e$ is imperfect. Moreover, we say that e is an *anticritical* edge of the complement $\overline{G}$: $e \notin E(\overline{G})$ holds and $\overline{G} + e$ is imperfect due to the Perfect Graph Theorem [11]. Whenever we delete (add) a critical (anticritical) edge e of a perfect graph G, we create in $G - e$ $(G + e)$ *minimally* imperfect subgraphs. In order to attack Problem 1, we study in Sect. 2 those subgraphs $G_e \subseteq G$ which yield the minimally imperfect subgraphs $G_e - e \subseteq G - e$ $(G_e + e \subseteq G + e)$. One example of a critical edge is a single *short chord* of a cycle with odd length ≥ 5 (that is a chord whose endnodes have distance two on the cycle), the deletion of e yields an odd hole. Hence, one rule in the sense of Problem 1 is "never omit a single short chord of an odd cycle". Analogously, adding the edge between the endnodes of a chordless path P_{2k+1} with $k \geq 2$ yields an odd hole, too. "Never add an edge connecting the endnodes of a chordless path of even length ≥ 4" is, therefore, another rule in the sense of Problem 1. Further results of that type are presented in Sect. 2. Sect. 3 investigates whether it is easier to solve Problem 1 if the graph in question belongs to a certain *subclass* of perfect graphs. Certain perfect graphs may not possess any critical or anticritical edge, then we could choose an arbitrary edge keeping perfectness.

Problem 2. In which perfect graphs do critical and anticritical edges occur at all?

One answer gives a new characterization of *Meyniel graphs*. They have been introduced in [12] as graphs all odd cycles of length ≥ 5 of which admit at least two chords. So critical edges in form of single short chords of odd cycles do obviously not occur in Meyniel graphs. Actually, removing any edge from a Meyniel graph keeps perfectness by [9] and we show: *a graph is Meyniel if and only if it does not admit any critical edge*. In particular, every perfect but non-Meyniel graph admits at least one critical edge. We study in Sect. 3 for such classes of perfect graphs which minimally imperfect subgraphs may occur after deleting (adding) a critical (anticritical) edge. The large abundance of classes of perfect graphs led us to mention only results for some "classical" classes: *Strongly perfect graphs* have been introduced in [2] to be graphs all of whose subgraphs $G' \subseteq G$ admit a stable set that has a non-empty intersection with all maximal cliques of G'. *Weakly triangulated graphs* are defined to have neither holes C_k nor antiholes $\overline{C}_k$ with $k \geq 5$ as induced subgraphs [6]. Meyniel [13] called a graph G *strict quasi parity* if each of its non-complete subgraphs has an even pair and *quasi parity* if G' or $\overline{G}'$ owns an even pair for each subgraph $G' \subseteq G$. In Sect. 4, we treat the following problem and present several classes of perfect graphs for which the answer is in the affirmative.

Problem 3. For a certain perfect graph, is there an order of all the edges to be deleted (added) so that we get a sequence of perfect graphs ending up with a stable set (clique)?

It turns out that it does not suffice to identify non-critical or non-anticritical edges: we can certainly remove an arbitrary edge from a Meyniel graph keeping perfectness but, at present, we do not know anything about critical edges of the resulting graph. Thus we have to look for edges the deletion or addition of which preserves membership within the corresponding *subclass* of perfect graphs. By this way, we present the studied ordering of edges to be deleted or added for, e.g., weakly triangulated graphs.

2   Critical and Anticritical Edges

For every critical (anticritical) edge e of a perfect graph G there is necessarily at least one subgraph $G_e \subseteq G$ such that $G_e - e$ ($G_e + e$) is *minimally* imperfect. We study those subgraphs G_e in order to give an answer to Problem 1. If $G_e - e \subseteq G - e$ is an odd hole, then G_e is isomorphic to an odd cycle of length ≥ 5 which admits precisely one chord, namely e. Moreover, e has to be a short chord of this cycle, forming a triangle with two of its edges, since G_e must not contain an odd hole (the case $G_e - e = C_7$ is depicted in Fig. 2(a) with $e = xy$). Thus, we have immediately:

Proposition 1. *If $G_e - e$ is an odd hole C_{2k+1} with $k \geq 2$, then $\omega(G_e) = 3$ and $\alpha(G_e) = k$ holds. G_e contains an even hole C_{2k}. A bull and the complement of a claw and of a diamond appear in G_e if $k \geq 3$.*

In the complementary graph $\overline{G}$, e is an anticritical edge and $\overline{G}_e + e$ an odd antihole (see Fig. 2(b)). Clearly, the complementary statement of Proposition 1 is true for $\overline{G}_e$.

Proposition 2. *If $G_e + e$ is an odd antihole $\overline{C}_{2k+1}$ with $k \geq 2$, then $\omega(G_e) = k$ and $\alpha(G_e) = 3$ holds. G_e contains an even antihole $\overline{C}_{2k}$. A bull, a claw, and a diamond appear in G_e if $k \geq 3$.*

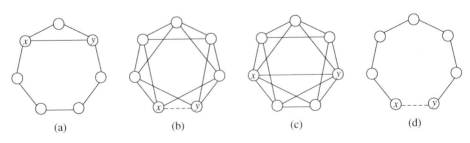

| (a) | (b) | (c) | (d) |

Fig. 2. The cases $|G_e| = 7$ with critical or anticritical edge $e = xy$.

If $G_e - e \subseteq G - e$ is an odd antihole, then G_e is the complement of an induced path P_{2k+1} with $k \geq 2$ (the case $G_e - e = \overline{C}_7$ is depicted in Fig. 2(c) with $e = xy$). We have, therefore:

Proposition 3. *If $G_e - e$ is an odd antihole $\overline{C}_{2k+1}$ with $k \geq 2$, then $\omega(G_e) = k + 1$ and $\alpha(G_e) = 2$ holds. G_e is isomorphic to $\overline{P}_{2k+1}$. A diamond appears in G_e if $k \geq 3$.*

In the complement $\overline{G}$, e is an anticritical edge, $\overline{G}_e + e$ an odd hole, and $\overline{G}_e$ an induced path (see Fig. 2(d)). The complementary assertion of Proposition 3 is true for $\overline{G}_e$.

Proposition 4. *If $G_e + e$ is an odd hole C_{2k+1} with $k \geq 2$, then $\omega(G_e) = 2$ and $\alpha(G_e) = k + 1$ holds. G_e is isomorphic to P_{2k+1}. The complement of a diamond appears in G_e if $k \geq 3$.*

Of course, such a description cannot be given if $G_e - e$ or $G_e + e$ is supposed to be a monster and, if the Strong Perfect Graph Conjecture is true, this case does not occur at all. With look at Problem 1, we list some properties of G_e if e is a critical edge and $G_e - e$ a generally minimal imperfect graph.

Lemma 1. *Let e be a critical edge of a perfect graph G and $G_e - e \subseteq G - e$ be minimally imperfect.*

(i) *The endnodes of e belong to the intersection of all maximum cliques of G_e.*
(ii) *The endnodes of e occur in a triangle of G_e.*
(iii) *An even hole is running through e in G_e.*
(iv) *If $G_e - e$ is a monster, then e belongs to a K_5 and to a diamond of G_e.*

Proof. Consider a critical edge e of a perfect graph G and let $G_e - e \subseteq G - e$ be minimally imperfect. We know $\omega(G_e) = \chi(G_e)$ but $\omega(G_e - e) < \chi(G_e - e)$. Since the removal of an edge cannot increase the chromatic number, $\omega(G_e) > \omega(G_e - e)$ follows. That means deleting e destroys all maximum cliques of G_e and (i) is true. In particular, $\omega(G_e) = 2$ and $\omega(G_e - e) = 1$ follows from (i) if the endnodes of e do not possess any common neighbor in G_e. Hence the endnodes of e occur in a triangle of G_e and (ii) is shown.

Due to the Even Pair Lemma [13] there is an odd induced path between every two non-adjacent nodes of $G_e - e$, in particular, between the endnodes of e. Thus an even hole is running through e in G_e and we obtain (iii).

In the special case if $G_e - e$ is supposed to be a monster, we need $\omega(G_e - e) \geq 4$ due to a result of Tucker [18]. Condition (i) implies $\omega(G_e - e) < 4$ if e does not belong to a K_5 in G_e. To complete the proof of (iv), consider a common neighbor z of the endnodes x and y of e which exists by (ii). The nodes x, z, y induce a P_3 in $G_e - e$, hence a hole C is running through x, z, y in $G_e - e$ due to a result of Hoàng [10]. If $G_e - e$ is a monster, C has to be an even hole of length 4 (otherwise $G_e - e$ or G_e contain an odd hole). $\qquad\square$

As immediate consequence, the complementary results hold for $\overline{G}_e$ if e is anticritical.

Lemma 2. *Suppose $e = xy$ to be a critical edge of the perfect graph $\overline{G}$.*

(i)   *The nodes x and y belong to a stable set of size 3 in G.*
(ii) *G admits an even antihole containing x and y.*
(iii) *If e is M-anticritical, then x and y appear in the complement of a diamond and are contained in a stable set of size 5.*

Lemma 1(ii),(iii) and Lemma 2(i),(ii) can be seen as rules which we ask for in Problem 1: given a perfect graph, how to identify a non-critical or non-anticritical edge? We are able to give further conditions when an edge of a perfect graph G cannot be critical.

Lemma 3. *Suppose $e = xy$ to be a critical edge of a perfect graph G. Then x and y do not form twins, antitwins, or a comparable pair of G. Neither x nor y is a simplicial node in G (that is a node with a clique as neighborhood).*

Proof. Consider a perfect graph G, a critical edge $e = xy$ of G, and a subgraph $G_e \subseteq G$ such that $G_e - e$ is minimally imperfect. Then x and y do not form a comparable pair, twins, and antitwins of G, else they would also be a comparable pair, twins, and antitwins in $G_e - e$ leading to a contradiction to a well-known fact, the Replacement Lemma [11], and the Antitwin Lemma [14], respectively. Now, neither x nor y can be a simplicial node, since then x and y are either twins or a comparable pair of $G_e - e$. □

If two nodes form twins, antitwins, or a comparable pair in G, they do so in the complementary graph $\overline{G}$. In addition, x and y form a *2-pair* if all induced paths connecting x and y have length two. We can also show that x and y must not form a 2-pair if $e \notin E(G)$ is supposed to be an anticritical edge of G.

Lemma 4. *Let $e = xy$ be an anticritical edge of the perfect graph G. Then x and y are neither twins, antitwins, a comparable pair, nor a 2-pair of G.*

Proof. Consider a perfect graph G, an anticritical edge $e = xy \notin E(G)$, and $G_e \subseteq G$ such that $G_e + e \subseteq G - e$ is minimally imperfect. Then Lemma 3 implies that x and y does not form a comparable pair, twins, and antitwins of G. Now assume x and y to form a 2-pair. In the case $N_{G_e}(x) - N_{G_e}(y) = \emptyset$ (all neighbors of x in G_e belong to the neighborhood of y), x and y would be a comparable pair in $G_e + e$, hence there is a node $x' \in N_{G_e}(x) - N_{G_e}(y)$. Since x and y are a 2-pair in G_e and x' and y are non-adjacent, all paths in $G_e + e$ connecting x' and y must contain x or a node in $N_{G_e}(x) \cap N_{G_e}(y)$ in contradiction to the Star-Cutset Lemma [3]. □

3   Which Graph Classes Admit Critical or Anticritical Edges?

This section is devoted to the investigation of Problem 2: In which perfect graphs do critical or anticritical edges occur at all? More precisely, are there classes of perfect graphs so that no graph in this class has a critical or anticritical edge? For some graph classes, this follows immediately using the results established in the previous section. E.g., every critical edge is contained in a triangle by Lemma 3(ii), so *bipartite graphs* do not possess any critical edge (since they contain no odd cycles). Lemma 3(iii) says that every critical edge has to occur in an even hole. Consequently, no *triangulated graph* has a critical edge (since triangulated graphs do not admit any hole of length ≥ 4). Both properties together yield the existence of an odd cycle of length ≥ 5 in the union of the triangle and the even hole. So *line-perfect graphs* cannot admit critical edges, too (since they are defined to contain no odd cycles of length ≥ 5).

The next theorem *characterizes the perfect graphs without any critical edge.* It also yields a new characterization of Meyniel graphs, defined to contain no odd cycles of length ≥ 5 with at most one chord. The proof of this theorem relies on the perfectness of a superclass of Meyniel graphs. Let $G = (V, E)$ be a Meyniel graph and $V' \subseteq V$. The *slim graph* $G(V')$ is obtained by deleting every edge of G with both endnodes in V' and $G(V')$ is perfect due to Hertz [9].

Theorem 1. (Hougardy, Wagler) *A perfect graph does not admit any critical edge if and only if it is Meyniel.*

Proof. Let G be a perfect graph without any critical edge. Assume G to be not a Meyniel graph then G admits an odd cycle C with length ≥ 5 which has at most one chord. Since G is perfect, C cannot be chordless, hence C possess precisely one chord e. If e were not a short chord of C, an odd hole would be contained in G. Thus e has to be a short chord of C and so it is an H-critical edge of G in contradiction to the precondition.

On the other hand, suppose $G = (V, E)$ to be a Meyniel graph. Then, for every subset $V' \subseteq V$, the graph $G(V')$ generated from G by deleting every edge of G with both endnodes in V' is a slim graph, still perfect by Hertz [9]. Thus, if we choose an arbitrary pair of adjacent nodes in a Meyniel graph and delete the edge connecting them, we get a perfect graph again, i.e., a Meyniel graph cannot admit any critical edge.   □

Making use of this new characterization of Meyniel graphs, a perfect graph does not admit any anticritical edge if and only if it is co-Meyniel, the complement of a Meyniel graph. Hence we obtain immediately:

Corollary 1. *A perfect graph does neither admit any critical nor anticritical edge if and only if it is both Meyniel and co-Meyniel.*

Consequently, critical (anticritical) edges occur in all perfect graphs G not contained in the class of Meyniel (co-Meyniel) graphs. Our aim is to find out which minimally imperfect subgraphs may occur in $G - e$ ($G + e$). The results listed in the next lemma are obtained by combining the knowledge on forbidden subgraphs in the respective graph classes and the results from the previous section.

Lemma 5. *Let G be a perfect graph, $e \in E(G)$ be a critical edge of G, and $e' \notin E(G)$ be an anticritical edge of G.*

(i) *If G is diamond-free Berge, then $G - e$ and $G + e'$ do not contain a $\overline{C}_{2k+1}$ with $k \geq 3$ by Proposition 3 and 2. Furthermore, $G - e$ cannot admit any monster by Lemma 1(iv).*

(ii) *For every K_4-free Berge graph G, Proposition 3 and Lemma 1(iv) imply that $G - e$ has odd holes as the only minimally imperfect subgraphs. $G + e'$ does not contain any $\overline{C}_{2k+1}$ with $k \geq 4$ by Proposition 4.*

(iii) *Is G weakly triangulated or bull-free Berge, then $G - e$ and $G + e'$ cannot contain C_{2k+1} and $\overline{C}_{2k+1}$ with $k \geq 3$ by Proposition 1 and 2, respectively.*

(iv) *Is G strict quasi parity, a slim graph, or strongly perfect, then no even antihole $\overline{C}_{2k}$ with $k \geq 3$ appears in G. Hence, $G + e$ does not admit a $\overline{C}_{2k+1}$ with $k \geq 3$ by Proposition 2.*

4 Perfect and Co-perfect Edge Orders

After considering the occurrence of critical edges in several classes of perfect graphs, we turn to Problem 3: we are interested whether it is possible, for graphs in a certain class $\mathcal{C}$ of perfect graphs, to successively delete or add edges keeping perfectness until a stable set or a clique is reached. Moreover, the existence of such edge orders for all graphs in $\mathcal{C}$ would provide a constructive method to generate the graphs in $\mathcal{C}$ by consecutively deleting or adding edges. For that, we use knowledge from the previous sections.

Let $G = (V, E)$ be a perfect graph. We call a numbering $(e_1, \ldots, e_m)$ of its edge set E a *perfect edge order* if, for $G = G_0$, all graphs $G_i := G_{i-1} - e_i$ are perfect for $1 \leq i \leq m$. Clearly, e_i has to be a non-critical edge of G_{i-1} for $1 \leq i \leq m$, and G_m is a stable set. Analogously, we say that a perfect graph G admits a *co-perfect edge order* iff its complement $\overline{G}$ has a perfect edge order. Here we simply use the numbering of the edges of $\overline{G}$ for the non-edges of G and get finally a clique.

Note that it does not suffice to identify non-critical or non-anticritical edges in the perfect graphs in question. E.g., we can certainly delete an arbitrary edge of a Meyniel graph keeping perfectness by [9], but we may obtain a slim graph that is not Meyniel and do not know anything about its non-critical edges so far. Hence, we cannot provide perfect edge orders of Meyniel graphs, although the graphs in this class are even characterized that they do not contain any critical edge due to Theorem 1. So we mainly have to look for edges such that their deletion or addition preserves the membership to the corresponding *subclass* of perfect graphs.

In general, we have to look for *critical (anticritical) graphs* with respect to the subclass $\mathcal{C}$ of perfect graphs under consideration: that are graphs which lose the studied property by deleting (adding) an arbitrary edge. If we can ensure that no (anti)critical graph with respect to $\mathcal{C}$ exists, then we know there is a (co-)perfect edge order for all graphs in $\mathcal{C}$: every graph G in $\mathcal{C}$ admits at least one edge e such that $G - e$ $(G + e)$ still belongs to $\mathcal{C}$. It is "only" left to find that edge e. In order to answer the question whether or not Meyniel graphs admit a perfect edge order we have, therefore, to solve the following problem:

Problem 4. Find a critical Meyniel graph or show that no such graph exists.

Looking for graph classes without critical graphs, we first observe that there are no critical graphs with respect to every *monotone* class (that is a class defined by some forbidden *partial* subgraphs). The simplest example is the class of bipartite graphs having no odd cycles. Obviously, deleting an *arbitrary* edge of a bipartite graph yields a bipartite graph again. Hence, bipartite graphs admit perfect edge orders and, in particular, *every* of their edge orders is perfect. A superclass of bipartite graphs consists of all line-perfect graphs having no odd cycles of length ≥ 5 as forbidden partial subgraphs. Again, there are no critical line-perfect graphs, hence every line-perfect graph admits a perfect edge order. Moreover, we have that they are *precisely* those perfect graphs such that *every* of their edge orders is perfect.

Theorem 2. *A graph is line-perfect iff all edge orders are perfect.*

Proof. A line-perfect graph G does not contain any odd cycle of length ≥ 5 as partial subgraph. Obviously, $G - e$ does also not contain any odd cycle of length ≥ 5 $\forall e \in E(G)$ and, therefore, is still line-perfect. Thus every ordering of $E(G)$ is perfect.

Now, assume G to be perfect but not line-perfect. Then G admits a cycle C of odd length at least 5 as partial subgraph. C cannot be chordless and every edge order of G that deletes all chords of C before an edge of C is not perfect.   □

Bipartite and line-perfect graphs are subclasses of Meyniel graphs. A further class of Meyniel graphs for which we know a perfect edge order consists off all triangulated graphs. For those graphs, we have a well-known structural result, namely, that a graph is triangulated iff every subgraph has a simplicial node. Consider a triangulated graph $G = (V, E)$ and let x be a simplicial node of G. Then no edge e incident to x is critical by Lemma 3. The graph $G - e$ is not only still perfect, but even still triangulated, since x is also simplicial in $G - e$. As a consequence, we obtain a perfect edge order for triangulated graphs:

Theorem 3. *Every triangulated graph G admits a perfect edge order $(e_1, \ldots, e_m)$ with $G = G_0$, $G_i = G_{i-1} - e_i$, and e_i incident to a simplicial node of G_{i-1} for $1 \leq i \leq m$.*

Clearly, the complementary classes of bipartite, line-perfect, and triangulated graphs admit the corresponding co-perfect edge orders. For one class of perfect (but not Meyniel) graphs, both a perfect and a co-perfect edge order are known, namely, for weakly triangulated graphs. Let G be non-complete and weakly triangulated. Then a 2-pair x, y occurs in G due to a characterization of weakly triangulated graphs given by Hayward, Hoàng, and Maffray in [8]. The graph $G + xy$ is not only perfect by Lemma 4 but still weakly triangulated by a result of Spinrad and Sritharan [17]. Consequently, we obtain a co-perfect edge order for weakly triangulated graphs by consecutively adding edges between 2-pairs (called 2-pair non-edge order in [17]).

Theorem 4. (Spinrad and Sritharan [17]) *Every weakly triangulated graph G admits a co-perfect edge order $(e_1, \ldots, e_{\overline{m}})$ with $G = G_0$, $G_i = G_{i-1} + e_i$ such that the endnodes of e_i form a 2-pair in G_{i-1} for $1 \leq i \leq \overline{m} = |E(\overline{G})|$.*

That the class of weakly triangulated graphs is closed under complementation yields particularly a perfect edge order for every weakly triangulated graph. Hayward proved in [7] that the following perfect edge order of a weakly triangulated graph G corresponds to the 2-pair non-edge order for $\overline{G}$ given by Spinrad and Sritharan [17].

Theorem 5. (Hayward [7]) *Every weakly triangulated graph G admits a perfect edge order $(e_1, \ldots, e_m)$ with $G = G_0$, $G_i = G_{i-1} - e_i$ such that e_i is not the middle edge of any P_4 in G_i for $1 \leq i \leq m = |E(G)|$.*

Weakly triangulated graphs are defined as a common generalization of triangulated and co-triangulated graphs. Hence, the above theorems provide also a co-perfect edge order for triangulated graphs and a perfect edge order for co-triangulated graphs. In addition, the 2-pair non-edge order for weakly triangulated graphs enables us to establish a co-perfect edge order for bipartite graphs G: either there are non-adjacent nodes a, b in different color classes of G and $G + ab$ is still bipartite, or G is as complete bipartite graph weakly triangulated.

At present, perfect or co-perfect edge orders are not known to the author for other subclasses of perfect graphs. However, we can ensure that no such edge order exists for some classes $\mathcal{C}$ of perfect graphs: If $\mathcal{C}$ contains a *critically (anticritically) perfect graph* G such that deleting (adding) an arbitrary edge yields an imperfect graph (see [20] for more details), then G is in particular critical (anticritical) with respect to $\mathcal{C}$ and there is no perfect (co-perfect) edge order for the graphs in $\mathcal{C}$.

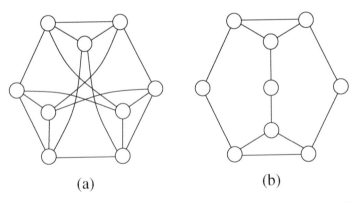

(a)   (b)

Fig. 3. (a) critically and anticritically perfect graph, (b) anticritically perfect graph.

The graph depicted in Fig. 3(a) is critically and anticritically perfect and belongs to the classes of F-free Berge graphs where F is a claw, a diamond, or a K_4. Hence, that graph is also (anti)critical with respect to those classes and there are, therefore, neither perfect nor co-perfect edge orders for those classes. The graph in Fig. 3(b) is anticritically perfect, strongly perfect, strict quasi parity and quasi parity. Consequently, there are no co-perfect edge orders for all strongly perfect, all strict quasi parity, or all quasi parity graphs. In particular, there is no way to construct all the graphs belonging to those classes of perfect graphs by consecutively adding edges.

References

1. Berge, C.: Färbungen von Graphen, deren sämtliche bzw. deren ungerade Kreise starr sind. Wiss. Zeitschrift der Martin-Luther-Universität Halle-Wittenberg (1961) 114–115
2. Berge, C., Duchet, P.: Strongly Perfect Graphs. In: Berge, C., Chvátal, V. (eds.): Topics on Perfect Graphs. North Holland, Amsterdam (1984) 57–61
3. Chvátal, V.: Star-Cutsets and Perfect Graphs. J. Combin. Theory (B) **39** (1985) 189–199
4. Chvátal, V., Sbihi, N.: Bull-Free Berge Graphs are Perfect. Graphs and Combinatorics **3** (1987) 127–139
5. Grötschel, M., Lovász, L., Schrijver, A.: The Ellipsoid Method and its Consequences in Combinatorial Optimization. Combinatorica **1** (1981) 169–197
6. Hayward, R.B.: Weakly Triangulated Graphs. J. Combin. Theory (B) **39** (1985) 200–209
7. Hayward, R.B.: Generating Weakly Triangulated Graphs. J. Graph Theory **21**, No. 1 (1996) 67–69
8. Hayward, R.B., Hoàng, C.T., Maffray, F.: Optimizing Weakly Triangulated Graphs. Graphs and Combinatorics **5** (1989) 339–349, erratum in **6** (1990) 33–35
9. Hertz, A.: Slim Graphs. Graphs and Combinatorics **5** (1989) 149–157
10. Hoàng, C.T.: Some Properties of Minimal Imperfect Graphs. Discrete Math. **160** (1996) 165–175
11. Lovász, L.: Normal Hypergraphs and the Weak Perfect Graph Conjecture. Discrete Math. **2** (1972) 253–267
12. Meyniel, H.: On The Perfect Graph Conjecture. Discrete Math. **16** (1976) 339–342
13. Meyniel, H.: A New Property of Critical Imperfect Graphs and some Consequences. Europ. J. Combinatorics **8** (1987) 313–316
14. Olariu, S.: No Antitwins in Minimal Imperfect Graphs. J. Combin. Theory (B) **45** (1988) 255–257
15. Padberg, M.W.: Perfect Zero-One Matrices. Math. Programming **6** (1974) 180–196
16. Parthasarathy, K.R., Ravindra, G.: The Strong Perfect Graph Conjecture is True for $K_{1,3}$-free Graphs. J. Combin. Theory (B) **21** (1976) 212–223
17. Spinrad, J., Sritharan, R.: Algorithms for Weakly Triangulated Graphs. Discrete Appl. Math. **59** (1995) 181–191
18. Tucker, A.: Critical Perfect Graphs and Perfect 3-chromatic Graphs. J. Combin. Theory (B) **23** (1977) 143–149
19. Tucker, A.: Coloring Perfect (K_4-e)-free Graphs. J. Combin. Theory (B) **42** (1987) 313–318
20. Wagler, A.: Critical Edges in Perfect Graphs. PhD thesis, TU Berlin (2000)

Author Index

Lecture Notes in Computer Science

For information about Vols. 1–2126
please contact your bookseller or Springer-Verlag

Vol. 2168: L. De Raedt, A. Siebes (Eds.), Principles of Data Mining and Knowledge Discovery. Proceedings, 2001. XVII, 510 pages. 2001. (Subseries LNAI).

Vol. 2170: S. Palazzo (Ed.), Evolutionary Trends of the Internet. Proceedings, 2001. XIII, 722 pages. 2001.

Vol. 2172: C. Batini, F. Giunchiglia, P. Giorgini, M. Mecella (Eds.), Cooperative Information Systems. Proceedings, 2001. XI, 450 pages. 2001.

Vol. 2173: T. Eiter, W. Faber, M. Truszczynski (Eds.), Logic Programming and Nonmonotonic Reasoning. Proceedings, 2001. XI, 444 pages. 2001. (Subseries LNAI).

Vol. 2174: F. Baader, G. Brewka, T. Eiter (Eds.), KI 2001: Advances in Artificial Intelligence. Proceedings, 2001. XIII, 471 pages. 2001. (Subseries LNAI).

Vol. 2175: F. Esposito (Ed.), AI*IA 2001: Advances in Artificial Intelligence. Proceedings, 2001. XII, 396 pages. 2001. (Subseries LNAI).

Vol. 2176: K.-D. Althoff, R.L. Feldmann, W. Müller (Eds.), Advances in Learning Software Organizations. Proceedings, 2001. XI, 241 pages. 2001.

Vol. 2177: G. Butler, S. Jarzabek (Eds.), Generative and Component-Based Software Engineering. Proceedings, 2001. X, 203 pages. 2001.

Vol. 2180: J. Welch (Ed.), Distributed Computing. Proceedings, 2001. X, 343 pages. 2001.

Vol. 2181: C. Y. Westort (Ed.), Digital Earth Moving. Proceedings, 2001. XII, 117 pages. 2001.

Vol. 2182: M. Klusch, F. Zambonelli (Eds.), Cooperative Information Agents V. Proceedings, 2001. XII, 288 pages. 2001. (Subseries LNAI).

Vol. 2183: R. Kahle, P. Schroeder-Heister, R. Stärk (Eds.), Proof Theory in Computer Science. Proceedings, 2001. IX, 239 pages. 2001.

Vol. 2184: M. Tucci (Ed.), Multimedia Databases and Image Communication. Proceedings, 2001. X, 225 pages. 2001.

Vol. 2185: M. Gogolla, C. Kobryn (Eds.), «UML» 2001 – The Unified Modeling Language. Proceedings, 2001. XIV, 510 pages. 2001.

Vol. 2186: J. Bosch (Ed.), Generative and Component-Based Software Engineering. Proceedings, 2001. VIII, 177 pages. 2001.

Vol. 2187: U. Voges (Ed.), Computer Safety, Reliability and Security. Proceedings, 2001. XVI, 261 pages. 2001.

Vol. 2188: F. Bomarius, S. Komi-Sirviö (Eds.), Product Focused Software Process Improvement. Proceedings, 2001. XI, 382 pages. 2001.

Vol. 2189: F. Hoffmann, D.J. Hand, N. Adams, D. Fisher, G. Guimaraes (Eds.), Advances in Intelligent Data Analysis. Proceedings, 2001. XII, 384 pages. 2001.

Vol. 2190: A. de Antonio, R. Aylett, D. Ballin (Eds.), Intelligent Virtual Agents. Proceedings, 2001. VIII, 245 pages. 2001. (Subseries LNAI).

Vol. 2191: B. Radig, S. Florczyk (Eds.), Pattern Recognition. Proceedings, 2001. XVI, 452 pages. 2001.

Vol. 2192: A. Yonezawa, S. Matsuoka (Eds.), Metalevel Architectures and Separation of Crosscutting Concerns. Proceedings, 2001. XI, 283 pages. 2001.

Vol. 2193: F. Casati, D. Georgakopoulos, M.-C. Shan (Eds.), Technologies for E-Services. Proceedings, 2001. X, 213 pages. 2001.

Vol. 2194: A.K. Datta, T. Herman (Eds.), Self-Stabilizing Systems. Proceedings, 2001. VII, 229 pages. 2001.

Vol. 2195: H.-Y. Shum, M. Liao, S.-F. Chang (Eds.), Advances in Multimedia Information Processing – PCM 2001. Proceedings, 2001. XX, 1149 pages. 2001.

Vol. 2196: W. Taha (Ed.), Semantics, Applications, and Implementation of Program Generation. Proceedings, 2001. X, 219 pages. 2001.

Vol. 2197: O. Balet, G. Subsol, P. Torguet (Eds.), Virtual Storytelling. Proceedings, 2001. XI, 213 pages. 2001.

Vol. 2198: N. Zhong, Y. Yao, J. Liu, S. Ohsuga (Eds.), Web Intelligence: Research and Development. Proceedings, 2001. XVI, 615 pages. 2001. (Subseries LNAI).

Vol. 2199: J. Crespo, V. Maojo, F. Martin (Eds.), Medical Data Analysis. Proceedings, 2001. X, 311 pages. 2001.

Vol. 2200: G.I. Davida, Y. Frankel (Eds.), Information Security. Proceedings, 2001. XIII, 554 pages. 2001.

Vol. 2201: G.D. Abowd, B. Brumitt, S. Shafer (Eds.), Ubicomp 2001: Ubiquitous Computing. Proceedings, 2001. XIII, 372 pages. 2001.

Vol. 2202: A. Restivo, S. Ronchi Della Rocca, L. Roversi (Eds.), Theoretical Computer Science. Proceedings, 2001. XI, 440 pages. 2001.

Vol. 2204: A. Brandstädt, V.B. Le (Eds.), Graph-Theoretic Concepts in Computer Science. Proceedings, 2001. X, 329 pages. 2001.

Vol. 2205: D.R. Montello (Ed.), Spatial Information Theory. Proceedings, 2001. XIV, 503 pages. 2001.

Vol. 2206: B. Reusch (Ed.), Computational Intelligence. Proceedings, 2001. XVII, 1003 pages. 2001. ·

Vol. 2207: I.W. Marshall, S. Nettles, N. Wakamiya (Eds.), Active Networks. Proceedings, 2001. IX, 165 pages. 2001.

Vol. 2208: W.J. Niessen, M.A. Viergever (Eds.), Medical Image Computing and Computer-Assisted Intervention – MICCAI 2001. Proceedings, 2001. XXXV, 1446 pages. 2001.

Vol. 2209: W. Jonker (Ed.), Databases in Telecommunications II. Proceedings, 2001. VII, 179 pages. 2001.

Vol. 2210: Y. Liu, K. Tanaka, M. Iwata, T. Higuchi, M. Yasunaga (Eds.), Evolvable Systems: From Biology to Hardware. Proceedings, 2001. XI, 341 pages. 2001.

Vol. 2211: T.A. Henzinger, C.M. Kirsch (Eds.), Embedded Software. Proceedings, 2001. IX, 504 pages. 2001.

Vol. 2212: W. Lee, L. Mé, A. Wespi (Eds.), Recent Advances in Intrusion Detection. Proceedings, 2001. X, 205 pages. 2001.

Vol. 2213: M.J. van Sinderen, L.J.M. Nieuwenhuis (Eds.), Protocols for Multimedia Systems. Proceedings, 2001. XII, 239 pages. 2001.

Vol. 2215: N. Kobayashi, B.C. Pierce (Eds.), Theoretical Aspects of Computer Software. Proceedings, 2001. XV, 561 pages. 2001.

Vol. 2217: T. Gomi (Ed.), Evolutionary Robotics. Proceedings, 2001. XI, 139 pages. 2001.

||| || ||▮||||||||||| ||||| ||| | || || ||| ||| |||